Solid State Physics

A primer

Solid State Physics

A primer

Luciano Colombo
University of Cagliari, Italy

IOP Publishing, Bristol, UK

ISBN   978-0-7503-2265-2 (ebook)
ISBN   978-0-7503-2262-1 (print)
ISBN   978-0-7503-2263-8 (myPrint)
ISBN   978-0-7503-2264-5 (mobi)

DOI   10.1088/978-0-7503-2265-2

Version: 20210301

IOP ebooks

British Library Cataloguing-in-Publication Data: A catalogue record for this book is available from the British Library.

Published by IOP Publishing, wholly owned by The Institute of Physics, London

IOP Publishing, Temple Circus, Temple Way, Bristol, BS1 6HG, UK

US Office: IOP Publishing, Inc., 190 North Independence Mall West, Suite 601, Philadelphia, PA 19106, USA

To my students.

Contents

Part II Vibrational, thermal, and elastic properties

Part III Electronic structure

8   The band theory   8-1

9   Semiconductors   9-1

Foreword

When Professor Colombo approached me to write the preface of this book, I was both surprised and honoured. Surprised because there are many eminent scholars out there that could 'do the job', and honoured because I have the greatest respect for Professor Colombo and his work. I have known Professor Colombo for many years—we interacted on several occasions (often at the MRS Fall Meeting in Boston), he visited me in Montréal, I visited him in Cagliari (invited to give a short course on models of laser ablation). I have always been impressed by the clarity and rigour of his mind, his knowledge of various topics in physics, his desire to not only do the 'best physics' but also to communicate it with clarity and passion. He has been a model for me and I tried my best, over the years, to imitate his style.

Professor Colombo told me once that writing a book was some sort of 'romantic' adventure. I was not too sure what he meant (and I did not dare to ask!), but the idea remained in my head. I suspect that he sees this as some sort of an idealised accomplishment, because as a professor, you teach and you need (and love!) the students to understand. This requires clarity and pedagogical aptitudes, and this is precisely what one finds in this excellent book: a text written for students. Having myself taught solid-state physics, in particular recently as an introductory course to undergraduate students (one semester, three credits), this book would have been perfect for my class, both with the material covered and with the level of knowledge in quantum physics, statistical physics and molecular physics expected from students. The amount of material covered is perhaps a bit large for a one-semester course, but some more advanced topics may be left for personal reading. This book fills a gap that is largely unoccupied. There are not many texts that address this topic in such a comprehensive way, and at the beginner level.

I admit that the title of the book, *Solid State Physics*, surprised me a bit at first, as it seems to me that 'condensed-matter physics' is perhaps more modern and inclusive. But I thought 'What's in a name? That which we call a rose, by any other name would smell as sweet' (William Shakespeare, *Romeo and Juliet*). We exchanged on this issue with Professor Colombo and I fully adhere to his views. The solid state is somehow a subset of the condensed state, one that is idealised and can be described using the fabulous mathematical toolbox that applies to periodic systems, including periodic boundary conditions. And this is relatively easy for students to understand and visualise. While the ideal solid state really does not exist in nature (just as a circle does not exist), it provides a reference for understanding more complex systems found in the real world. From a pedagogical viewpoint, further, understanding the basics of the discipline, i.e. solid-state physics, is a prerequisite to dealing with more complex situations, i.e. condensed-matter physics, e.g. with such problems as amorphous systems (lack of periodic order), nano-materials (importance of boundary effects), etc. To cite Professor Colombo, this primer and the others in this series 'are my personal battle to promote the need to train students, especially at undergraduate level, by passing a strong, robust, general and universally applicable set of fundamentals'. This statement, I would say, is the clear act of faith of a remarkable professor, fully dedicated to his art: kudos to Professor Colombo.

The book is organised in three parts, each divided in several chapters. The first part, 'Preliminary concepts', presents the basic basics. Chapter 1, 'The overall picture', is clever: it answers first the necessary question 'what are we talking about?' (crystalline solids, which are approached from the point of view of diffusion—very nice!), then moves on to discuss, upfront, concepts that percolate throughout the various topics which will be addressed in subsequent chapters. The late George K Horton (Rutgers), who was an authority in the area of dynamical properties of solids, told me once that writing a paper (or, for that matter, a book chapter or a book) is not like writing a novel: one should not keep the 'punch' for the end but say early where one is going, and this will serve to light the route that one follows. This is precisely what this first chapter does (and I wish I had followed George Horton's advice more often myself!). The connection between the atomic structure and the complicated many-body electronic structure is introduced right at the outset (semi-classical approximation, frozen-core approximation, Born–Oppenheimer approximation, etc). Chapter 2 goes into the 'heart of the matter' and discusses in detail the crystalline structure, both in direct and in reciprocal space (including an introduction to the Bloch theorem and one-electron wavefunctions), and inevitable deviations from the ideal picture (defects). The stage is set.

The second part, 'Vibrational, thermal, and elastic properties', emerges naturally from chapter 2. Once again, Professor Colombo excels in communicating his message effectively. As it should, the first chapter in this section, chapter 3, deals with lattice dynamics: modes of vibration, dispersion relations (acoustic and optic modes), dynamical matrix and, evidently, phonons. The presentation of phonons as a gas of pseudo-particles within the framework of the quantum theory of the harmonic crystal, is simple, direct and effective. From a communications viewpoint, and this diffuses throughout the book, the author is clearly 'speaking' to the reader —the student—just as if he was standing in front of the class. I would have loved to have been one of his students (I have been on a few occasions)! Chapter 3 continues with a clever discussion of the experimental determination of the dispersion relations —an exercise that is not *a priori* simple and one that is, in my view, essential (and sometimes overlooked)—and of the vibrational density of states, which saves one from having to count the modes one by one! Chapter 4 is concerned with the thermal properties of solids—heat capacity (including the Debye model and the quantum framework that underlies the theory) and the very important anharmonic effects, including a very nice section on phonon–phonon interactions for the most avid readers (with an excellent discussion of the crystal momentum and normal/Umklapp processes), and thermal transport (Boltzmann equation). The second part closes with chapter 5 on a discussion of elastic properties (strain, stress, elastic constants), moving from a discrete, atomic description of the crystal to a continuous description of it. This topic allows one to connect to the real world: you *know* that solids are made of atoms, but you *see* how they react to pressure or temperature, for example. I would not normally cover this subject in my undergraduate course—but perhaps I should!

The third part, 'Electronic structure', is where 'the beef is'. It consists of five chapters and begins with a general discussion of electrons in crystals (chapter 6), more precisely methods for constructing the electronic structure, namely free-electron, tight-binding and density-functional theory, all of which are developed in detail in subsequent chapters. This chapter also reviews Fermi–Dirac statistics and revisits in detail the Bloch theorem and Bloch wavefunctions. It ends with a thorough discussion of the Kronig–Penney model which provides a simple, clear introduction to the concept of band structure with its allowed regions and forbidden regions (gaps). Chapter 7 is devoted to the free-electron theory of metals, and includes a very nice discussion of the Drude model (and the reasons of its failure) and of the Sommerfeld theory of conduction. This is one occasion to introduce the electronic density of states, both at zero (ground-state) and at finite temperature, showing very nicely the contributions of electrons and phonons to the specific heat, and how the latter dominates at high temperatures. The chapter closes on a discussion of relaxation times and scattering, and limitations of the Sommerfeld theory. The latter topic provides a natural transition to chapter 8 ('The band theory') that conveys many important concepts, notably the nearly-free electron gas (weak potential) which permits the author to introduce such important ideas as energy bands and energy gaps, hence explaining the difference between insulators, semi-conductors and metals. Further to this, Professor Colombo discusses in detail the tight-binding approach (and he is famous for his work in this area, among others) and its success in describing the electronic structure. A thorough description of the band structure follows—with such topics as electron dynamics, field effects, electrons and holes, etc—and the chapter closes with a short overview of other approaches to the band structure. Chapter 9, 'Semiconductors', is dedicated to the theoretical framework underlying the physics of semiconductors—important for applications—and, in particular, to electron transport. It introduces such important concepts as doping, charge transport, carrier statistics and populations, optical (photon) absorption and related carrier transitions and excitons. Finally, chapter 10 introduces the now ubiquitous density-functional theory—Hohenberg–Kohn and Kohn–Sham theorems and practical implementation of the method; this is a great idea, not common in solid-state physics books, as it allows students to connect to topical research and shows them that quantum mechanics is a 'practical tool'!

A few more notes; in line with the logic of 'tell them what you are going to tell them', each chapter starts by a summarising 'syllabus'. This is extremely useful and helps give a direction to the reader. Further, the book includes several appendices where specific and useful topics are discussed (like, for instance, 'Essential thermo-dynamics', 'The tight-binding theory'). Finally, the text contains many illustrations that are very well done and very informative; graphical representations of complex (or abstract) concepts that describe real-world objects is extremely useful for students to develop their intuition.

This book is a 'primer' and, in that respect, cannot cover the subject matter in an exhaustive manner. Yet, I find it to be remarkably complete, in particular for a book intended to undergraduate students. And all of this in 250 pages is a genuine *tour de force*. True, there are some topics that could have been covered (e.g. superconductivity), but there is already a lot of material for a one-semester course. I would have loved to have this book for my class. Thanks to Professor Colombo for this excellent addition to the literature. I predict that it will become a 'classic'.

Laurent J Lewis
Département de physique
Université de Montréal
September 2020

Presentation of the 'primer series'

This is the second volume of a series of three books that, as a whole, account for an introduction to the huge field usually referred to as 'condensed matter physics': they are respectively addressed to atomic and molecular physics, to solid state physics, and to statistical methods for the description of classical or quantum ensembles of particles. They are based on my 20 year experience of teaching undergraduate courses on these topics for bachelor-level programs in physical and engineering sciences at the University of Cagliari (Italy).

The volumes are called 'Primers' to underline that the pedagogical aspects have been privileged over those of completeness. In particular, I selected the contents of each volume so as to keep limited its number of pages and so that the topics actually covered correspond to the *syllabus* of a typical one-semester course.

More important, however, was the choice of the style of presentation: I wanted to avoid an excessively formal treatment, preferring instead the exploration of the underlying physical features and always placing phenomenology at the centre of the discussion. More specifically, the main characteristics of this book series are:

- emphasis is always given to the physical content, rather than to formal proofs, i.e. mathematics is kept at the minimum level possible, without affecting rigour or clear thinking;
- an in-depth analysis is presented about the merits and faults of any approximation used, incorporating also a thorough discussion of the conceptual framework supporting any adopted physical model;
- prominence is always on the underlying physical basis or principle, rather than to applications;
- when discussing the proposed experiments, the focus is given to their conceptual background, rather than to the details of the instrumental setup.

Despite the tutorial approach, I nevertheless wanted to follow the Italian academic tradition, which provides even the elementary introduction to condensed matter physics at a quantum level. I hope that my efforts have optimally combined ease-of-access and rigour, especially conceptually.

The intentionally non-encyclopaedic content and the tutorial character of these 'Primers' should facilitate their use even for students not specifically enrolled in a university *curriculum* in physics. I hope, in particular, that my textbooks could be accessible to students in chemistry, materials science and also of many engineering branches. In view of this, I have included a brief outline of non-relativistic quantum mechanics in the first 'Primer', a subject that does not appear in the typical engineering *curricula*. For the rest, classical mechanics, elementary thermodynamics and Maxwell theory of electromagnetism are used, to which all students of natural and engineering sciences are normally exposed.

Each 'Primer' is organised in parts, divided into chapters. This structure is tailored to facilitate the planning of a one-semester course: these volumes aim at being their main teaching tool. More specifically, each part identifies an independent teaching module, while each chapter corresponds to about two weeks of lecturing.

I cannot conclude this general introduction without thanking the many students who, over the years, have attended my courses in condensed matter physics at the University of Cagliari. Through the continuous exchange of ideas with them I have gradually understood how best to organise my teaching and the corresponding study material. As a matter of fact, the contents that I have collected in these volumes were born from this very fruitful dialogue.

Luciano Colombo

Cagliari, June 2019

Acknowledgements

I am really indebted to Ms C Mitchell—Commissioning Editor of IOP Publishing— and Mr Daniel Heatley—ebooks Editorial Assistant of IOP Publishing—for their great enthusiasm in accepting my initial proposal and for shaping it into a suitable editorial product. I also warmly acknowledge Ms Emily Tapp— Commissioning Editor of IOP Publishing—and Mr R Trevelyan—Editorial Assistant of IOP Publishing—for supporting my writing efforts of the second Primer with great professionalism, always promptly replying to my queries and clarifying my doubts. Overall, their assistance has been really precious.

Introduction to: 'Solid state physics: a primer'

In my view the 'solid state' is an idealised state of condensed matter where translational invariance rules most physical phenomena. By 'idealised' I simply mean that real solid materials are in fact not at all perfectly periodic: they contain defects, either point-like and extended, as imposed by thermodynamics; they are contaminated or doped, as result of natural or intentional processes randomly altering their pristine chemistry; they have a finite size, that is they have surfaces. While all these phenomena are indeed fascinating topics of condensed matter physics, a first pedagogical approach to solids is in my opinion better framed within the idealisation of a surface-free and periodically invariant system, namely a crystal. This allows one to develop a simple formalism—in several important cases analytically workable—which leads to quite a few important achievements like, for instance, understanding the difference between metals and insulators or describing the thermal transport in terms of the corpuscular phonon language.

A modern approach to the fundamental physics of crystalline solids does require the use of quantum mechanics, which is the language adopted in any chapter, although classical physics is often used in order to set up the first approach to a problem or, through the manifestation of its inadequacy, to remark the real need of a full quantum treatment. Furthermore, despite the tutorial approach, some robust basic notion on atomic physics is nevertheless needed in order to fully appreciate the physics discussed in this book.

A possible subtitle for this Primer could have been: 'a first introduction to the physics of phonons and electrons in crystalline systems'. As a matter of fact, these are the only two topics—treated separately in view of the fundamental adiabatic approximation—here discussed, among many other important ones which are simply omitted, since this textbook is primarily intended as the study material for a first undergraduate course in condensed matter physics. Despite this somewhat narrow perspective, I believe that the focus on the physics of phonons and electrons in periodic crystals is enough to define the minimal-complexity framework needed to understand the fundamentals of condensed matter physics. It is understood that, in order to complete the student's background in this field, a more advanced course on dielectric, optical, magnetic ordering, and many-body properties is mostly needed, including topics like the physics of defects and surfaces, the mechanical properties of solids, and superconductivity.

This Primer is divided in three parts: the first one provides, at first, a careful account of the hierarchy of approximations adopted to treat the solid state and, next, the formal tools to describe the crystalline atomic structures; the second part deals with the vibrational, thermal, and elastic properties of crystal solids; the third part is addressed to their electronic structure. A short presentation of the density functional theory—which is nowadays considered the 'standard model' to treat condensed matter systems—is offered in the last chapter. Nine appendices are added to the text, each focussed on some technical development which, at first reading, can be skipped without compromising the general understanding of the arguments developed in the main text. A bibliography is added to each chapter as a guideline for further reading.

The volume contains many figures, most of which are 'conceptual', that is: they are basically intended to provide a graphical representation of the main ideas and concepts developed in the written part. Tables with numerical values of important physical properties are included as well, in the attempt to provide the reader with information useful to 'quantify' the physical results presented. Finally, a list of all the mathematical symbols used in the volume is given at the beginning as an orientation guide while reading.

Luciano Colombo
Cagliari, February 2021

Acknowledgements

I am really indebted with many friends who helped me by critically reading (in part or totally) the pre-editorial version of the book (for this special thanks go to Dr G Malloci and Professor C Melis, both at the Department of Physics, University of Cagliari, Italy, and to Dr S Giordano, CNRS and University of Lille, France) or by providing me with the output of their calculations used to plot some figures (Dr G Fugallo, CNRS and University of Nantes, France and Dr A Antidormi, ICN2, Barcelona, Spain) or for suggesting updated bibliographic entries. Finally, Professor C Melis is once again warmly thanked for assisting me in generating several figures reporting experimental data taken from literature. The care these friends have taken in checking my original manuscript has corrected several unclear passages and many misprints, thus greatly improving my presentation. If there are still errors or omissions they should be attributed solely to me.

Author biography

Luciano Colombo

Luciano Colombo received his doctoral degree in physics from the University of Pavia (I) in 1989 and then he was a post-doc at the École polytechnique fédérale de Lausanne (CH) and at the International School for Advanced Studies (I). He became assistant professor (tenured) at the University of Milano (I) in 1990, next moving to the University of Milano-Bicocca (I) in 1996 for an equivalent position. Since 2002 he has been full professor of theoretical condensed matter physics at the University of Cagliari (I) and since 2015 fellow of the 'Istituto Lombardo—Accademia di Scienze e Lettere' (Milano, I). He has been the principal investigator of several research projects addressed to solid-state and materials physics problems, the supervisor of more than 80 students (at bachelor, master, and PhD level), and the mentor of about 20 post-docs. He is the author, or coauthor, of more than 276 scientific articles and 9 books (this included). More about him can be found at: http://people.unica.it/lucianocolombo.

Symbols

α	optical absorption coefficient
α_T	isothermal compressibility
β	thermal expansion coefficient
γ	weighted Grüneisen parameter
$\gamma_{s\mathbf{q}}$	Grüneisen parameter for $(s\mathbf{q})$ phonon mode
$\Delta^2 R(t)$	atomic mean square displacement
ϵ_0	vacuum permittivity
ϵ_{ij}	ijth component of the second-rank strain tensor
$\epsilon_{\mathrm{r}}(\omega)$	frequency-dependent relative permittivity of a material
κ_{e}	thermal conductivity of electrons (generic)
$\kappa_{\mathrm{e}}^{\mathrm{Drude}}$	thermal conductivity of electrons, calculated according to the Drude theory
$\kappa_{\mathrm{e}}^{\mathrm{Sommerfeld}}$	thermal conductivity of electrons, calculated according to the Sommerfeld theory
κ_{l}	lattice thermal conductivity
κ_{tot}	total thermal conductivity
λ	first Lame' coefficient
λ_{e}	electron mean free path
$\lambda_{s\mathbf{q}}$	mean free path of the $(s\mathbf{q})$ phonon mode
μ	second Lame' coefficient
μ_{c}	chemical potential
μ_{e}	electron mobility
μ_{h}	hole mobility
μ_{B}	electron Bohr magneton
μ_{N}	nuclear magneton
μ_{x}	exciton effective mass
$\rho(\mathbf{r})$	electron density
$\rho_{\mathrm{GS}}(\mathbf{r})$	ground-state electron density
ρ^{HEG}	electron density in a homogeneous electron gas
σ_{e}	direct-current electron conductivity
σ_{h}	direct-current hole conductivity
σ_{tot}	direct-current total conductivity
τ_{e}	electron relaxation time
τ_{h}	hole relaxation time
$\tau_{s\mathbf{q}}$	relaxation time (lifetime) of the $(s\mathbf{q})$ phonon mode
$\varphi_{\mathrm{KS}}(\mathbf{r})$	Kohn–Sham orbital
$\Phi(\mathbf{r}, \mathbf{R})$	total crystalline wavefunction
Φ_{e}	flux of electrons
Φ_{h}	flux of holes
Φ_{ph}	flux of photons
$\Psi_{\mathrm{e}}^{(\mathbf{R})}(\mathbf{r})$	total electron wavefunction (when ions are clamped in the $\mathbf{R}$ configuration)
$\Psi_{\mathrm{n}}(\mathbf{r})$	total ion wavefunction
ζ	atomic valence
ω_{p}	plasma frequency
ω_{E}	Einstein frequency
ω_{D}	Debey frequency

a_{CB}	deformation potential for the conduction band
a_{H}	Bohr radius
a_{VB}	deformation potential for the valence band
a_0	lattice constant of cubic crystals
$\{\mathbf{a}_1, \mathbf{a}_2, \mathbf{a}_3\}$	translation vectors of the direct lattice
A	atomic mass number
$\{\mathbf{b}_1, \mathbf{b}_2, \mathbf{b}_3\}$	translation vectors of the reciprocal lattice
B	bulk modulus
$\mathbf{B}$	externally applied magnetic field
$\mathbf{B}_{\mathrm{d}}$	Burgers vector of a dislocation
c	speed of light
c_V^{e}	constant-volume specific heat (electron contribution)
c_V^{l}	constant-volume specific heat (lattice contribution)
c_V^{tot}	total constant-volume specific heat
C_{ij}	ijth component of the elastic stiffness tensor in the compact Voigt notation
C_{ijkh}	$ijkh$th component of the fourth rank elastic stiffness tensor
$C_V^{\mathrm{Drude}}(T)$	constant-volume heat capacity (Drude interpolation scheme)
$C_V^{\mathrm{Einstein}}(T)$	constant-volume heat capacity (Einstein model)
$C_V^{\mathrm{quantum}}(T)$	constant-volume heat capacity (full quantum expression)
$\mathbb{C}^{\mathrm{Voigt}}$	matrix representation of the elastic stiffness tensor in the compact Voigt notation
D_{e}	electron diffusivity
D_{h}	hole diffusivity
D_{ij}	ijth component of the elastic compliance tensor in the compact Voigt notation
D_{ijkh}	$ijkh$th component of the fourth rank elastic compliance tensor
e	electron charge (absolute value)
e_{cohesive}	cohesive energy per atom of a crystal
$\mathbf{E}$	externally applied electric field
E_{c}	energy at the bottom of the conduction band
$E_{\mathrm{c}}^{\mathrm{HEG}}$	correlation energy of a homogeneous electron gas
$E_{\mathrm{c}}^{\mathrm{LDA}}$	correlation energy in the local density approximation
E_{cohesive}	cohesive energy of a crystal
$E_{\mathrm{e}}^{(\mathbf{R})}$	total electron energy (when ions are clamped in the configuration $\mathbf{R}$)
E_{v}	energy at the top of the valence band
$E_{\mathrm{x}}^{\mathrm{HEG}}$	exchange energy of a homogeneous electron gas
$E_{\mathrm{x}}^{\mathrm{LDA}}$	exchange energy in the local density approximation
E_{F}	Fermi energy
$E_{\mathrm{CB}}(\mathbf{k})$	dispersion relation for the conduction band
$E_{\mathrm{VB}}(\mathbf{k})$	dispersion relation for the valence band
E_T	total energy of the crystal
E_{x}	exciton binding energy
$E_{\mathrm{x},n}$	exciton discrete energy spectrum
$E_{\mathrm{xc}}[\rho(\mathbf{r})]$	exchange-correlation energy functional
$E_{\mathrm{xc}}^{\mathrm{LDA}}$	exchange-correlation energy in the local density approximation
$F[\rho(\mathbf{r})]$	electron total energy functional

$g(r)$	pair correlation function
$g_n(E)$	electronic density of states per unit volume for the band n of a semiconductor
$\bar{g}_{CB}(E)$	$g_n(E)$ for the conduction band in parabolic bands approximation
$\bar{g}_{VB}(E)$	$g_n(E)$ for the valence band in parabolic bands approximation
$\mathbf{G}$	lattice vector (reciprocal lattice)
$\mathcal{G}$	Gibbs free energy
G	carrier generation rate
$G(E)$	electronic density of states (eDOS) of a metal
$G(\omega)$	vibrational density of states (vDOS) or phonon density of states
$G_n(E)$	electronic density of states for the band n of a semiconductor
$G_D(\omega)$	vibrational density of states (Drude interpolation scheme)
h	Planck constant
$\hbar$	$\hbar = h/2\pi$
$\mathcal{H}$	enthalpy
$\hat{i}$	unit vector of the x-axis
$\hat{j}$	unit vector of the y-axis
$\mathbf{J}_{CB}$	electron charge density current (conduction band)
$\mathbf{J}_{VB}$	hole charge density current (valence band)
$\mathbf{J}_{q,tot}$	total charge density current
$\mathbf{J}_h$	heat density current
$\hat{k}$	unit vector of the z-axis
$\mathbf{k}$	electron wavevector
k_B	Boltzmann constant
k_F	Fermi wavevector
$\mathbf{L}_d$	line vector of a dislocation
m_e	electron rest mass
m_e^*	electron effective mass
m_h^*	hole effective mass
m_n	neutron rest mass
m_p	proton rest mass
$\hat{n} = (n_1, n_2, n_3)$	unit vector of a generic direction in space
$\mathcal{N}_A$	Avogadro number
$n_{BE}(s\mathbf{q}, T)$	Bose–Einstein distribution for (population of) the $(s\mathbf{q})$ phonon mode
$n_{FD}(E, T)$	Fermi–Dirac distribution for (population of) the electron level with energy E
$\bar{n}(s\mathbf{q}, T)$	non-equilibrium population of the $(s\mathbf{q})$ phonon mode
N_c	effective density of states in conduction band
N_v	effective density of states in valence band
P	pressure
$\mathbf{q}$	phonon wavevector (or wavevector of a lattice vibrational wave)
$\mathbf{r}$	generic electron position within a crystal
r_x	electron–hole orbital radius (exciton size)
$\mathbf{R}$	generic ion position within a crystal
R	carrier recombination rate
$\mathbf{R}_b$	ion position within the basis
$\mathcal{R}_H$	Rydberg constant for hydrogen
$\mathbf{R}_l$	lattice vector (direct lattice)
S	entropy

T	temperature
T_{D}	Debye temperature
T_{F}	Fermi temperature
T_{ij}	ijth component of the second rank stress tensor
$\hat{T}_{\mathbf{R}_{\mathrm{l}}}$	translation operator such that $\hat{T}_{\mathbf{R}_{\mathrm{l}}} f(\mathbf{r}) = f(\mathbf{r} + \mathbf{R}_{\mathrm{l}})$
u	elastic energy density
$\mathbf{u}$	ionic displacement vector
$\mathcal{U}$	internal energy
$U(\mathbf{R})$	crystalline ion energy (for ions in configuration $\mathbf{R}$)
$\mathbf{v}_{\mathrm{d}}$	electron drift velocity in a weak field regime
$\mathbf{v}_{\mathrm{e}}^{\mathrm{th}}$	electron thermal velocity
$\mathbf{v}_{\mathrm{F}}$	electron Fermi velocity
$\mathbf{v}_{\mathrm{sat}}$	electron saturation velocity in a strong field regime
$\mathbf{v}_{\mathrm{CB}}(\mathbf{k})$	velocity of an electron in the conduction band
$\mathbf{v}_{\mathrm{VB}}(\mathbf{k})$	velocity of an electron in the valence band
V	volume (generic)
V_{c}	volume of the primitive unit cell (direct lattice)
V_{cfp}	crystal field potential
V_{ee}	electron–electron Coulomb interaction potential
V_{Hartree}	Hartree potential
V_{ne}	nucleus–electron Coulomb interaction potential
V_{nn}	nucleus–nucleus Coulomb interaction potential
V_{xc}	exchange-correlation potential
w	atomic weight
Z	atomic number

Part I

Preliminary concepts

IOP Publishing

Solid State Physics
A primer
Luciano Colombo

Chapter 1

The overall picture

Syllabus—After defining what condensed matter (and, in particular, the crystalline solid state) is, we build the theoretical framework of minimum complexity able to deal with the corresponding atomic-scale physics. Proceeding from a limited set of basic definitions, a hierarchy of approximations is introduced, like, e.g. the separation between the nuclear and the electronic degrees of freedom and the single-particle picture.

1.1 Basic definitions

The first task we must accomplish is defining the physical system we are actually interested in. This semantic exercise is quite important, since it will properly define the topic treated in this textbook: *the physics of crystalline solids.*

A fruitful constitutive hypothesis to start from is embodied by the *atomistic picture* [1, 2] which, relying on robust experimental evidence, states that *ordinary matter is made by elementary constituents known to be atoms*[1]. We agree that *condensed matter* forms whenever a *very large number of such atoms* (belonging to just one or more chemical species) *tightly bind together* by electrostatic interactions. Both features are indeed necessary in order to sharply define the state of aggregation we are interested in: (i) the fact that the number of atoms is very large allows us to exclude single molecules[2] from the horizon of our interest, while (ii) the strong character of their mutual interactions allows us to neglect the case of gaseous systems.

The definition just given is actually very generic and it does not allow us to distinguish between two paradigmatically different situations. In order to clarify and resolve this ambiguity, let us consider a sample of condensed matter and let us label

---

[1] By 'ordinary' we mean matter as typically organised in materials found on Earth.

[2] A single molecule may have as many as thousands of atoms, like e.g. RNA or DNA nucleic acids: this is still a comparatively smaller number than found in a material specimen with any dimension from the micro- to the macro-scale.

doi:10.1088/978-0-7503-2265-2ch1   1-1

by $\mathbf{R}_\alpha(t)$ the position of its αth atom at time t. We define the *mean square atomic displacement* $\Delta^2 R(t)$ as

$$\Delta^2 R(t) = \frac{1}{N} \sum_{\alpha=1}^{N} |\mathbf{R}_\alpha(t) - \mathbf{R}_\alpha(0)|^2, \tag{1.1}$$

where $\mathbf{R}_\alpha(0)$ represents the initial position of the αth atom and N is the total number of particles in the system. It is understood that the system is in equilibrium at temperature T. The calculation of $\Delta^2 R(t)$ for a silicon sample is reported in figure 1.1 at two different temperatures, respectively, above and below its melting temperature $T_m^{Si} = 1685$ K. If, according to atomic diffusion theory [3], we now link such a quantity to the corresponding diffusion coefficient $D(T)$

$$D(T) = \lim_{t \to +\infty} \frac{1}{6} \frac{\Delta^2 R(t)}{t}, \tag{1.2}$$

we immediately realise that below T_m^{Si} the sample does not show any self-diffusion characteristics[3] while above T_m^{Si} it flows. In other words, the definition of condensed matter given above allows *both solid and liquid systems* to be called *condensates*, despite their physics being largely different. Therefore, we make it clear that from now on we will focus our attention only on the *solid state*, i.e. only on *condensed matter systems that do not show any diffusive behaviour* (an introduction to the fascinating physics of liquids can be found elsewhere [4]).

The description just given of solid state is still inconveniently too general, since it is compatible with atomic arrangements showing quite unalike structural order

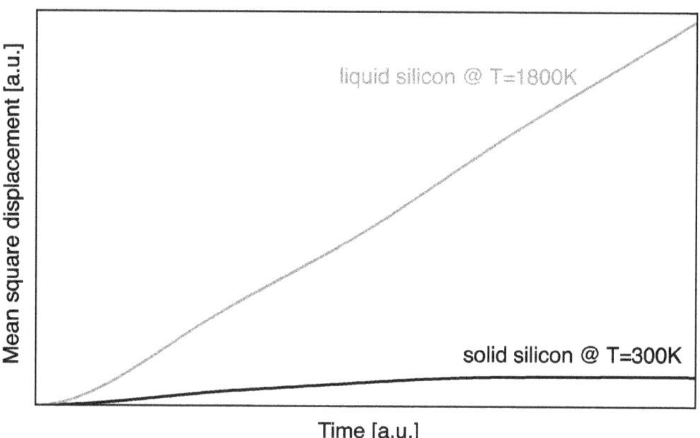

Figure 1.1. The mean square displacement of silicon at $T = 300$ K (black) and at $T = 1800$ K (red). Data are calculated by molecular dynamics simulations. Arbitrary units (a.u.) are used in both axes.

---

[3] The phenomenological Fick law of diffusion $\mathbf{J}(T) = -D(T)\nabla C$ links the amount of matter $\mathbf{J}$ flowing through a unit area in a unit time interval to the concentration gradient ∇C across that unit area. Accordingly, zero-diffusivity systems do not flow.

features: we must duly refine this notion. To this aim let us consider a monoatomic solid state system and let us introduce its *pair correlation function*[4] $g(r)$ which at a generic position $\mathbf{r}$ is defined as

$$g(r) = \frac{1}{N\rho} \left\langle \sum_{\alpha \neq \beta = 1}^{N} \delta(\mathbf{r} - \mathbf{R}_{\alpha\beta}) \right\rangle, \tag{1.3}$$

where $\rho = N/V$ is the average number density in the system and $\mathbf{R}_{\alpha\beta} = \mathbf{R}_{\beta} - \mathbf{R}_{\alpha}$ is the distance vector between atoms α and β. The $\langle \cdots \rangle$ brackets indicate that the sum is time- and direction-averaged by keeping constant the number of particles in the system as well as its volume and temperature, while the δ-symbol is defined usually as: it is worth one if its argument is null, otherwise it is worth zero.

Equation (1.3) describes how the number density varies within the system: more specifically, it provides the probability of finding a pair of (α, β) particles at a distance $r = R_{\beta} - R_{\alpha}$ apart, relative to the corresponding probability calculated for a uniform (random) distribution of similar atoms at the same density. In figure 1.2 we report the pair correlation function for two solid state silicon systems, both at room temperature. The peaks of the $g(r)$ function indicate *local densifications* of matter, while its minima indicate *local rarefactions*, in both cases referred to the

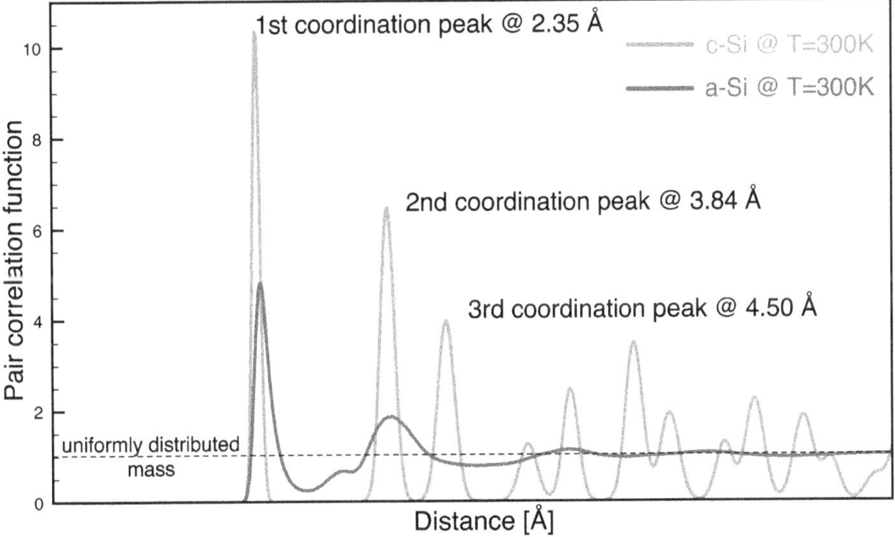

Figure 1.2. Pair correlation function for a crystalline (c-Si, red) and an amorphous (a-Si, blue) silicon sample with same density and both kept at room temperature. The actual position of the first-, second-, and third-coordination peak is reported in the case of c-Si. The dashed line represents the case of a uniformly distributed mass with the very same density. Data are calculated by molecular dynamics simulations.

---

[4] A thorough discussion of the pair correlation function is found elsewhere [3, 4].

normalised value $g(r) = 1$ corresponding, as anticipated above, to a uniformly distributed mass. These peaks *are precise markings of the discrete distribution of the atoms forming the solid*. For instance, the first peak reveals the position of the first coordination shell, i.e. the average distance at which the first nearest neighbours are found. Furthermore, because of the very definition of pair correlation function, we argue that the integral of each peak represents the *average number of neighbours found at that distance*.

It is apparent from figure 1.2 that there is a striking difference between the two samples, namely: while a persistent pattern is found in one case, no structure at all is observed beyond ~4.5 Å in the other one. We conclude that figure 1.2 allows us to distinguish between a discrete distribution of atoms characterised by a *regular repeated pattern* and a discrete distribution of atoms showing *no regularity above the shell of second-next-neighbours*. Solid state systems corresponding to the first situation are characterised by *long-range structural order* and are named *crystals*; on the other hand, solids showing just a *short-range structural order* are referred to as *amorphous systems*. We state that from now on we will focus our attention only on *solid state crystalline materials*.

In order to conclude the task of unambiguously defining the material systems to be studied, we still have to address a last important issue. It is trivial to observe that any object has a size, possibly even very large (especially when compared with typical interatomic spacings or even molecular sizes), but in any case finite. In short: *a material system has boundary surfaces*. This implies that the set of atoms forming any condensed matter system can be divided into two categories: *bulk atoms* and *surface atoms*. They feel a vastly different physics, as intuitively understood by considering that surface atoms (i) are under-coordinated with respect to bulk ones and that (ii) they are exposed to external perturbations (or, equivalently, they interact with the environment surrounding the system). A primary consequence is that *surfaces usually show a lower degree of crystallinity than the bulk regions*: this is an annoying issue, in view of the above comfortable definition of crystalline state. Furthermore, *a large number of phenomena only occur at surfaces*, like growth, catalysis, or chemical bond reconstruction to name just a few. Surface physics is, therefore, an industry with many specifics and it represents a very lively and rapidly expanding sub-discipline of modern condensed matter physics [5–9]; nevertheless, we will not deal with it because it largely falls beyond the scope of this primer. Rather, *we are primarily interested in bulk properties* and we will focus on them by invoking the following justification: by assuming a nearly constant number density ρ, the number of bulk and surface atoms, respectively, scales as $\sim\rho V$ and $\sim\rho V^{2/3}$ for a system of volume V. This implies that *for a 'large enough' specimen, the number of surface atoms becomes negligibly small as compared to bulk ones*.

Now, the point is: what does 'large enough' mean? There is no rigorous answer to such a seemingly innocent question. For instance, we could be tempted to provide a simple pragmatic reply: such a condition is reached whenever bulk-like properties become unaffected by surface phenomena. Unfortunately, this apparently reassuring definition is not accurate enough to deal with a number of important issues like, e.g. the proper definition of the work required to extract an electron from a solid[5] or the accurate description of polarisation effects in ionic crystals[6].

The conclusion is that the present introduction to solid state physics, basically focussed on the bulk properties of crystalline materials, is made conceptually clean only by introducing an abstraction, namely by *introducing suitable periodic boundary conditions* that allow for dealing with *ideal surface-free systems*[7]. The procedure is illustrated in figure 1.3 in the case of a one-dimensional crystalline solid[8]: a number N of atoms is considered so that they are arranged with the proper crystalline regularity (that is: atoms are equally spaced by an amount a hereafter referred to as the lattice spacing); next, we join the two terminal ends of the crystalline chain so that the first and last atom actually overlap, their positions coincide, and they are further treated as one particle. The formalisation of this procedure is

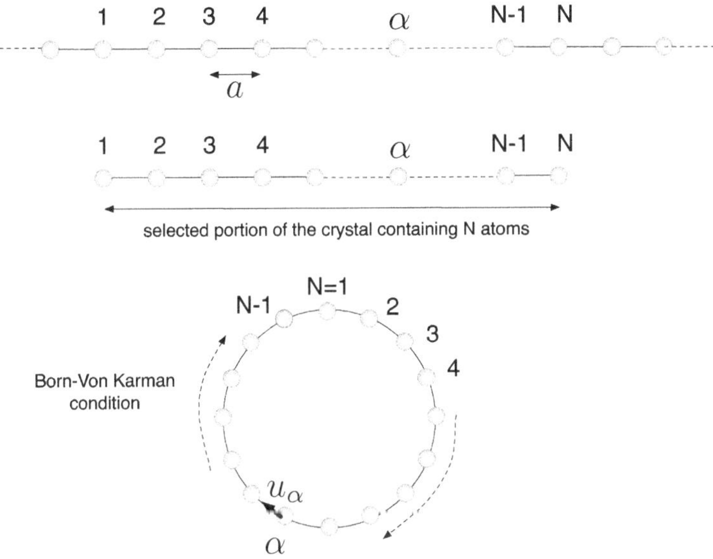

Figure 1.3. Graphical representation of the Born–von Karman periodic boundary condition for a one-dimensional crystalline solid with lattice spacing a.

---

[5] Such an energy does depend on some surface-specific features like the distortion of the electronic charge distribution with respect to the inner bulk.

[6] Any arbitrary truncation of a solid of such a kind generates a different charge distribution at the surface which, because of the long-range character of Coulomb interactions, differently affects the physics in the bulk region.

[7] We remark that selecting a solid state condensate with crystal order, among many other possible atomic gatherings, was just a matter of choice. On the other hand, the use of conditions that free us from the existence of something as real as a surface corresponds to *adopting a model*. Accordingly, the physics we will elaborate within its framework will be paradigmatically representative, conceptually relevant, and predictive under many concerns, but unavoidably incomplete.

[8] Of course, the choice of a one-dimensional system is made just for convenience of graphical representation: as a matter of fact, boundary conditions can be applied in one, two, or three dimensions, as needed.

straightforward: let us indicate by $u(R_\alpha)$ the displacement of the αth atom along the chain[9]. Imposing the above periodic boundary conditions is equivalent to placing

$$u(R_\alpha) = u(R_\alpha + (N-1)a) \qquad \text{and} \qquad \psi(r_i) = \psi(r_i + (N-1)a), \qquad (1.4)$$

where we have completed the picture by specifying as well how such conditions act on the wavefunction $\psi(r_i)$ of the ith crystalline electron[10]. We take this opportunity to specify that throughout this book the positions of the atoms and electrons will be indicated by upper and lower case letters, respectively, and labelled by Greek and Latin indices, respectively. Equation (1.4), known as the *Born–von Karman condition*, completes the reasoning aimed at removing surfaces from our problem. We finally remark that following this construction we effectively reduce the number of atoms of the system from N to $(N-1)$. However, provided that N is 'large enough' we can set $(N-1) \sim N$. It is easy to understand that the expression 'large enough' in this case is used in a less dense way of physical consequences than previously discussed: it is just an allowable mathematical convenience.

In conclusion, we have eventually set all terms unequivocally defining what 'a bulk property of a crystalline system' indeed means. After this semantic introduction we are ready to start with the true solid state physics.

1.2 Synopsis of atomic physics

In order to fully exploit the atomistic picture that we are going to use as the constitutive hypothesis to describe solid state materials, we need to recall some basic notion of atomic physics [1, 2, 10], define the notation adopted throughout the present volume, and set the value of some fundamental physical constants.

1.2.1 Atomic structure

We know that an atom is a *bound system* consisting of a *nucleus* with a positive charge $+N_p e$, where N_p is the *atomic number*, that is the number of *protons*, and a set of Z *electrons*, each carrying a charge $-e$. We recall that $e = 1.602\,19 \times 10^{-19}$ C is the elementary electric charge. If $Z = N_p$ then the atom is in a neutral configuration, while if $Z \neq N_p$ then we say that the atom has been ionised (either positively or negatively provided that Z is smaller or larger than N_p, respectively). The nucleus also contains a number N_n of *neutrons*, carrying no electric charge. While all electrons have the same mass $m_e = 9.109 \times 10^{-31}$ kg, the nucleus of each chemical species has instead a specific mass M determined as: $M = (N_p + N_n)m_p$, where $m_p = 1.672 \times 10^{-27}$ kg is the proton mass[11]. We remark that $A = N_p + N_n$ is referred

---

[9] The notion that atoms oscillate around their equilibrium position is here assumed on a purely intuitive basis, but it will be extensively discussed in the next chapters.

[10] We have anticipated an important twofold concept, namely that under some suitable approximations we can treat electrons as single particles and we can separate atomic motions from electron dynamics. The next section is devoted to this issue.

[11] For our purposes it is sufficient to set the same mass to both protons and neutrons.

to as the *atomic mass number*. Atoms with the same number of protons, but a different number of neutrons are referred to as *isotopes*.

As for nuclei, we will further neglect their inner structure by treating them as point-like, massive, and charged objects[12]. This is indeed a very good approximation for any situation described in this volume and, therefore, protons and neutrons will no longer enter as single objects in our theory. On the other hand, electrons will be individually addressed. Nuclei and electrons are inherently non-classical objects and, therefore, they must be duly described in quantum mechanical terms.

1.2.2 Angular and magnetic momenta

In addition to their charge and mass, electrons are further characterised by their *spin* [2, 10]: an *intrinsic angular momentum* $\mathbf{S}$, whose square modulus S^2 and z-component obey the following quantisation rules

$$S^2 = s(s + 1)\hbar^2$$
$$S_z = m_s\hbar,$$

(1.5)

with $s = 1/2$ and $m_s = \pm 1/2$ (spin 'up' or 'down') known as the spin quantum numbers and $\hbar = h/2\pi = 1.054\,46 \times 10^{-34}$ J s is the reduced Planck constant. An *intrinsic spin magnetic moment* $\mathbf{M}_S$, similarly quantised, is attributed to each electron according to

$$\mathbf{M}_S = -g_S \frac{\mu_B}{\hbar}\mathbf{S},$$

(1.6)

where $\mu_B = e\hbar/2m_e = 9.273\,2 \times 10^{-24}$ J T^{-1} is the Bohr magneton, and $g_S \sim 2$ is the spin g-factor.

Similarly, each nucleus, in addition to being charged, also carries a *magnetic moment* $\mathbf{M}_N$ [11] which for our purposes is conveniently defined as

$$\mathbf{M}_N = g_N \frac{\mu_N}{\hbar}\mathbf{N},$$

(1.7)

where g_N is the *nuclear g-factor* (a dimensionless constant), μ_N is the *nuclear magneton*

$$\mu_N = \frac{m_e}{m_p} \mu_B = 5.050\,82 \times 10^{-27}$ J T$^{-1},$$

(1.8)

and $\mathbf{N}$ is the *nuclear spin* or, equivalently, the total nuclear angular momentum.

Electrons are also characterised by an *orbital magnetic moment*, since their orbital motion around the nucleus corresponds to a current or, equivalently, to a magnetic moment $\mathbf{M}_L$ defined as [1, 2, 10]

---

[12] Actually, nuclei are neither really point-like nor is their charge spherically distributed. Because of this, nuclear electric quadrupole moment effects can be observed in high-resolution atomic spectroscopy measurements [10].

$$\mathbf{M}_L = -g_L \frac{\mu_B}{\hbar} \mathbf{L}, \tag{1.9}$$

where g_L is the *orbital g-factor* and $\mathbf{L}$ is the electron *orbital angular momentum* obeying the quantisation rules

$$L^2 = l(l + 1)\hbar^2$$
$$L_z = m_l\hbar, \tag{1.10}$$

cast in terms of the *orbital quantum number* $l = 0, 1, 2, \ldots$ and of the *magnetic quantum number* $m_l = 0, \pm1, \pm2, \ldots, \pm l$. The spectroscopic notation is widely adopted to label quantum states differing by l: we will set $l = 0 \rightarrow$ s-states, $l = 1 \rightarrow$ p-states, $l = 2 \rightarrow$ d-states, and so on [2].

1.2.3 Electronic configuration

The central problem of the physics of atoms is to determine their *ground-state configuration*, that is: the distribution of their electrons, among all available quantum states, corresponding to the minimum total energy. For a multi-electron atom this task is accomplished by following a rather complex procedure, qualitatively summarised below. A full account can be found elsewhere [1, 2, 10].

The first step consists in solving the complete Schrödinger equation[13] for the atom: a formidable indeed many-body quantum problem. The full scenario contains *electrostatic interactions* (among electrons and between the nucleus and each electron) as well as *magnetic interactions* (among all existing magnetic dipoles). Coulomb interactions are by far the strongest ones and they determine the main features of the atomic energy spectrum which can be calculated, for instance, within the central field approximation (CFA)[14]. Here each electron is treated as a *single-particle* undergoing an *average central field* due to the nucleus and the remaining electrons. In this way the many-body problem is reduced to Z single-particle ones, each separately solved by ordinary methods of atomic physics. The resulting CFA electron wavefunctions $\psi_{nlm_l}^{\mathrm{CFA}}(\mathbf{r}) = \bar{R}_{nl}(r)\, Y_{lm_l}(\theta, \phi)$ are written in polar coordinates (the central field has by construction a spherical symmetry!) as the product between a radial function $\bar{R}$ and a spherical harmonic function Y. Accordingly, each quantum state is labelled by three quantum numbers, namely: the *principal quantum number* $n = 1, 2, 3, \ldots$ and the l and m_l ones already introduced in section 1.2.2 where their values have been assigned[15]. A twofold picture emerges that (i) the *energy spectrum is discrete* and (ii) *allowed atomic quantum states are organised in*

---

[13] For brevity we will hereafter adopt a widely used notation abuse: what in the text has been called the 'Schrödinger equation' should be more rigorously indicated as the eigenvalue equation for the Hamiltonian operator describing the system of our interest. In short: time dependence has been already eliminated from our problem.

[14] Although some important many-body features are overlooked in this theory, it nevertheless contains all the essential features needed for the present discussion.

[15] We remark that the emerging organisation of quantum states is justified straightforwardly within the central field approximation, but it is also fruitfully adopted in more refined quantum approaches like the Hartree, the Hartree–Fock or the configuration-interaction theory.

shells and sub-shells, respectively, corresponding to a given value of the n and of $l = 0, 1, 2, \ldots, (n-1)$.

Once the energy spectrum has been determined by only considering Coulomb couplings, we can add magnetic interactions as twofold perturbations of electronic and nuclear nature. The semi-classical *vector model of the atom* provides a very direct and pedagogical treatment of magnetic interactions among electrons: each electron pair i, j undergoes the following interactions

- *orbit–orbit coupling*, corresponding to an energy $E_{oo} = \xi_{LL}\, \mathbf{L}_i \cdot \mathbf{L}_j$
- *spin–spin coupling*, corresponding to an energy $E_{ss} = \xi_{SS}\, \mathbf{S}_i \cdot \mathbf{S}_j$
- *spin–orbit coupling*, corresponding to an energy $E_{so} = \xi_{LS}\, \mathbf{L}_i \cdot \mathbf{S}_j$

where ξ_{LL}, ξ_{SS} and ξ_{LS} are the *magnetic coupling constants*. A major task of atomic physics is providing their accurate determination. While electrostatic interactions determine the main features of the atomic energy spectrum, magnetic interactions provide its *fine structure*. Next, we eventually consider the nucleus by observing that its dipole moment given in equation (1.7) feels both the magnetic field $\mathbf{B}_{\text{electrons}}$ generated by the orbital motion of the electrons and the dipole–dipole interactions between electron and nuclear spins. Accordingly, a new energy term E_{hp} comes into play

$$E_{\text{hp}} = -\mathbf{M}_{\text{N}} \cdot \mathbf{B}_{\text{electrons}}, \tag{1.11}$$

providing the ultimate *hyperfine structure* of the atomic energy spectrum.

The next step eventually provides the electronic configuration we are interested in. It basically consists in placing electrons on each allowed state, starting from the lowest-energy one, by fulfilling both the *Pauli principle* and the *Hund rules*. According to the Pauli principle, the maximum number of electrons that can be accommodated in a sub-shell with orbital quantum number l is $2(2l + 1)$ and, therefore, we can place up to $2n^2$ electrons in a shell with principal quantum number n. A sub-shell is said to be incomplete if the number of electrons there accommodated is smaller than $2(2l + 1)$. Following a standard picture, fully exploited in the periodic table of elements, we will refer to electrons belonging to the highest-energy and incomplete sub-shell as *valence electrons*, while electrons belonging to low-energy and complete sub-shells will be named *core electrons*.

1.3 Setting up the atomistic model for a solid state system

Trying to plug the full atomistic picture into condensed matter physics is a hopeless enterprise and an unreasonable choice as well: the resulting mathematical problem would be too complicated to be solved by any analytical or numerical tool and, furthermore, several details specific to the single atomic system are actually marginal when matter is organised in condensates. In order to proceed, *we need approxima-tions*. Far from being a fallback choice, this way of proceeding will allow us to bring out the most salient physical aspects of the solid state, avoiding an excess of detail that, in reality, would not translate into new meaningful knowledge. We are therefore going to develop a *hierarchy of approximations* that will actually constitute the backbone of our *working model* for crystalline solids. In the following chapters these approximations will be critically readdressed whenever some phenomenology questions their validity.

1.3.1 Semi-classical approximation

In general, we will treat electric, charge current, and magnetic effects according to the *classical Maxwell electromagnetism*; on the other hand, ion and electron physics will be described according to *quantum mechanics*. However, there will be some exceptions to this general choice.

First of all, we remark that the process of emission or absorption of electromagnetic energy by any material system will be described through the concept of *photon*. This approach represents the most basic way to include the quantum nature of electromagnetic radiation into our elementary theory. We will not go any further because any improvement of this picture, admittedly simplified, would fall beyond the scope of this tutorial introduction to solid state physics.

Finally, the dynamics of crystal lattices will be firstly treated by classical mechanics in order to easily catch the phenomenology of ionic vibrations. Next, a fully quantum picture will be developed through the concept of *phonon*.

1.3.2 Frozen-core approximation

To a large extent, the chemical properties of an atom are dictated by its valence electrons [2]. In particular, valence electrons rule over the formation mechanism of interatomic bonds, so ultimately affecting most of the physical properties in a condensed matter system. This suggests that *core electrons are expected to play a minor role in determining most of solid state properties*. We can exploit this observation by introducing the *frozen-core approximation* which will greatly simplify the picture. This approximation can be cast in a very simple form according to the scheme

$$\text{atom} = \underbrace{\text{nucleus} + \text{core electrons}}_{\text{ion}} + \text{valence electrons}$$
$$= \text{ion} + \text{valence electrons},$$

which suggests the following: we will implement the atomistic description of a crystalline solid assuming that it consists of a *collection of ions and valence electrons*. The former will be described as point-like objects with a nuclear mass specific to their chemical species[16] and carrying a positive charge. If there are in total $Z = Z_c + Z_v$ electrons (where Z_c and Z_v are the number of core and valence electrons, respectively), then the *ionic charge* Q will be assigned the value $Q = +Z_v$, in units of the elementary charge e.

The main advantage of the frozen-core approximation is a *dramatic reduction of the number of electronic degrees of freedom to deal with*: for instance, the main features of the electronic structure in a silicon crystal will be studied by considering

---

[16] The total mass of the core electrons is definitely negligible with respect to that of the bare nucleus.

just four valence electrons for each ion, instead of the full 14 electron set found in a silicon atom. In conclusion, hereafter *when referring to 'electrons' we will actually mean 'valence electrons'*, while core ones will never be addressed since they are attached to nuclei forming ions.

1.3.3 Non-magnetic and non-relativistic approximations

A detailed calculation of the three coupling constants ξ_{LL}, ξ_{SS} and ξ_{LS} providing the atomic fine structure leads to important information, namely: *magnetic interactions are comparatively much smaller than electrostatic ones* [10]. This is confirmed experimentally by observing that the splitting of the energy levels caused by the above coupling terms is always some good order-of-magnitude smaller than their separation predicted by only considering Coulomb interactions [2]. Therefore, our choice will be to *completely neglect magnetic interactions* among electrons. This choice is straightforwardly extended to the nucleus, following a simple argument. The hyperfine energy terms given in equation (1.11) depend on $\mathbf{M}_N$, that is, on the nuclear magneton. Now, it turns out that $\mu_N \ll \mu_B$ since the proton mass is much larger than the electron one and, therefore, *hyperfine interactions are also negligibly small if compared to Coulomb ones*. In summary, *no magnetic couplings of whatsoever origin will be in the first instance considered* when developing our solid state theory[17].

It is, however, important to remark the above *non-magnetic approximation does not imply that we are neglecting the true existence of the electron spin*: as a matter of fact, this cannot be done because we would lose the fundamental notion that electrons are fermions and, therefore, they obey (i) the Fermi–Dirac statistics (see section 6.2) and (ii) the Pauli principle. Rather, we reconcile the simplified non-magnetic treatment with the need to take into account the spin by multiplying the space part of the wavefunction by a spin function which is in charge of just characterising the actual spin state of the electron. The resulting total wavefunction will be labelled[18] by a fourfold set of quantum numbers $\{n, l, m_l, s\}$ so that, although no spin-related features are present in this simplified picture, it will not be possible for two electrons to have the same set of four quantum numbers: if they lie on the same energy level, then they must differ at least for their spin orientation.

Since spin is a relativistic feature [10, 12], neglecting any corresponding coupling term is tantamount to developing a theory in the *non-relativistic approximation*. We will take this in the widest meaning and, accordingly, we will overlook not only spin-mediated interactions but also any other relativistic effect[19]. In particular, we will use anywhere the *rest mass* m_e for electrons.

In conclusion, ions and electrons will interact only via Coulomb coupling: indeed a major simplification for our developing theory.

---

[17] Their effects are usually treated as a perturbation on the quantum energies predicted by only considering Coulomb interactions. This is, for instance, the case of the spin-splitting of the energy bands.

[18] This is strictly true only within the central field approximation.

[19] They correct the electron kinetic energy and the nuclear Coulomb potential energy [10].

1.3.4 Adiabatic approximation

It is now time to formulate the full quantum mechanical problem describing the physics of a solid; to this aim we are exploiting the set of approximations discussed in the previous sections and using the labelling notation for ions and electrons anticipated in section 1.1. In particular, the full set of electronic and ionic coordinates will be indicated by the shortcuts $\mathbf{r}$ and $\mathbf{R}$, respectively, while the *total crystalline wavefunction* will be written as $\Phi(\mathbf{r}, \mathbf{R})$.

The classical expressions for the Coulomb interactions at work within the crystal (under the frozen-core, non-magnetic, and non-relativistic approximations) are

$$V_{\mathrm{ne}}(\mathbf{r}, \mathbf{R}) = -\frac{e^2}{4\pi\epsilon_0} \sum_{i,\alpha} \frac{Q_\alpha}{|\,\mathbf{r}_i - \mathbf{R}_\alpha\,|}$$

$$V_{\mathrm{nn}}(\mathbf{r}, \mathbf{R}) = +\frac{e^2}{4\pi\epsilon_0} \sum_{\alpha>\beta} \frac{Q_\alpha Q_\beta}{|\,\mathbf{R}_\alpha - \mathbf{R}_\beta\,|} \qquad (1.12)$$

$$V_{\mathrm{ee}}(\mathbf{r}, \mathbf{R}) = +\frac{e^2}{4\pi\epsilon_0} \sum_{i>j} \frac{1}{|\,\mathbf{r}_i - \mathbf{r}_j\,|},$$

describing, respectively, the ion–electron, the ion–ion, and the electron–electron interactions. The corresponding Schrödinger equation is written as

$$\left[-\frac{\hbar^2}{2} \sum_\alpha \frac{1}{M_\alpha} \nabla_\alpha^2 - \frac{\hbar^2}{2m_{\mathrm{e}}} \sum_i \nabla_i^2 + \hat{V}_{\mathrm{ne}}(\mathbf{r}, \mathbf{R}) + \hat{V}_{\mathrm{nn}}(\mathbf{R}) + \hat{V}_{\mathrm{ee}}(\mathbf{r}) \right] \Phi(\mathbf{r}, \mathbf{R}) = E_T \Phi(\mathbf{r}, \mathbf{R}), \quad (1.13)$$

where E_T is the *total energy of the crystal* including any electronic or ionic contribution, while the symbols ∇_α^2 and ∇_i^2 imply derivatives with respect to the αth ion and ith electron coordinates, respectively. We remark that quantum operators will always be indicated with the same symbol used for their classical counterpart, but topped with the $\wedge$ hat symbol.

The quantum problem given in equation (1.13) is still unsolvable by any analytical or numerical tool, in spite of the many drastic approximations we have worked out to formulate it. This implies that in order to continue we need to further simplify our theory. Good for us, we can take profit from a very basic fact: ions are always much more massive than electrons and, therefore, their dynamics is expected to be much slower. In other words, *at any moment electrons feel ions as if they were stationary in their instantaneous positions*. The opposite view is equally valid: *at any moment electrons will be in their ground-state for the specific ionic configuration occurring at that time*. This conclusion is better understood by taking into account that the maximum ionic oscillation frequency is about 10^{13} s^{-1}, while the collective excitations of the electron system[20] have a typical 10^{16} s^{-1} frequency. Consequently, electrons almost immediately respond to a perturbation due to ionic motion,

---

[20] They are called *plasma oscillations* and will be studied in section 7.2.2.

re-adjusting in their new ground-state. In other words, *electrons instantaneously follow the atomic vibrations.* The separation between vibrational and electronic excitations is reflected in spectroscopic measurements: while emission/absorption of photons caused by the oscillations of the ions around their equilibrium position typically occurs in the infrared region, transitions occurring between electronic states fall in the visible and ultraviolet regions of the electromagnetic spectrum. In short: *the physics of electrons and nuclei unfolds on quite different energy scales.*

We exploit these phenomenological arguments by formally *decoupling electronic and vibrational degrees of freedom* or, equivalently, by separating the $\mathbf{r}$- and the $\mathbf{R}$-dependence of the total wavefunction Φ as

$$\Phi(\mathbf{r}, \mathbf{R}) = \Psi_n(\mathbf{R}) \, \Psi_e^{(\mathbf{R})}(\mathbf{r}), \tag{1.14}$$

where $\Psi_n(\mathbf{R})$ and $\Psi_e^{(\mathbf{R})}(\mathbf{r})$ are the *total ionic wavefunction* and the *total electronic wavefunction*, respectively. More specifically, we state that $\Psi_e^{(\mathbf{R})}(\mathbf{r})$ describes the crystalline electronic structure corresponding to the clamped-ion configuration $\mathbf{R}$ and, therefore, it must be understood that Ψ_e has an analytical dependence of the $\mathbf{r}$-coordinate set, while it has a parametric dependence on the $\mathbf{R}$-coordinate set. Equation (1.14) is usually referred to as the *adiabatic approximation* because *it implies that there is no exchange of energy between the system of electrons and the ion lattice.* Although we will also deal with non-adiabatic phenomena in the following chapters, let us for the moment proceed below this approximation.

By inserting equation (1.14) into equation (1.13) we straightforwardly obtain the following *eigenvalue problem for the total electron wavefunction*

$$\left[-\frac{\hbar^2}{2m_e} \sum_i \nabla_i^2 + \hat{V}_{ne}(\mathbf{r}, \mathbf{R}) + \hat{V}_{ee}(\mathbf{r}) \right] \Psi_e^{(\mathbf{R})}(\mathbf{r}) = E_e^{(\mathbf{R})} \Psi_e^{(\mathbf{R})}(\mathbf{r}), \tag{1.15}$$

where $E_e^{(\mathbf{R})}$ has the meaning of *total electron energy when ions are clamped in the configuration* $\mathbf{R}$. This is an important achievement: setting the problem as shown in equation (1.15) implies that we need to separately calculate $E_e^{(\mathbf{R})}$ for each possible ion displacement pattern. By combining equations (1.13) and (1.15), the corresponding *eigenvalue problem for the total ion wavefunction* is immediately obtained as

$$\left[-\frac{\hbar^2}{2} \sum_\alpha \frac{1}{M_\alpha} \nabla_\alpha^2 + \hat{V}_{nn}(\mathbf{R}) + E_e^{(\mathbf{R})} \right] \Psi_n(\mathbf{R}) = E_T \Psi_n(\mathbf{R}), \tag{1.16}$$

where in this case the difference $E_T - E_e^{(\mathbf{R})} = U(\mathbf{R})$ must be understood as the crystalline ion energy. In practice, the main conceptual consequence of the adiabatic approximation is that *solid state physics is traced back to the separate solution of equations (1.15) and (1.16);* any non-adiabatic effect will be treated as a perturbation to their solutions.

Although we have based the adiabatic approximation on very robust phenomenological evidence, we still need to rigorously establish its accuracy. This task is accomplished by a non trivial argument that we are developing in a simplified form, just focussing on its main conceptual steps[21]. The $\Psi_{\mathrm{e}}^{(\mathbf{R})}(\mathbf{r})$ solutions of equation (1.15) represent a complete set and, with no loss of generality, we can further assume that they have been properly orthonormalised. Therefore, we can express the total wavefunctions Φ as the following linear combination[22]

$$\Phi(\mathbf{r}, \mathbf{R}) = \sum_m \Xi_m(\mathbf{R}) \, \Psi_{\mathrm{e}, m}^{(\mathbf{R})}(\mathbf{r}), \tag{1.17}$$

where $\Xi_m(\mathbf{R})$ are the expansion coefficients and in the present context the label m is a shortcut for the full set of quantum numbers describing the states of the electron system, found by solving equation (1.15). By inserting this representation into equation (1.13) and using the orthonormality of the $\Psi_{\mathrm{e}}^{(\mathbf{R})}(\mathbf{r})$ wavefunctions, after some algebra we obtain the eigenvalue equation for the $\Xi_m(\mathbf{R})$'s

$$\left[-\frac{\hbar^2}{2} \sum_\alpha \frac{1}{M_\alpha} \nabla_\alpha^2 + \hat{V}_{\mathrm{nn}}(\mathbf{R}) + E_{\mathrm{e}}^{(\mathbf{R})} + \sum_m (\hat{A}_{mm'} + \hat{B}_{mm'}) \right] \Xi_m(\mathbf{R}) = E_T \, \Xi_m(\mathbf{R}), \tag{1.18}$$

where[23]

$$\hat{A}_{mm'} = -\hbar^2 \sum_\alpha \frac{1}{M_\alpha} \left[\int d\mathbf{r} \, \Psi_{\mathrm{e}, m}^{(\mathbf{R})*}(\mathbf{r}) \nabla_\alpha \Psi_{\mathrm{e}, m'}^{(\mathbf{R})}(\mathbf{r})) \right] \cdot \nabla_\alpha$$

$$\hat{B}_{mm'} = -\frac{\hbar^2}{2} \sum_\alpha \frac{1}{M_\alpha} \int d\mathbf{r} \, \Psi_{\mathrm{e}, m}^{(\mathbf{R})*}(\mathbf{r}) \nabla_\alpha^2 \Psi_{\mathrm{e}, m'}^{(\mathbf{R})}(\mathbf{r}). \tag{1.19}$$

We easily recognise that the adiabatic approximation is formally equivalent to dropping off the terms $\sum_m (\hat{A}_{mm'} + \hat{B}_{mm'}) \Xi_n(\mathbf{R})$ from equation (1.18): under this condition, the $\Xi_m(\mathbf{R})$ functions are nothing other than the total ion wavefunctions $\Psi_n(\mathbf{R})$ calculated when the electronic contribution to the total energy is provided by the mth state. Now, the point is to assess how really accurate is the choice of neglecting these terms in our theory.

To this aim we write the position of the αth atom in the form $\mathbf{R}_\alpha = \mathbf{R}_\alpha^{(0)} + \kappa \, \mathbf{u}_\alpha$, where $\mathbf{R}_\alpha^{(0)}$ is its equilibrium position, $\mathbf{u}_\alpha$ its displacement, and κ a suitable expansion parameter which allows us to write the total crystalline Hamiltonian appearing in equation (1.13) in the form

$$\begin{aligned} \hat{H} &= -\frac{\hbar^2}{2} \sum_\alpha \frac{1}{M_\alpha} \nabla_\alpha^2 - \frac{\hbar^2}{2m_{\mathrm{e}}} \sum_i \nabla_i^2 + \hat{V}_{\mathrm{ne}} + \hat{V}_{\mathrm{nn}} + \hat{V}_{\mathrm{ee}} \\ &= -\frac{\hbar^2}{2} \sum_\alpha \frac{1}{M_\alpha} \nabla_\alpha^2 - \frac{\hbar^2}{2m_{\mathrm{e}}} \sum_i \nabla_i^2 + \hat{V}_{\mathrm{ee}} + \hat{V}_{\mathrm{ne}}^{(0)} + \hat{V}_{\mathrm{nn}}^{(0)} \\ &\quad + [\kappa \hat{V}_{\mathrm{ne}}^{(1)} + \kappa^2 \hat{V}_{\mathrm{ne}}^{(2)} + \kappa^3 \hat{V}_{\mathrm{ne}}^{(3)} + \cdots] + [\kappa \hat{V}_{\mathrm{nn}}^{(1)} + \kappa^2 \hat{V}_{\mathrm{nn}}^{(2)} + \kappa^3 \hat{V}_{\mathrm{nn}}^{(3)} + \cdots], \end{aligned} \tag{1.20}$$

---

[21] The complete formal development is found in [13, 14].

[22] In general, the eigenfunctions of an Hermitean operator represent a complete set and, therefore, they can be used to expand any other wavefunction defined in the same volume of space and obeying the same boundary conditions as a linear combination of them.

[23] The asterisk * indicates the complex conjugate function.

where $\hat{V}_{\text{ne}}^{(0)}$ and $\hat{V}_{\text{nn}}^{(0)}$ are the nucleus–electron and the nucleus–nucleus terms calculated for ions at their equilibrium positions, respectively, while any term written as $\hat{V}^{(\lambda)}$ is λth order in the ionic displacements $\mathbf{U}_\alpha$ (the derivatives appearing in these terms are evaluated with all ions in their equilibrium positions). Equation (1.20) is known as the *Born–Oppenheimer expansion*. In the limiting case of $M_\alpha \to +\infty$ the total Hamiltonian operator reduces to

$$\lim_{M_\alpha \to 0} \hat{H} = -\frac{\hbar^2}{2m_{\text{e}}} \sum_i \nabla_i^2 + \hat{V}_{\text{ee}} + \hat{V}_{\text{ne}}^{(0)} + \hat{V}_{\text{nn}}^{(0)}, \tag{1.21}$$

indicating that $\kappa \to 0$ as well. Since such an expansion parameter is adimensional, this conclusion suggests that we can set $\kappa = (m_{\text{e}}/\langle M \rangle)^\zeta$ where $\langle M \rangle$ is the average ion mass and $\zeta > 0$ a real number[24].

By using the Born–Oppenheimer expansion for the $\hat{V}_{\text{ne}}$ term, it is possible to get a solution of equation (1.15) through a standard perturbative quantum mechanical calculation: this provides the $\Psi_{\text{e}}^{(\mathbf{R})}(\mathbf{r})$ functions. Next, by repeating a similar expansion procedure to equation (1.18), the $\hat{A}_{mm'}$ and $\hat{B}_{mm'}$ operators are, after a non trivial calculation, eventually found to be $\hat{A}_{mm'} \sim \mathcal{O}(\kappa^3)$ and $\hat{B}_{mm'} \sim \mathcal{O}(\kappa^4)$. In conclusion, *non-adiabatic terms appear at high order in the expansion parameter κ.* Since for any possible real crystal we have $m_{\text{e}}/\langle M \rangle \ll 1$, the conclusion is twofold: (i) the Born–Oppenheimer expansion is quickly convergent and (ii) dropping off the terms $\sum_m (\hat{A}_{mm'} + \hat{B}_{mm'}) \, \Xi_m(\mathbf{R})$ from equation (1.18) is really a more than acceptable approximation.

It is worth remarking that the formal argument we outlined above is centred on the expansion parameter $\kappa = (m_{\text{e}}/\langle M \rangle)^\zeta$ and, therefore, it is fully consistent with the phenomenological argument used to set the discussion on the adiabatic approximation: *the very large difference in dynamical inertia between electrons and ions supports a theoretical approach where their degrees of freedom are separately taken into account.*

1.4 Mastering many-body features

Equations (1.15) and (1.16) are the *constitutive equations of solid state theory*. Although they have been derived under many (and sometimes strong) approximations, they both are a formidable many-body quantum problem complicated to the extent that the present theoretical and computational knowledge does not allow for their exact solution. In order to overcome this impasse two rather different strategies will be adopted, as presented below.

---

[24] Its is proved [13, 14] that by setting $\zeta = 1/4$, the total Hamiltonian expanded up to the second order corresponds to the so-called harmonic approximation. While we will not provide the formal proof of this, we will make an extensive use of this approximation when studying the lattice dynamics, although we will formulate it in a much more phenomenological way.

1.4.1 Managing the electron problem: single-particle approximation

As far as the electronic problem given in equation (1.15) is concerned, we choose to adopt another very drastic approximation, namely: we will handle the electron problem within the *single-particle approximation.*

In principle, by exploiting a key result of atomic physics [2, 10], we could represent the total electron wavefunction $\Psi_e^{(R)}(r)$ as an anti-symmetrised product of one-electron wavefunctions. In this context, each electron is described by a *spin-orbital* function given by the product between a space- and a spin-wavefunction (this latter is added to correct the limitations of the non-relativistic approximation and thus enforcing the Pauli principle). Next, the best set of spin-orbitals is obtained by a quantum mechanical variational procedure, where the total energy of the ground-state of the crystalline system we are interested in is usually minimised. This leads to rather complicated integro-differential equations, known as Hartree–Fock (HF) equations [15], which are typically solved through a self-consistent procedure implemented by a numerical calculation. We remark that the HF ground-state energy is an excess approximation to the true value since it is obtained through a variational principle. Furthermore, we remark that, while the HF method correctly accounts for exchange effects (which are a truly many-body attribute), it is unable to deal with quantum correlation features. The HF method [16] has been widely used since it allows for qualitatively understanding many solid state physics problems. Nevertheless, its deficiencies in fully catching all the many-body features and the fact that a single anti-symmetrised product of spin-orbitals is in general found to be a poor representation of the true $\Psi_e^{(R)}(r)$ many-particle wavefunction have motivated the development of more sophisticated many-body theories which are overviewed in [15, 17] and fully developed in [18, 19].

A more accessible approach consists in adopting the *single-particle approxima-tion*[25]: the full set of electron–ion and electron–electron interactions are represented by *an effective one-electron potential with the same periodicity of the underlying crystal lattice*, hereafter referred to as '*crystal field*'. In other words, we will assume that *each electron independently moves under the action of a local crystal field potential* describing its embedding into a crystalline environment made by all ions and by all the remaining electrons. We will indicate such a crystal field potential (cfp) as $V_{cfp}(r)$.

The first feature of $V_{cfp}(r)$ to account for is its periodicity. At this stage we have not yet developed the mathematical background to rigorously treat the geometrical characteristics of a crystal. We can nevertheless justify our assumption of periodic potential by looking at the simple graphical rendering of figure 1.4: by embedding an electron into a crystal, it will experience a net potential closely resembling that of an isolated bare ion when closely approaching it; on the other hand, the net potential will flatten off in the interstitial regions. Next, we remark that the crystal field potential is local: its value only depends on the instantaneous position actually

---

[25] This picture is equivalently referred to as *independent electron approximation* or *one-electron approximation.*

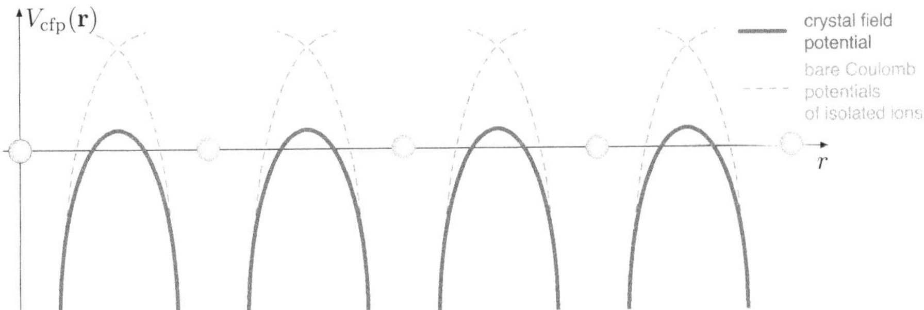

Figure 1.4. Graphical representation of the potential acting on an electron in the case of a one-dimension solid. Full line (blue): actual crystal field potential; dashed line (red): superposition of bare Coulomb potentials due to isolated ions.

occupied by the single electron. This is another very convenient practical advantage offered by the one-electron approximation with respect to many-body theories.

The resulting formalism is clean and conceptually simple: we are reduced to solving the one-electron Schrödinger equation for each ith electron

$$\left[-\frac{\hbar^2}{2m_{\mathrm{e}}}\nabla^2 + \hat{V}_{\mathrm{cfp}}(\mathbf{r}_i) \right]\psi_{m_i}(\mathbf{r}_i) = E_{m_i}^{(\mathbf{R})}\psi_{m_i}(\mathbf{r}_i), \tag{1.22}$$

where $\psi_{m_i}(\mathbf{r}_i)$ and $E_{m_i}^{(\mathbf{R})}$ are its wavefunction and energy, while m_i stands for a suitable set of quantum numbers describing the crystalline quantum state of the ith electron. The adopted labelling once again makes it clear that the equation is solved for ions clamped in the $\mathbf{R}$ configuration. The total electron energy approximating the eigenvalue of equation (1.15) is calculated as

$$E_{\mathrm{e}}^{(\mathbf{R})} = \sum_i E_{m_i}^{(\mathbf{R})}. \tag{1.23}$$

It is manifest that equation (1.22) greatly simplifies the actual many-body problem describing the electron system. However, we remark that *we are not treating electrons as non-interacting particles*. Rather, we are assuming that a suitable choice of $V_{\mathrm{cfp}}(\mathbf{r})$ will allow us to include in this simplified picture the most relevant features of the electron–electron interactions.

Before even addressing the resulting description of the crystalline electronic structure, we need to clarify, at least conceptually, how $V_{\mathrm{cfp}}(\mathbf{r}_i)$ can be defined. The most effective procedure is implemented by the *self-consistent-field method*, which is nowadays the most widely used approach in modern solid state physics since it naturally translates into efficient computational algorithms allowing for its numerical determination. Let us assume that a clever initial guess for the single-particle wavefunctions $\psi_{m_i}^{(0)}(\mathbf{r}_i)$ is possible. For instance, these functions could be atomic orbitals, superpositions of plane waves, or any other orthogonalised set of suitable wavefunctions. Then, the zero-order approximation of the electron–electron Coulomb potential energy for the ith electron can be written as

$$V_{\text{ee}}^{(0)}(\mathbf{r}_i) = \frac{e^2}{4\pi\epsilon_0} \sum_{j \neq i} \int \left[\psi_{m_j}^{(0)}(\mathbf{r}_j) \right]^* \frac{1}{\left| \mathbf{r}_j - \mathbf{r}_i \right|} \, \psi_{m_j}^{(0)}(\mathbf{r}_j) \, d\mathbf{r}_j, \tag{1.24}$$

so that the classical total potential energy on such a particle is

$$V_{\text{cfp}}^{(0)}(\mathbf{r}_i) = V_{\text{ee}}^{(0)}(\mathbf{r}_i) + V_{\text{ne}}(\mathbf{r}_i, \mathbf{R}). \tag{1.25}$$

By inserting the corresponding operator $\hat{V}_{\text{cfp}}^{(0)}(\mathbf{r}_i)$ into equation (1.22) and solving it, we get a new set of wavefunctions $\psi_{m_i}^{(1)}(\mathbf{r}_i)$: they allow us to refine our initial guess on electron–electron Coulomb potential energy, now calculated as

$$V_{\text{ee}}^{(1)}(\mathbf{r}_i) = \frac{e^2}{4\pi\epsilon_0} \sum_{j \neq i} \int \left[\psi_{m_j}^{(1)}(\mathbf{r}_j) \right]^* \frac{1}{\left| \mathbf{r}_j - \mathbf{r}_i \right|} \, \psi_{m_j}^{(1)}(\mathbf{r}_j) \, d\mathbf{r}_j. \tag{1.26}$$

By repeating the previous calculation, we solve again equation (1.22) for the new estimate of the potential energy term. This will return a set of second-order eigenfunctions $\psi_{m_i}^{(2)}(\mathbf{r}_i)$ which we can further use to generate the second-order approximation to the potential energy, and so on. The procedure is iterated until the $(n - 1)$ th order wavefunctions are found to differ from the nth order ones less than an agreed degree of numerical accuracy. Full convergence is at this stage proclaimed: we have both the 'true' effective crystal field potential and one-electron wavefunctions.

In the next chapters the electronic structure of crystalline solids will be studied under assumption that the one-electron local crystal field potential is known: this will lead to the so-called electronic band theory. A rather different approach will be outlined as well, named *density functional theory* (DFT). In this case, the key quantities of the theory are the density (instead of wavefunctions) and the total energy of the electron system. DFT is nowadays considered the 'standard model' for the ab *initio* prediction of the physical properties of condensed matter systems.

1.4.2 Managing the ion problem: classical approximation

Let us now turn to considering equation (1.16). Even in this case the quantum mechanical problem is inherently many-body and, therefore, quite difficult to solve. However, we can adopt a simplification procedure based on the same arguments put forward in support of the adiabatic approximation: ions are comparatively very massive objects and, accordingly, *we can in first approximation treat them classically.*

In practice, we will at first elaborate some suitable classical model for the force field describing the interactions among ions. Next, we will investigate the ionic displacements by solving their Newton equations of motion. Any relevant quantum feature will be eventually added *ex post*, by a quantisation procedure of the classical displacement field. As we will see, this will replace the classical description based on lattice waves with the notion of quantum lattice vibrational field, whose quanta will be referred to as *phonons*.

References

[1] Demtröder W 2010 *Atoms, Molecules and Photons* (Berlin: Springer)
[2] Colombo L 2019 *Atomic and Molecular Physics: A Primer* (Bristol, UK: IOP Publishing)
[3] Haile J M 1997 *Molecular Dynamics Simulations: Elementary Methods* (New York: Wiley)
[4] Hansen J-P and McDonald I R 2006 *Theory of Simple Liquids* 3rd edn (Oxford: Academic)
[5] Prutton M 1983 *Surface Physics* (Oxford: Oxford University Press)
[6] Zangwill A 1988 *Physics at Surfaces* (Cambridge: Cambridge University Press)
[7] Desjonqueres M-C and Spanjaard D 1996 *Concepts in Surface Physics* (Berlin: Springer)
[8] Bechstedt F 2003 *Principles of Surface Physics* (Berlin: Springer)
[9] Luth H 2014 *Solid Surfaces, Interfaces and Thin Films* 6th edn (Berlin: Springer)
[10] Bransden B H and Joachain C J 1983 *Physics of Atoms and Molecules* (Harlow: Addison-Wesley)
[11] Povh B, Rith K, Scholz C and Zetsche F 2009 *Particles and Nuclei* 6th edn (Berlin: Springer)
[12] Greiner W 1990 *Relativistic Quantum Mechanics* (Berlin: Springer)
[13] Born M and Huang K 1954 *Dynamical Theory of Crystal Lattices* (Oxford: Oxford University Press)
[14] Böttger H 1983 *Principles of the Theory of Lattice Dynamics* (Berlin: Akademie)
[15] Grosso G and Pastori Parravicini G 2014 *Solid State Physics* 2nd edn (Oxford: Academic)
[16] Pisani C, Dovesi R and Roetti C 1988 *Hartree–Fock Ab Initio Treatment of Crystalline Systems* (New York: Springer)
[17] Ashcroft N W and Mermin N D 1976 *Solid State Physics* (London: Holt-Saunders)
[18] Inkson C J 1984 *Many-body Theory of Solids–An Introduction* (New York: Springer)
[19] Bruus H and Flensberg K 2004 *Many-body Quantum Theory in Condensed Matter Physics: An Introduction* (Oxford: Oxford University Press)

IOP Publishing

Solid State Physics
A primer
Luciano Colombo

Chapter 2

The crystalline atomic architecture

Syllabus—*The basic notions of crystallography are here outlined, setting the mathematical background to treat the physics of translationally invariant solid state systems. The concept of direct lattice is developed, together with a number of formal tools useful to describe the periodicity and symmetry of a crystalline array of atoms. By discussing the main features of x-rays scattering, we further introduce the concept of reciprocal lattice; then, we formally prove the Bloch theorem setting the general mathematical form of any one-electron crystalline wavefunction. Thermodynamics is then invoked to prove that the concept of perfect crystal is just an idealisation: solids do contain both point- and extended-defects, whose classification and characteristics are outlined. A qualitative classification of crystalline solids and a rudimental model for their cohesive energy are eventually discussed.*

2.1 Translational invariance, symmetry, and defects

In section 1.1 we have qualitatively defined a *crystal as a periodic arrangement of atoms*, displaying long-range order. We have developed this notion in contrast to the case of an amorphous solid, where just short-range coordination is observed, while no regular repeated pattern is found beyond the shell of first- or second-nearest neighbours. In addition, in disordered solids even the local environment surrounding each atom is different from place to place. This definition of crystal is correct, but definitely insufficient to provide us with all the mathematical tools needed to characterise the *translational invariance* and the *symmetry* imposed by the periodicity. The discipline in charge of elaborating these tools is called *crystallography* [1–3]; it makes use of the mathematical methods of the group theory which can be developed at a tutorial [4, 5] or advanced level [6, 7]. In this formal environment it is possible to characterise any possible existing crystal structure, unambiguously defining its periodicity and symmetry properties which, in turn, can be fruitfully used to describe any *physical property* of the crystal. Only the most simple crystal structures and their basic physics will be here addressed and, therefore, we will not

doi:10.1088/978-0-7503-2265-2ch2   2-1

need the full theoretical machinery of group theory. As a matter of fact, we will make use of just the basic notions of crystallography, such as the concept of *direct* and *reciprocal lattice*, together with their most important geometrical characteristics. These are the topics of sections 2.2 and 2.4, respectively, while in section 2.3 a detailed classification of the crystals structures is reported.

In defining the concept of lattice we will explicitly resort to the *idealisation of perfect crystal*, namely: we will assume that *the periodic atomic architecture forming a crystalline solid is completely defect-free*. While useful under many circumstances and physically sound in a large number of prototypical cases, this picture does not faithfully correspond to the real situation: *thermodynamics dictates that at any finite temperature a crystal lattice must contain defects*. Therefore, in section 2.5 we will describe the most common lattice defects, stressing the difference between point and extended ones. The chapter is eventually concluded by a phenomenological classification of solids (Section 2.6) and a tutorial introduction to cohesive theory (section 2.7).

2.2 The direct lattice

2.2.1 Basic definitions

The existence of a discrete regular distribution of matter within a crystal suggests that its organisation consists in a *space periodic repetition of identical structural units*, each of which may contain one or more atoms[1]. In this latter case, the fundamental structural unit could be formed by atoms belonging to the same or to different chemical species: we will refer to *elemental* or *compound* systems in the two cases, respectively.

By some formal rules discussed below we can distinguish between the *geometry* and the *structure* of a crystal. Its geometry is described by introducing a suitable *discrete grid of points in the space* fulfilling the fundamental requirement of *translational invariance*. In doing that, we have in general many different options and, in particular, we could define the grid such that all its points are equivalent in any respect to the origin of the frame of reference set to define positions. This requirement imposes that *the arrangement and the orientation of the points on the grid must appear the same from whichever site is selected to look at their distribution*.

In general, not all atoms of a crystal occupy a position with these characteristics, as illustrated in figure 2.1 for graphene[2]: a single atomic plane of carbon atoms placed at the corners of regular hexagons. It is clear from figure 2.1(left) that atoms labelled by capital letters A, B, C, ⋯ lie at positions fully equivalent (under both the arrangement and orientation criteria) to the position of the origin O, while atoms labelled by lower letters a, b, c, ⋯ do not have this property. The grid of points A, B, C, ⋯ shown in figure 2.1(right) is named the *lattice* of graphene.

---

[1] Or molecules in the case of molecular crystals [8, 9] which, however, we will never consider.

[2] It will be sometimes convenient to illustrate new key concepts by discussing the case of two-dimensional lattices, where pictures are more easy to understand. Conclusions will be straightforwardly generalised in three dimensions.

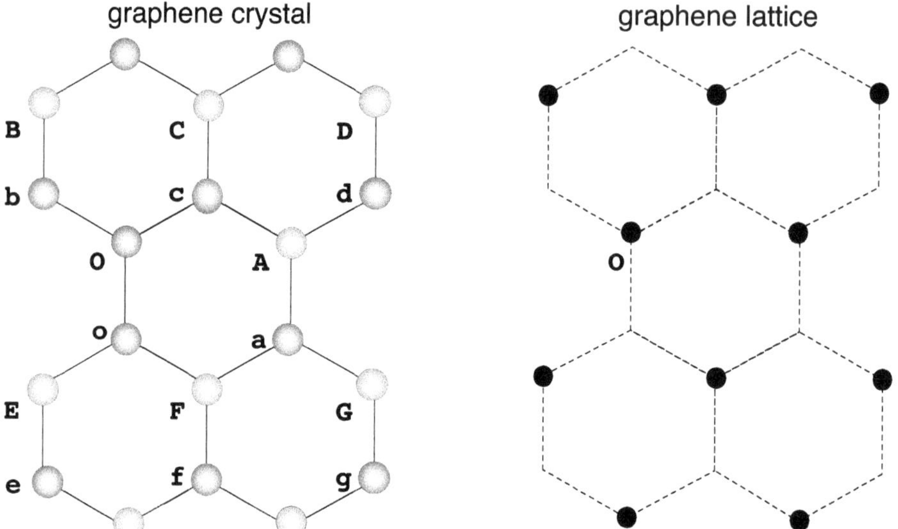

Figure 2.1. The graphene crystal (left) and its two-dimensional lattice (right). The red (blue) atoms are equivalent (non-equivalent) to that at the origin (grey).

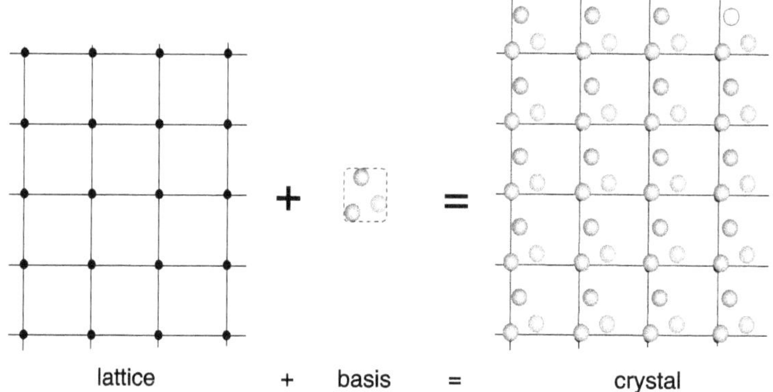

Figure 2.2. Generating a two-dimensional crystal by combining a square lattice with a three-atom basis.

Once a lattice is defined, we can decorate it by *associating each of its points with a set of atoms*, possibly differing in chemical nature. This set is referred to as the crystal *basis*. The procedure is shown in figure 2.2 in the hypothetical case of a two-dimensional square lattice with a basis formed by three unlike atoms, while in the previously discussed case of graphene, the basis consists in just two atoms like the pairs (O,c), (A,d), (E,o), and so on. This picture should make it clear that *in general the geometry and the structure of the crystal are not the same.*

In summary, the lattice and the basis carry different, but equally important, information: geometrical and chemical, respectively. Therefore, *a crystal can be defined as the sum of a lattice and a basis.* This synthesis not only provides us with an operational definition of 'crystal', but it also defines the sequence of the topics addressed in the rest of this section. We remark that, for reasons that will be clear later in this chapter, the lattice is better defined as *direct lattice*, meaning that it is hosted in the space where distances are measured in units of meter.

2.2.2 Direct lattice vectors

The vector positions $\mathbf{R}_l$ of the lattice points are defined as

$$\mathbf{R}_l = n_1 \mathbf{a}_1 + n_2 \mathbf{a}_2 + n_3 \mathbf{a}_3, \tag{2.1}$$

where $\{\mathbf{a}_1, \mathbf{a}_2, \mathbf{a}_3\}$ are named *translation vectors* and $n_1, n_2, n_3 = 0, \pm1, \pm2, \pm3, \ldots.$ Translation vectors must not all lie on the same plane. Through equation (2.1) an infinite lattice is generated (for this reason $\mathbf{R}_l$ is also referred to as *lattice vector*), with translational invariance: the geometrical situation is just the same if viewed from any two positions $\mathbf{r}$ and $\mathbf{r}'$ such that $\mathbf{r}' = \mathbf{r} + \mathbf{R}_l$ as illustrated in figure 2.3 in the case of a two-dimensional square lattice.

The choice of translation vectors is not unique, as shown in figure 2.4: the same lattice can be equivalently spanned by different sets of translation vectors. We accordingly distinguish between *primitive translation vectors* and *conventional translation vectors* following a very simple criterion: if lattice points are found only at the corners of the parallelepiped whose edges are defined by $\{\mathbf{a}_1, \mathbf{a}_2, \mathbf{a}_3\}$, then the translation vectors are primitive. This is the case of the red and blue sets of vectors in figure 2.4; conversely, the magenta vectors represent a conventional set. Lattices generated by primitive translation vectors are referred to as *Bravais lattices*. In this case, lattice points closest to a given point are named its *nearest neighbours*. Their number (necessarily equal for each lattice point because of the translational invariance property) is a characteristic of the specific Bravais lattice: it is called the *coordination number*.

The volume V_c of the parallelepiped defined by the translation vectors $\{\mathbf{a}_1, \mathbf{a}_2, \mathbf{a}_3\}$ is

$$V_c = |\, \mathbf{a}_1 \cdot \mathbf{a}_2 \times \mathbf{a}_3 \,|, \tag{2.2}$$

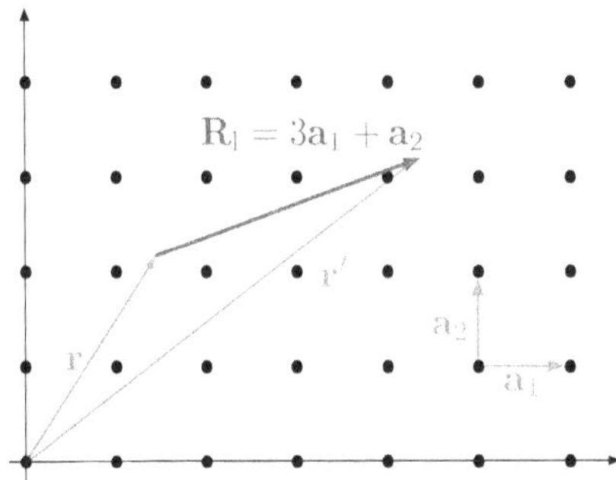

Figure 2.3. Pictorial representation of the translational invariance in the case of a two-dimensional square lattice. Since $\mathbf{R}_l$ is assigned according to equation (2.1), the physics looks the same if observed in $\mathbf{r}$ or in $\mathbf{r}' = \mathbf{r} + \mathbf{R}_l$.

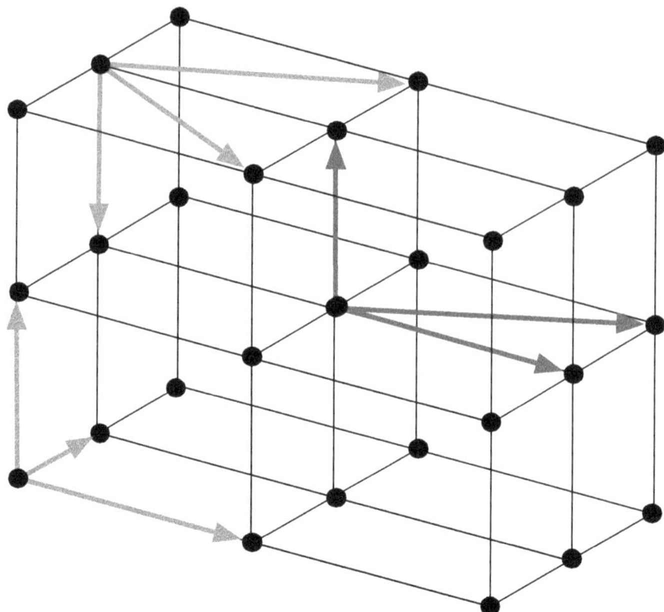

Figure 2.4. Some different choices of translation vectors in a three-dimensional lattice. The red, blue, and magenta sets of vectors equally span the lattice.

as imposed by vector algebra, and the corresponding portion of space is referred to as the *unit cell* of the crystal. In the case where it is defined by primitive translation vectors, it is more precisely labelled as a *primitive unit cell*; otherwise it is named a *conventional unit cell*. Such a volume has the remarkable property that it fills all space without overlapping and without leaving voids when translated through equation (2.1).

A conventional unit cell is typically larger than the primitive one, but it could have the exact symmetry of the lattice it represents. This is illustrated in figure 2.5 in the case of the three-dimensional face-centred cubic crystal[3]. In this case the conventional cell has four times the volume of the unit one, but it fully reflects the leading cubic symmetry of the lattice. The use of conventional cells explains some phenomenological results, like e.g., the remarkable similarities between different specimens of the same material or even the empirical law originally formulated by N Steno in 1761, still used in mineralogy, according to which the faces of crystals made by the same substance form identical angles.

In generating the primitive unit cell of a crystal we can add an additional request, namely that *it defines the region of space closer to a given lattice point than to any other one*. This special kind of primitive cell is called *Wigner–Seitz cell*. It is not difficult to generate such a cell: just draw a line connecting the selected lattice point with its first next neighbours; then, at the midpoint of each segment draw a plane

---

[3] Once again, we are anticipating some information discussed later in this chapter; here it is sufficient to say that in such a lattice the points are placed at the corners of a cube, as well as at the centre of its faces.

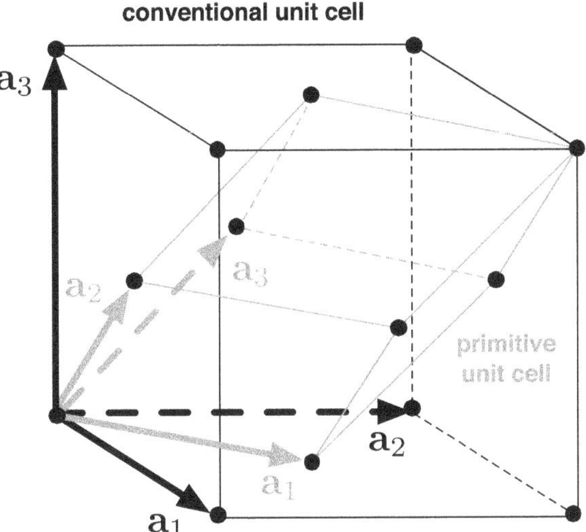

Figure 2.5. Primitive (red) and conventional (black) unit cell for the face-centred cubic lattice. The corresponding translation vectors are indicated with the same colour code.

normal to the segment itself; the portion of space contained within these planes has by construction the required property. The Wigner–Seitz cell is a particular case of primitive unit cell and, like all the others, contains only one lattice point.

2.2.3 Bravais lattices

Translational invariance represents the dominant structural feature of any crystal, largely dictating its physics. Nevertheless, it is not the only operation taking the lattice in itself. For instance, let us consider the face-centred cubic lattice shown in figure 2.5: it is easy to recognise that any rotation of a $\pi/2$ angle about a line normal to a face and passing through its centre leaves the lattice unchanged. Similarly, a reflection in any plane defined by the cube faces takes the lattice in itself. These are just simple examples of *non-translational symmetry operations*: their full description is the core business of crystallography [4–7]. Here we limit ourselves to defining some general features allowing for the classification of the Bravais lattices.

First of all, we understand that all the operations we are dealing with are rigid, that is, *they do not change the distance between lattice points*. In other words, we are not considering deformations. Under this constraint, we can distinguish between *pure translations* and other *operations that leave just one lattice point fixed*. For example, imagine a two-dimensional square lattice and a rotation of a $\pi/2$ angle about a line normal to the plane and passing through a lattice point. It is a key result of crystallography that by combining a translation with an action leaving just one lattice point fixed we get a symmetry operation for the selected lattice. We do not formally prove this result, but the graphical example shown in figure 2.6 makes it plausible. In summary, *all operations taking a lattice in itself are either pure translations or leave a particular lattice point fixed or are a combination of the two.*

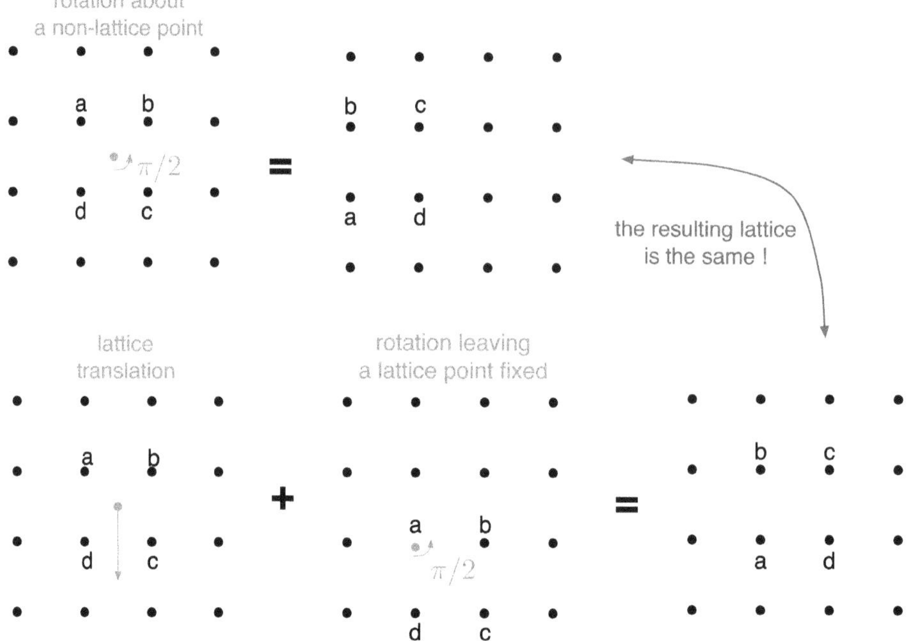

Figure 2.6. Graphical proof that a symmetry rotation about a line normal to the plane containing a two-dimensional square lattice and passing through a non-lattice point (top) is equivalently represented by a sequence of a lattice translation and a rotation leaving fixed a lattice point. The labelling of lattice points is just a guide to figure out the action of any operation: in fact, they are all equivalent.

The full set of symmetry operations of the three kinds is known as the *space group* or *symmetry group* of the Bravais lattice. On the other hand, the subset of symmetry operations that leave a particular lattice point fixed is referred to as the *point group* of the Bravais lattice; their elements are called *point operations*.

Crystallography classifies the existing Bravais lattices according to the space group or to the point group they belong to. The seminal work by M L Frankheim and A Bravais, developed in the middle of the XIXth century, has proved that if we follow the space group criterion we get *14 Bravais lattices,* while if we follow the point group criterion we get *seven crystal systems.* These results hold for three-dimensional lattices, while two-dimensional ones present a much simpler situation: there exist just five two-dimensional Bravais lattices. More precisely, they are known as the oblique, the square, the hexagonal, the rectangular, and the centred rectangular lattice: their graphical representation is straightforward. Since Bravais lattices are more numerous than crystal classes, we understand that each of the latter can be attributed one or more lattice(s). The key concept is however that *Bravais lattices belonging to the same crystal class have the same point group* or, equivalently, they are characterised by the same set of point operations.

The seven three-dimensional crystal systems are classified by using the lengths of the edges of their conventional cell and the angles they form. We will use the

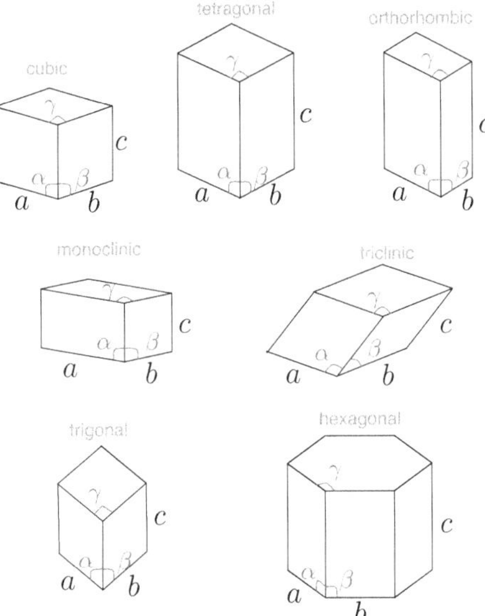

Figure 2.7. The conventional unit cell of the seven crystal systems, each one characterised by a different point group symmetry. Black Latin symbols (a, b, c) indicate the length of the cell sides while blue Greek symbols (α, β, γ) indicate the angles between cell edges.

standard notation to indicate with (a, b, c) the edge lengths and with (α, β, γ) the angles. Also, (a, b, c) will be referred to as the *lattice constants*. The crystal systems are shown in figure 2.7 and named:

1. **Cubic**: it is characterised by $a = b = c$ and $\alpha = \beta = \gamma = \pi/2$. Its point group contains all the symmetry operations of a cube. There are *three Bravais lattices for this system*, namely the *simple cubic* (lattice points at the corners of the cube), the *body-centred cubic* (lattice points at the corners of the cube and at its centre), and the *face-centred cubic* (lattice points at the corners of the cube and at the centre of its faces). We will refer to them by using the acronyms sc, bcc, and fcc, respectively. The sc lattice is the only one for which the conventional cell shown in figure 2.8 is a primitive cell.

2. **Tetragonal**: it is characterised by $a = b \neq c$ and $\alpha = \beta = \gamma = \pi/2$. Its point group contains all the symmetry operations of an orthogonal prism with a square basis and an unequal height. There are *two Bravais lattices for this system*, namely the *simple tetragonal* (lattice points at the corners of the orthogonal prism) and the *centred tetragonal* (lattice points are the corners of the orthogonal prism and at its centre).

3. **Orthorhombic**: it is characterised by $a \neq b \neq c$ and $\alpha = \beta = \gamma = \pi/2$. Its point group contains all the symmetry operations of an orthogonal prism with three unequal lengths. There are *four Bravais lattices for this system*, namely the *simple orthorhombic* (lattice points at the corners of the prism), the *base-centred orthorhombic* (lattice points at the corners of the prism and at the centre of the two bases), the *body-centred orthorhombic* (lattice points

at the corners of the prism and at its centre), and the *face-centred orthorhombic* (lattice points at the corners of the prism and at the centre of its faces).

4. **Monoclinic**: it is characterised by $a \neq b \neq c$ and $\alpha = \beta = \pi/2 \neq \gamma$. Its point group contains all the symmetry operations of a prism with three unequal lengths and a non-rectangular basis. There are *two Bravais lattices for this system,* namely the *simple monoclinic* (lattice points at the corners of the prism) and the *centred monoclinic* (lattice points at the corners of the prism and at its centre).

5. **Triclinic**: it is characterised by $a \neq b \neq c$ and $\alpha \neq \beta \neq \gamma$. Its point group contains all the symmetry operations of a three-dimensional object with parallel opposite faces, but no restrictions both on the side lengths and on the angles between edges. There is *just one Bravais lattice for this system* which is simply known as *triclinic*.

6. **Trigonal**: it is characterised by $a = b = c$ and $\alpha = \beta = \gamma < 2\pi/3$ (all angles not equal to $\pi/2$). Its point group contains all the symmetry operations of a rhombohedron (i.e. a cube which has been stretched along its diagonal). There is *just one Bravais lattice for this system* which is equally known as *trigonal* or *rhombohedral*.

7. **Hexagonal**: it is characterised by $a = b \neq c$ and $\alpha = \beta = \pi/2$, $\gamma = 2\pi/3$. Its point group contains all the symmetry operations of an orthogonal prism with a regular hexagon as basis. There is *just one Bravais lattice for this system* which is simply known as *hexagonal*.

In figure 2.8 a set of primitive translation vectors for the three cubic lattices are shown for illustration purposes.

We conclude this section by introducing the concept of *crystallographic axes*: they are imaginary lines drawn within the lattice, defining a useful frame of reference in the crystal since their intersection point will be used as the origin. They are not all coplanar; for cubic lattices, they are parallel to the edges of the conventional unit cell. This implies that in most lattices three crystallographic axes are found, corresponding to the (a, b, c) edges of figure 2.7. On the other hand, the hexagonal lattice needs four crystallographic axes: three of them fall in the same basal plane of the hexagonal prism, intersecting with a $2\pi/3$ angle at its centre; the fourth axis passes through the intersection point at a right angle to the plane formed by them.

2.2.4 Lattice planes and directions

Within a crystal lattice we can identify sets of planes with the threefold property of (i) containing lattice points, and being (ii) parallel and (iii) equally spaced. Such *lattice planes* play an important role in determining the diffractions of whatever waves are travelling within the crystal.

Since any plane is determined by three non collinear points in the space, the conventional procedure to identify a family of parallel lattice planes is based on

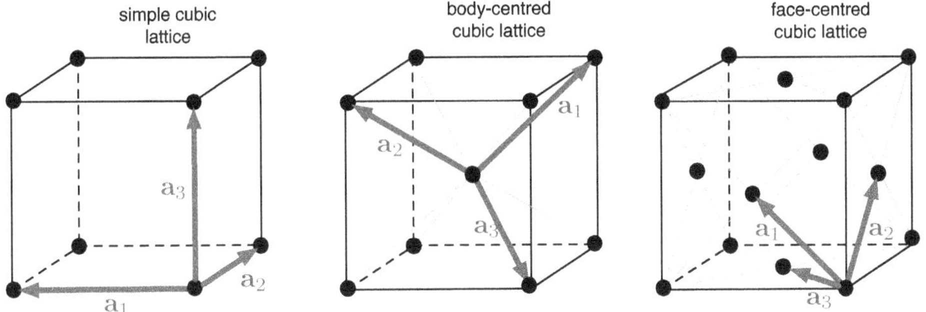

Figure 2.8. Primitive translation vectors for the sc, bcc, and fcc lattices. Black lines represent the underlying symmetry of the lattice; magenta lines represent just a guide to the eye helping to figure out the space relation among lattice point positions.

the intercepts made on the crystallographic axes by the nearest plane to the origin (not considering the plane that possibly contains the origin for the reason that will be immediately clear). The first step consists in finding such intercepts in terms of the lattice constants (a, b, c): this defines a set of three integer numbers; next, the reciprocal of these numbers is taken; finally, these reciprocals are reduced to the smallest three integer numbers (h, k, l) having the same ratio. For instance, let us consider the case shown in figure 2.9: the plane has intercepts with the crystallographic axes given by 3, 1, and 2 in units of the (a, b, c) lattice constants. Their reciprocal are 1/3, 1 ad 1/2. The smallest three integers with the same ratio are (263): they label the plane. The three numbers (h, k, l) are known as the *Miller indices* of the plane (in fact, they identify the family of its parallel lattice planes). In figure 2.10 some important planes in Bravais lattices of the cubic crystal system are shown. We finally remark that whenever a plane intercepts an axis on its negative side with respect to the origin, the corresponding Miller index is negative: in order to keep this information, this index will be labelled by placing a bar on it.

Similarly to planes, *even directions can be identified within a lattice*. In this case, a set of three integers is used and put in square parenthesis as $[u, v, w]$: they represent the set of smallest integers with the same ratio as the components of a vector pointing along the selected direction, referred to the crystallographic axes. In figure 2.11 we show some important directions in Bravais lattices of the cubic crystal system. We remark that only in this special case is the $[u, v, w]$ direction always normal to the (u, v, w) plane.

2.3 Crystal structures

2.3.1 The basis

Once the lattice has been determined, we can generate the actual *crystal structure* of a solid material by simply assigning to each lattice point the very same basis, whose definition was provided in section 2. In this case the actual position $\mathbf{R}$ of an ion is

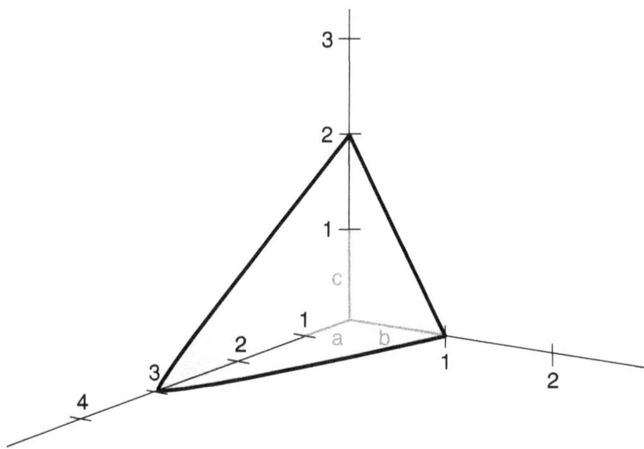

Figure 2.9. The shaded lattice plane is identified by the Miller indices (263). The (a, b, c) lattice constants are shown by red segments. In their units, the plane intercepts are (3, 1, 2).

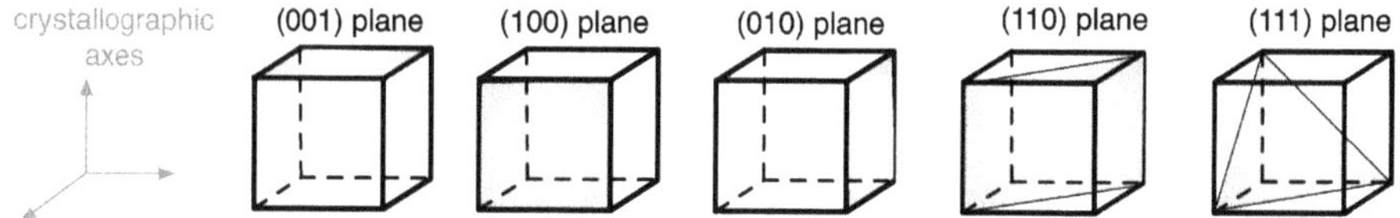

Figure 2.10. Some important planes and corresponding Miller indices for lattices of the cubic crystal system.

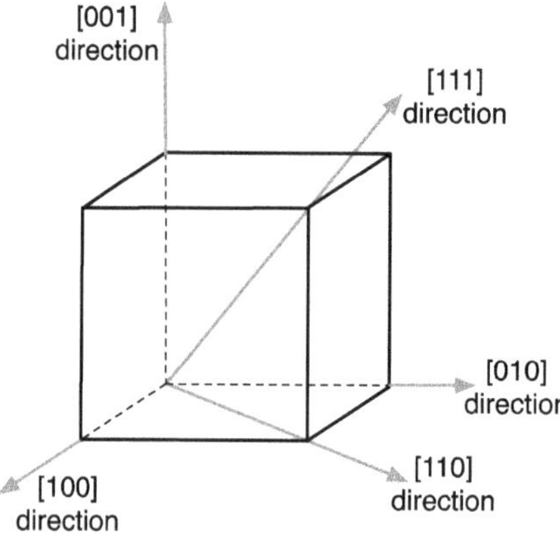

Figure 2.11. Some important directions for lattices of the cubic crystal system.

written by summing the lattice vector $\mathbf{R}_l$ to the vector $\mathbf{R}_b$ providing the relative position of the atom within the basis

$$\mathbf{R} = \mathbf{R}_l + \mathbf{R}_b, \tag{2.3}$$

where we have placed the origin of the frame of reference for each basis on the lattice points $\mathbf{R}_l$, without any rotation from place to place (in the case of a monoatomic

Bravais crystal, we obviously have $\mathbf{R}_b = 0$). The resulting periodic crystal structure eventually corresponds to a *physical object*, while the lattice was just an abstract geometrical entity. We nevertheless remark that such a crystal structure is a still idealised physical system since, at variance with real solids, (i) it is infinite and (ii) it does not contain any imperfection.

In the most general case, a basis will contain two or more atoms (or molecules), whose relative positions must be the same everywhere[4] in order to ensure the translational invariance. On the other hand, when the basis contains just one single atom, the resulting structure is referred to as a *monoatomic Bravais crystal*. This is indeed a special case of crystal structure, since it can be equivalently described by using a lattice with a basis, provided that a non-primitive unit cell has been chosen[5].

2.3.2 Classification of the crystal structures

A large variety of solids (mostly metallic) crystallise as Bravais crystals, mainly in the bcc and fcc form[6]. More specifically (standard chemical symbols hereafter appearing are defined in the periodic table reported in appendix A) we have:

1. The **bcc monoatomic Bravais crystal structure**: it is typically assumed by alkali metals (Li, Na, K, Rb, Cs), by some transition metals (V, Cr, Fe, Nb, Mo, Ta, W), by some alkaline-earth metals (Ba) or by other metals (Tl).
2. The **fcc monoatomic Bravais crystal structure**: it is typically assumed by noble elements (Ne, Ar, Kr, Xe), by some transition metals (Co, Ni, Cu, Rh, Pd, Ag, Ir, Pt, Au), by some alkaline-earth metals (Ca, Sr), by other metals (Al, Pb), by some lanthanides (La, Pr, Yb) or actinides (Th).

The set of non-Bravais crystals is of course much more rich, including elemental as well as compound solids. Some important crystal structures with a basis are:

3. The **sodium chloride structure**: it consists of an equal number of atoms of two A and B chemical species, placed at alternate points of an sc lattice. The resulting crystal structure is described as an fcc lattice with a basis of two different atoms in position $(0, 0, 0)a$ and $(1/2, 1/2, 1/2)a$, where a is the cubic lattice constant. This crystal structure is typical of AB ionic solids, where A is any alkali metal (Li, Na, K, Rb, Cs) and B any halogen atom (F, Br, Cl, I). Similarly, A–B compounds crystallise in this form, where A is a metal (Ag, Mg, Ca, Sr or Ba) and B is a non-metal (O, S, Se or Te) element.
4. The **cesium chloride structure**: it consists in bcc lattice whose sites are occupied by atoms of two A and B chemical species so that each A-atom has eight nearest neighbouring B-atoms (and vice versa). The resulting

---

[4] Meaning: the same in any basis assigned to each lattice point.

[5] A bcc Bravais crystal with lattice constant a can be equivalently described as an sc lattice with a two-atom basis by adopting a cubic conventional unit cell with side a and a basis of two atoms in positions $(0, 0, 0)a$ and $(1/2, 1/2, 1/2)a$. The crystallographic axes provide the frame of reference.

[6] The sc form is instead really very rare in normal conditions of temperature and pressure.

crystal structure is described as an sc lattice with a basis of two different atoms in position $(0, 0, 0)a$ and $(1/2, 1/2, 1/2)a$, where a is the cubic lattice constant. We find in this crystal structure Cs-based materials, like CsCl, CsBr, and CsI.

5. The **diamond structure**: it consists of two inter-penetrating fcc monoatomic Bravais lattices, displaced along the diagonal of the cubic conventional unit cell by $a/4$, where a is the cubic lattice constant. The two sub-lattices are occupied by atoms of the same chemical species. The resulting crystal structure is described as an fcc lattice with a basis of two identical atoms in position $(0, 0, 0)a$ and $(1/4, 1/4, 1/4)a$. Elemental semiconductors (Si, Ge, α-Sn) and diamond (C) crystallise in this structure.

6. The **zincblende structure**: it is similar to the diamond structure, but the two sub-lattices are occupied by atoms of two different chemical species. Compound semiconductors crystallise in this structure, in any IV–IV (SiC) or III–V (where III = Al, Ga, In and V = P, As, Sb) or II–VI (where II = Zn, Cd, Hg and VI = S, Se, Te) combination. Other crystals assuming this structure are CuA (where A = F, Cl, Br, I), BeB (where B = S, Se, Te), MnC (where C = S, Se).

7. The **hexagonal structure**: it consists of two inter-penetrating hexagonal Bravais lattices, shifted from one another by a displacement vector $(a/3, b/3, c/2)$, where $a = b$ and c are defined in figure 2.7. The 'ideal' hexagonal structure has a $c/a = \sqrt{8/3}$ ratio, while actual hexagonal crystals deviate from this value. The ideal c/a ratio is calculated by assuming that each lattice point is occupied by a hard sphere, a situation which is referred to as *hexagonal close packing* (see next section for more detail). However, atoms are not rigid spheres and, therefore, in real materials this ratio can assume different values. We find in crystal structure many elemental solids like those made by Cd, Mg, Nd, Os, Sc, Ti, Zn, and Zr with a c/a ratio of 1.89, 1.62, 1.61, 1.58, 1.59, 1,59, 1.86, and 1.59, respectively.

For all the formal developments discussed in the next chapters it is important to visualise the seven crystal structures, which are therefore shown in figure 2.12.

We conclude this section with an important remark. A basis is a group of atoms which, in principle, is not required to have full spherical symmetry like, instead, the single atom (in fact a totally symmetric object) assigned to each point of Bravais lattices. The consequence is that *the number of symmetry groups existing for general crystal structures is greatly increased* with respect to the case of monoatomic Bravais lattices. More specifically, crystallography has classified *230 space groups* and *32 crystallographic point groups* for general crystal structures (these numbers should be compared to 14 and 7, respectively, found in monoatomic Bravais lattices). While more details are found in [7], here we limit ourselves to mentioning the various kinds of symmetry operations

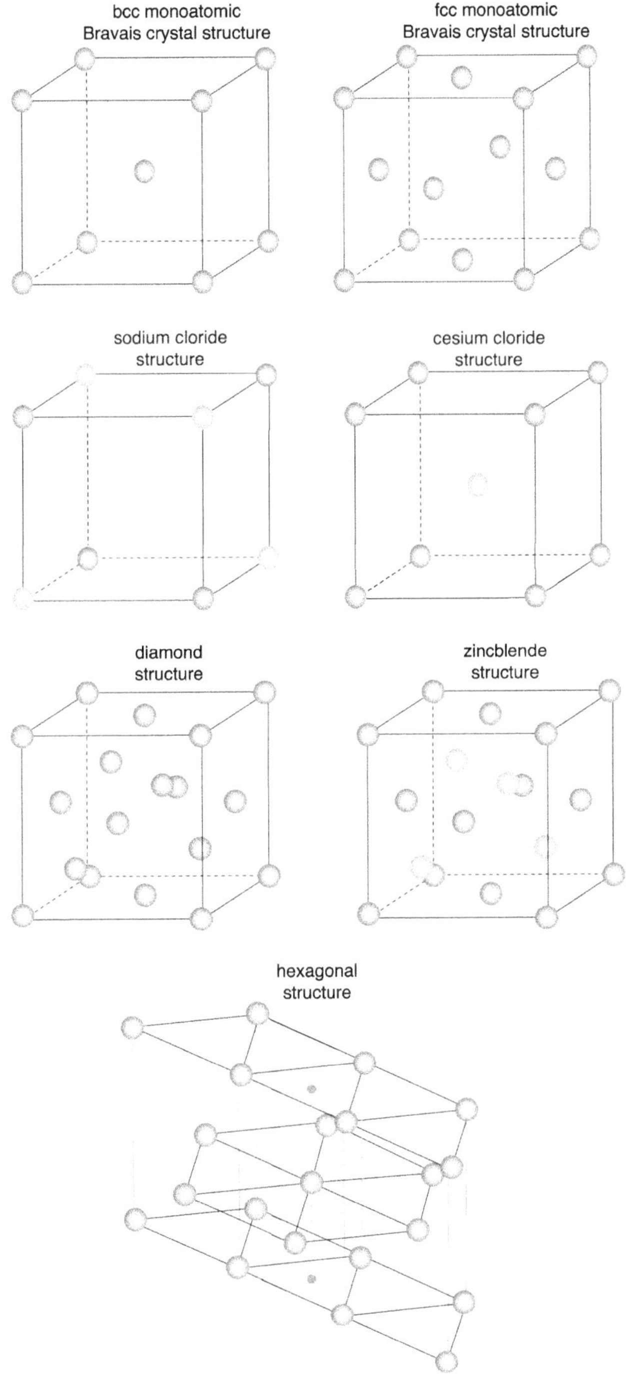

Figure 2.12. The seven crystal structures discussed in the text. Atoms are shown in blue or red colour. Black lines represent the underlying symmetry of the lattice; magenta lines represent just a guide to the eye helping one to figure out the space relation among atom positions.

found in crystallographic point groups. Basically, we can distinguish among: (i) *rotations* through integer multiples of $2\pi/n$ angles about some given axis, with n limited to the set $\{2, 3, 4, 6\}$ of values; (ii) *reflections* in a mirror plane; (iii) *inversions* through a fixed point; (iv) *roto-reflections* where a rotation is followed by a reflection in a plane, thus providing a symmetry through a net rotation of an angle not equal to $2\pi/n$; (v) *roto-inversions* where a rotation through a $2\pi/n$ angle is followed by an inversion in a point found on the rotation axis.

Rotational symmetry deserves a clarification. In listing all possible rotations by $2\pi/n$ angles we restricted n to just the four values $\{2, 3, 4, 6\}$. More specifically, we did not cover the case with $n = 5$ or $n = 7$. In order to figure out the fundamental reason for this limitation, it is useful to start by considering the full set of two-dimensional Bravais lattices (which have been listed in section 2): by very intuitive arguments it is easy to prove that they can have neither a five-fold nor a seven-fold rotation axis (we cannot exactly cover a floor with pentagonal or heptagonal tiles!). Then, it is really easy to prove that an n-fold rotation axis cannot exist in three-dimensional Bravais lattices if it does not exist in some two-dimensional one. In conclusion, even in three dimensions we cannot have five-fold or seven-fold rotational symmetry. There is, however, a notable exception to this general rule which we discuss in appendix B where the concept of quasi-crystal is outlined.

Two main classification schemes have been elaborated to label the crystallographic point groups [7, 10], among which the *Schoenflies nomenclature system* is the most widely used; here symmetry groups are put in the following categories:

- C_n: groups containing only a n-fold rotation axis;
- C_{nv}: groups containing a n-fold rotation axis and mirror planes containing it;
- C_{nh}: groups containing a n-fold rotation axis and just one mirror plane normal to it;
- S_n: groups containing only n-fold roto-reflection axis;
- D_n: groups containing a n-fold rotation axis and other two-fold rotation axes normal to it;
- D_{nh}: groups containing the same operations of D_n and an additional mirror plane normal to the n-fold axis;
- D_{nd}: groups containing the same operations of D_n and mirror planes containing the n-fold axis and bisecting the angles between the two-fold axes.

The second classification scheme is named *international notation*. Its correspondence with the Schoenflies nomenclature system is found elsewhere [7].

2.3.3 Packing

There is still a remaining criterion for classifying atomic architectures to be discussed. It is based on the assumption to treat *atoms as attracting hard spheres*. While this is clearly a very crude approximation, it is reasonably well satisfied by metals and this represents the phenomenological foundation for its applicability.

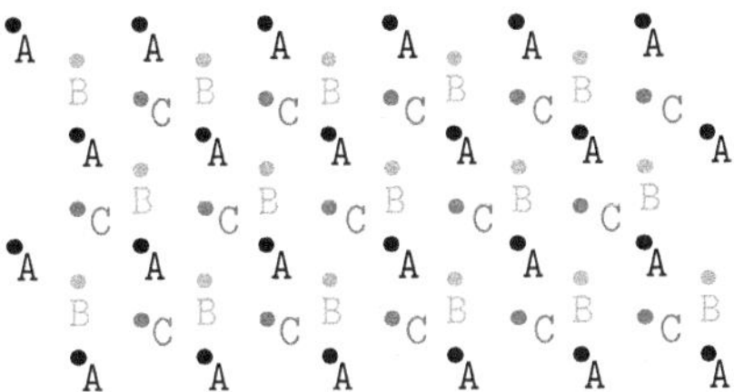

Figure 2.13. A close-packed sequence of hard-sphere layers. The first layer is marked grey (A lattice), the second in red (B lattice). The third layer can be placed on the A lattice or, alternatively, on the blue one (C lattice).

Since atoms are looked at as 'hard' spheres, they cannot overlap; however, because of their mutual attraction, they tend to assume an arrangement that minimises the total energy of the system. This implies that *they tend to pack as closely as possible*[7].

Let us start by arranging identical hard spheres on a plane: in the closest packing configuration the centres of the spheres lie on a two-dimensional triangular lattice, in positions marked by A letters in figure 2.13. By looking at this configuration from the top, we can identify for the second layer the new triangular lattice (lying on a plane parallel to the first layer) marked by B letters. While this choice is unique, when adding a third layer we can add spheres on the triangular lattice marked by C letters or, alternatively, on the triangular lattice once again marked by A letters (in both cases the third layer lies on a plane parallel to the two previous ones). The corresponding *stacking sequence* is ABCABCABC···and ABABABAB···, respectively: they are named *fcc structure* and *hexagonal close-packed (hpc) structure*. The atomic layers so generated correspond to (111) planes of the fcc lattice or to the basal plane of the hexagonal lattice. In both configurations each atom has 12 nearest neighbours: any model according to which the total energy of a crystal only depends on the number of nearest neighbours must necessarily predict the very same energy for fcc and hpc structures.

The number of different ways to pack hard spheres in arrangement other than the close-packed one is actually infinite. In table 2.1 we summarise some properties of the packing in cubic lattices. The *packing fraction* is the volume fraction occupied by the hard spheres: the labelling 'close packing' for fcc is justified by the fact that its packing fraction is maximum.

2.4 The reciprocal lattice

2.4.1 Fundamentals of x-ray diffraction by a lattice

The construction of crystal structures according to the formal rules developed in the previous section finds full experimental evidence by means of *x-ray crystallography*

---

[7] In essence this corresponds to the practical exercise of stacking cannonballs on top of each other.

Table 2.1. Some features of cubic lattices (volumes and distances in units of the a lattice constant).

	sc	bcc	fcc
Volume of the conventional unit cell	a^3	a^3	a^3
Volume of the primitive unit cell	a^3	$a^3/2$	$a^3/4$
Number of nearest neighbours	6	8	12
Distance of the nearest neighbours	a	$\sqrt{3}\,a/2$	$a/\sqrt{2}$
Packing fraction	$\pi/6$	$\sqrt{3}\,\pi/8$	$\sqrt{2}\,\pi/6$

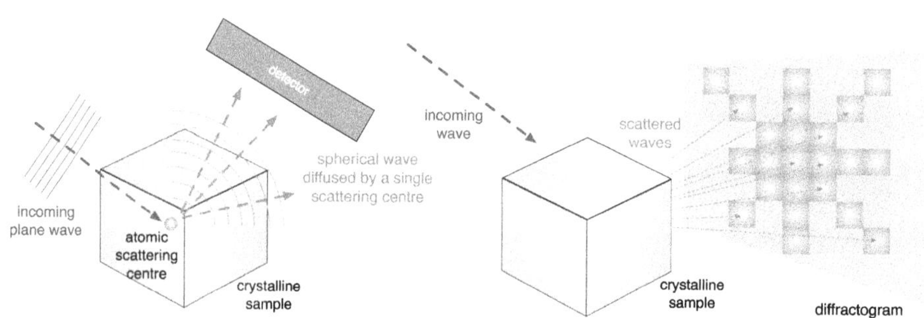

Figure 2.14. Pictorial representation of x-ray diffraction by a crystalline sample. Left: an incoming plane wave is scattered by each single atom forming the crystalline sample. Right: constructive and destructive interference phenomena between the diffused spherical waves provide light points and low brightness regions in the diffractogram collected by the detector.

[10–12], where an electromagnetic radiation with a typical wavelength in the range 0.01–10nm is made incident on and then diffracted[8] by a solid state crystalline sample.

The experimental situation is qualitatively summarised in figure 2.14: an incoming x-ray beam is collimated on a crystalline sample and the corresponding diffracted beam is collected by a detector as a distribution of light points, separated by regions of low brightness. They correspond, respectively, to the positions of maximum and minimum intensity of the diffracted beam; this picture is known as diffractogram. Assuming that the sample consists of a *discrete distribution of atomic scattering centres* and analysing by the laws of optics [13] all the angles formed between the direction of the incident and diffracted beams (as well as their intensities), it is possible to reconstruct the space distribution of the centres. In other words, from the distribution of the light points on the diffractogram it is eventually possible to specify the atomic architecture of the investigated sample.

---

[8] Diffraction [13] is the phenomenon of elastic diffusion of a wave by an ordered array of scattering centres. In the physics we are discussing, we will accordingly assume that the scattering centres are crystalline atoms.

2.4.2 Von Laue scattering conditions

We now perform a detailed analysis of the scattering events occurring in x-ray diffraction. Let $\mathbf{k}_{in}$ be the wavevector of the incoming monochromatic plane wave with amplitude

$$\mathcal{A}_{in}(\mathbf{r}, t) = \mathcal{A}_0 \exp\left[i(\mathbf{k}_{in} \cdot \mathbf{r} - \omega t)\right], \tag{2.4}$$

where ω is its angular frequency, while $\mathbf{r}$ and t indicate the position in space and time, respectively. Our goal is to predict the *amplitude $\mathcal{A}_{out}$ of the scattered waves*. To this aim we adopt a model originally developed by M von Laue and based on two simplifying assumptions: (i) *the incoming beam is weak* enough that its interaction with the sample does not affect the underlying crystal structure and (ii) *the scattering events are elastic*, that is, x-rays do not lose energy by diffusion (i.e. their intensity is unaffected by scattering).

We now remember that, according to the Huygens–Fresnel principle of elementary optics [13], any point-like object invested by a plane wave becomes the source of a scattered spherical wave. Accordingly, with reference to figure 2.15, we can write the amplitude $\mathcal{A}_{out}$ of the spherical wave emerging from the atom at position $\mathbf{R}$ and revealed by a detector at a distance $\mathbf{D}$ from the origin of the adopted frame of reference as

$$\mathcal{A}_{out} = \underbrace{\mathcal{A}_0 \exp\left[i(\mathbf{k}_{in} \cdot \mathbf{R} - \omega t)\right]}_{\text{incoming plane wave}} \cdot \underbrace{f_{\mathbf{R}}}_{\text{atomic form factor}} \cdot \underbrace{\frac{\exp\left[ik_{in}|\mathbf{D} - \mathbf{R}|\right]}{|\mathbf{D} - \mathbf{R}|}}_{\text{scattered spherical wave}}, \tag{2.5}$$

where the last term on the right-hand side accounts for phase change and amplitude decrease of the scattered wave. Such changes are due to the complex of quantum phenomena occurring in the interaction between the electromagnetic wave and the atom at position $\mathbf{R}$: their overall effect is summarised by the *atomic form factor $f_{\mathbf{R}}$* which depends on the atomic number Z of the atomic scatterer, as well as by its electron charge density $\rho(\mathbf{r})$ through the general expression

$$f_{\mathbf{R}} = \int \rho(\mathbf{r}) \exp[i\mathbf{K} \cdot (\mathbf{r} - \mathbf{R})]d\mathbf{r}, \tag{2.6}$$

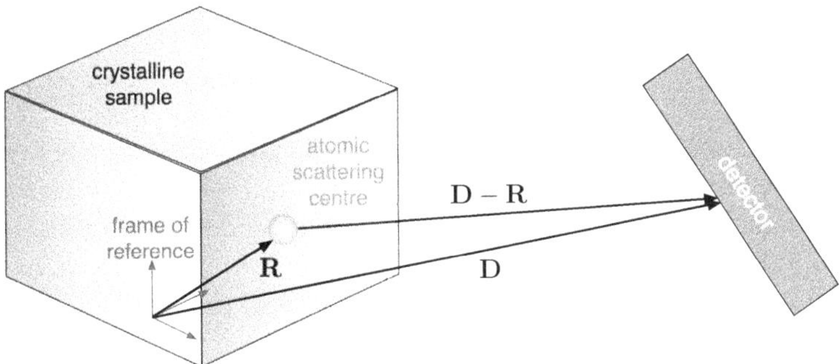

Figure 2.15. Geometry of the M von Laue model for x-ray scattering.

where $\mathbf{K} = \mathbf{k}_{\text{out}} - \mathbf{k}_{\text{in}}$ is the *scattering vector* (we have named $\mathbf{k}_{\text{out}}$ the wavevector of the scattered wave). The atomic form factor is calculated by the atomic theory of elastic scattering [14] and we will consider it as known.

Given a typical experimental setup for x-ray diffraction measurements, with no loss of generality we can assume $\mathbf{D} \gg \mathbf{R}$, since the detector is usually placed at a comparatively very large distance from the sample which, instead, is near the origin O of the laboratory frame of reference. Furthermore, at such a large distance the diffused spherical wave can be approximated to a very good extent by a plane wave with wavevector $\mathbf{k}_{\text{out}}$; under the assumption of elastic scattering, we have $k_{\text{in}} = k_{\text{out}}$. The geometry sketched in figure 2.15 suggests that $\mathbf{K}$ is nearly parallel both to $\mathbf{D}$ (measuring the sample–detector distance) and to $\mathbf{D} - \mathbf{R}$ (measuring the distance from the atom at position $\mathbf{R}$ and the detector). By combining the geometrical information collected so far, we can write

$$k_{\text{in}}|\mathbf{D} - \mathbf{R}| \sim \mathbf{k}_{\text{out}} \cdot (\mathbf{D} - \mathbf{R}) = k_{\text{out}}D - \mathbf{k}_{\text{out}} \cdot \mathbf{R} = k_{\text{in}}D - \mathbf{k}_{\text{out}} \cdot \mathbf{R}, \qquad (2.7)$$

which leads to the following expression for the amplitude of the scattered wave

$$\mathcal{A}_{\text{out}} = \frac{A_0 \exp\left[i(k_{\text{in}}D - \omega t)\right]}{D} \cdot f_{\mathbf{R}} \cdot \exp[-i\mathbf{K} \cdot \mathbf{R}]. \qquad (2.8)$$

We note that the first term on the right-hand side is easily calculated once the characteristics of the incoming wave (k_{in} and ω) and the geometry of the experiment (D) are known. We accordingly set $\Omega = A_0 \exp\left[i(k_{\text{in}}D - \omega t)\right]/D = \text{constant}$.

The total diffused wave has an amplitude $\mathcal{A}_{\text{tot}}$ resulting from the superposition of all waves scattered by single atoms. We therefore write

$$\mathcal{A}_{\text{tot}} = \Omega \sum_{\mathbf{R}} f_{\mathbf{R}} \exp[-i\mathbf{K} \cdot \mathbf{R}], \qquad (2.9)$$

where the lattice sum spans all the atoms in the crystal. Following the notation introduced in section 2.3.1, their position is $\mathbf{R} = \mathbf{R}_l + \mathbf{R}_b$. Equation (2.9) is then written as

$$\mathcal{A}_{\text{tot}} = \Omega \left[\sum_{\mathbf{R}_b} f_{\mathbf{R}_b} \exp(-i\mathbf{K} \cdot \mathbf{R}_b)\right]\left[\sum_{\mathbf{R}_l} \exp(-i\mathbf{K} \cdot \mathbf{R}_l)\right], \qquad (2.10)$$

where we have exploited the fact that each basis atom with position $\mathbf{R}_b$ has exactly the same form factor in any basis[9].

The maxima of the diffracted beam (corresponding to the bright light points in figure 2.14) correspond to the maxima of $\mathcal{A}_{\text{tot}}$, which we are now going to determine. We already know that Ω is a constant; the same is also true for the second term on the right-hand side of equation (2.10): it is calculated once for all, as soon as the chemical nature and the geometrical structure of the basis has been specified (the $f_{\mathbf{R}_b}$ form factors are provided by atomic physics). We conclude that the maxima of $\mathcal{A}_{\text{tot}}$

---

[9] The sum in the second term on the right-hand side of this equation is known as the *structure factor* [15].

correspond to *the maxima of the last term on the right-hand side of equation (2.10)*. Since $\mathbf{R}_l$ vectors are given by equation (2.1), the maximum condition occurs provided that

$$\mathbf{K} \cdot \mathbf{a}_1 = 2\pi l \qquad \mathbf{K} \cdot \mathbf{a}_2 = 2\pi m \qquad \mathbf{K} \cdot \mathbf{a}_3 = 2\pi n, \tag{2.11}$$

where l, m, n are integer numbers. Equations (2.11) are known as *Laue conditions*: they are remarkably important since they state that *allowed scattering vectors $\mathbf{K}$ lie on a discrete lattice.*

This is indeed quite a different kind of lattice with respect to the direct one, since here distances are measured in units of $(\text{meter})^{-1}$: for this reason it is referred as the *reciprocal lattice*. While the direct lattice spans the space where ionic positions are framed and their displacements take place, *the reciprocal lattice is the natural environment where the wavevectors of whatever wave-like phenomena within a crystal are defined.* This will affect the whole solid state physics.

2.4.3 Reciprocal lattice vectors

The reciprocal lattice is formally described by the same concepts developed in section 2 for the direct one. More specifically, its points are given by

$$\mathbf{G} = m_1\mathbf{b}_1 + m_2\mathbf{b}_2 + m_3\mathbf{b}_3, \tag{2.12}$$

where $\{\mathbf{b}_1, \mathbf{b}_2, \mathbf{b}_3\}$ are named *reciprocal translation vectors* and $m_1, m_2, m_3 = \pm 1, \pm 2, \pm 3, \ldots$. The maximum scattering vectors $\mathbf{K}$ entering equation (2.10) lie on this reciprocal lattice and, therefore, they must fulfil equation (2.12); accordingly, by setting $\mathbf{K} = \mathbf{G}$, after some little algebra we obtain that the Laue conditions are satisfied if

$$\mathbf{b}_1 = 2\pi \frac{\mathbf{a}_2 \times \mathbf{a}_3}{\mathbf{a}_1 \cdot \mathbf{a}_2 \times \mathbf{a}_3} \qquad \mathbf{b}_2 = 2\pi \frac{\mathbf{a}_3 \times \mathbf{a}_1}{\mathbf{a}_1 \cdot \mathbf{a}_2 \times \mathbf{a}_3} \qquad \mathbf{b}_3 = 2\pi \frac{\mathbf{a}_1 \times \mathbf{a}_2}{\mathbf{a}_1 \cdot \mathbf{a}_2 \times \mathbf{a}_3}. \tag{2.13}$$

This result provides the formal definition of $\mathbf{b}$-vectors and stresses the fact that *the reciprocal lattice is closely related to its direct counterpart.* A number of formal relations hold, including the following most important ones:

$$\begin{aligned}
\mathbf{a}_i \cdot \mathbf{b}_j &= 2\pi\delta_{ij} & \text{with} \quad i, j = 1, 2, 3 \\
\mathbf{G} \cdot \mathbf{a}_i &= 2\pi m_i & \text{with} \quad i = 1, 2, 3 \\
\exp(i\,\mathbf{G} \cdot \mathbf{R}_l) &= 1 & \text{for any pair of direct/reciprocal vectors,}
\end{aligned} \tag{2.14}$$

where δ_{ij} is Kroenecker delta-symbol. Furthermore, we observe that $\mathbf{b}_i$ and $\mathbf{b}_j$ with $i, j \neq k$ are normal to $\mathbf{a}_k$, while $\mathbf{b}_i$ is not in general parallel to $\mathbf{a}_i$: this holds only in crystals with orthogonal axes.

The reciprocal lattice has a number of important formal properties which are very easy to prove by applying the above definitions:

- the reciprocal lattice is a kind of Bravais lattice;
- through equation (2.12) an infinite lattice is generated, once again characterised by *translational invariance*;

- it is possible to identify volumes such that, upon translations by a $\mathbf{G}$ vector, they entirely fill the reciprocal space, without overlapping or leaving empty spaces: such volumes, therefore, play the same role as unit cells in the direct lattice;
- the reciprocal of the reciprocal lattice is nothing other than the original direct lattice;
- interestingly enough, two-dimensional direct lattices have a remarkable property, namely they are self-reciprocal: the direct and the reciprocal lattice have the very same symmetry.

In order to complete the discussion of cubic and hexagonal direct lattices developed in section 2, we discuss here the generation of their reciprocal lattice vectors. Let us consider at first sc, bcc, and fcc direct lattices: for them it is natural to adopt a Cartesian frame of reference whose three axes coincide with the edges of the cubit conventional unit cell; this set of Cartesian axes will be identified by the set of mutually normal unit vectors $\{\hat{\imath}, \hat{\jmath}, \hat{k}\}$. For the hexagonal lattice just the c axis is collinear with a Cartesian one. The corresponding sets of direct and reciprocal vectors are reported in table 2.2 for the four cases of interest.

Another interesting feature is that *any reciprocal lattice vector $\mathbf{G}$ is normal to a family of planes in the direct lattice*. It is easy to prove it. Let $\exp(i\mathbf{G} \cdot \mathbf{r})$ be a plane wave whose maximum (unity) amplitude values occur on parallel planes in real space, separated by a distance $d = 2\pi/G$. We know that $\mathbf{G}$ is normal to such a family of planes; we also know (see equation (2.14)) that $\exp(i\mathbf{G} \cdot \mathbf{R}_l) = 1$ for any $\mathbf{R}_l$ lattice vector. Therefore, we conclude that the family of parallel planes in real space must necessarily contain all those lattice planes which are spanned by suitable $\mathbf{R}_l$ vectors. Since the reciprocal lattice vector $\mathbf{G}$ is normal to them, we can use its three integers

Table 2.2. Direct $\{\mathbf{a}_1, \mathbf{a}_2, \mathbf{a}_3\}$ and reciprocal $\{\mathbf{b}_1, \mathbf{b}_2, \mathbf{b}_3\}$ translation vectors of cubic (sc, bcc, fcc) and hexagonal (hex) lattices. Just one lattice constant a is used in the cubic case, while two lattice constants (a, c) are needed for the hexagonal lattice (see figure 2.7).

	sc	bcc	fcc	hex
$\mathbf{a}_1$	$a\hat{\imath}$	$\frac{a}{2}(\hat{\jmath} + \hat{k} - \hat{\imath})$	$\frac{a}{2}(\hat{\jmath} + \hat{k})$	$a\hat{\imath}$
$\mathbf{a}_2$	$a\hat{\jmath}$	$\frac{a}{2}(\hat{k} + \hat{\imath} - \hat{\jmath})$	$\frac{a}{2}(\hat{k} + \hat{\imath})$	$a\left(-\frac{1}{2}\hat{\imath} + \frac{\sqrt{3}}{2}\hat{\jmath}\right)$
$\mathbf{a}_3$	$a\hat{k}$	$\frac{a}{2}(\hat{\imath} + \hat{\jmath} - \hat{k})$	$\frac{a}{2}(\hat{\imath} + \hat{\jmath})$	$c\hat{k}$
$\mathbf{b}_1$	$\frac{2\pi}{a}\hat{\imath}$	$\frac{2\pi}{a}(\hat{\jmath} + \hat{k})$	$\frac{2\pi}{a}(\hat{\jmath} + \hat{k} - \hat{\imath})$	$\frac{4\pi}{\sqrt{3}a}\left(\frac{\sqrt{3}}{2}\hat{\imath} + \frac{1}{2}\hat{\jmath}\right)$
$\mathbf{b}_2$	$\frac{2\pi}{a}\hat{\jmath}$	$\frac{2\pi}{a}(\hat{k} + \hat{\imath})$	$\frac{2\pi}{a}(\hat{k} + \hat{\imath} - \hat{\jmath})$	$\frac{4\pi}{\sqrt{3}a}\hat{\jmath}$
$\mathbf{b}_3$	$\frac{2\pi}{a}\hat{k}$	$\frac{2\pi}{a}(\hat{\imath} + \hat{\jmath})$	$\frac{2\pi}{a}(\hat{\imath} + \hat{\jmath} - \hat{k})$	$\frac{2\pi}{c}\hat{k}$

$\{m_1, m_2, m_3\}$ to label this family of direct lattice planes. In conclusion, *reciprocal lattice vectors provide an alternative definition for Miller indices*[10].

2.4.4 The Brillouin zone

The conventional way to generate the *reciprocal primitive unit cell* is by following the Wigner–Seitz construction: the resulting cell is referred to as the *first Brillouin zone*. By construction, it contains *all the wavevectors that are not linked by a reciprocal translational vector* **G**. Its volume is $(2\pi)^3/V_c$, where V_c is the volume of the primitive unit cell of the corresponding direct lattice defined in equation (2.2). We remark that the use of the adjective 'first' will be clear when discussing the vibrational and electronic properties of crystalline solids. We will hereafter make use of the acronym 1BZ to indicate the first Brillouin zone.

The boundaries of the 1BZ are given by planes which, as explained in the previous section, are in turn defined by means of reciprocal lattice vectors. The general principle is that *the 1BZ is the smallest volume in the reciprocal space which is enclosed by planes normally bisecting reciprocal lattice vectors drawn for the origin*. With reference to table 2.2, we can calculate that the boundary planes of the 1BZ for the three cubic lattices are defined as follows (the 1BZ of the hexagonal lattice is added for completeness):

- **simple cubic lattice**: take the six planes normal to the vectors $\pm 2\pi\hat{i}/a$, $\pm 2\pi\hat{j}/a$, and $\pm 2\pi\hat{k}/a$ at their midpoints: they define a *cubic volume*;
- **body-centred cubic lattice**: take the 12 planes normal to the vectors $2\pi(\pm\hat{j}\pm k)/a$, $2\pi(\pm\hat{k}\pm\hat{i})/a$, and $2\pi(\pm\hat{i}\pm\hat{j})/a$ at their midpoints: they define a *rhombic dodecahedron volume*;
- **face-centred cubic lattice**: take the eight planes normal to the vectors $2\pi(\pm\hat{i}\pm\hat{j}\pm\hat{k})/a$ at their midpoints and further cut them by another set of six planes bisecting the reciprocal lattice vectors $\pm 4\pi\hat{i}/a$, $\pm 4\pi\hat{j}/a$, and $\pm 4\pi\hat{k}/a$: the resulting volume is a *truncated octahedron*.

A number of *high symmetry lines and points* can be identified as shown in figure 2.16 by red lines and black dots, respectively. High-symmetry points of the 1BZ typically lie at the centre of the zone, edges, and faces, as well as at the corner points. They play an important role in solid state physics: whenever we need to visualise a crystalline physical property depending upon a wavevector[11], the conventional choice is to follow the path marked in red colour in figure 2.16, corresponding to the edges of the so called *irreducible part of the 1BZ*. For further reference, we report a standard labelling used for fcc crystals to indicate some high-symmetry directions: the three directions connecting the Γ zone-centre to the X, K, and L point are indicated as Δ, Σ, and Λ, respectively.

---

[10] In order to avoid any ambiguity, the widely adopted convention is that the shortest reciprocal vector parallel to **G** is used to label the family of lattice planes normal to it.

[11] Anticipating the topics of the next chapters, the physical property could be the frequency of a vibrational normal mode of the lattice or the energy of a crystalline electron.

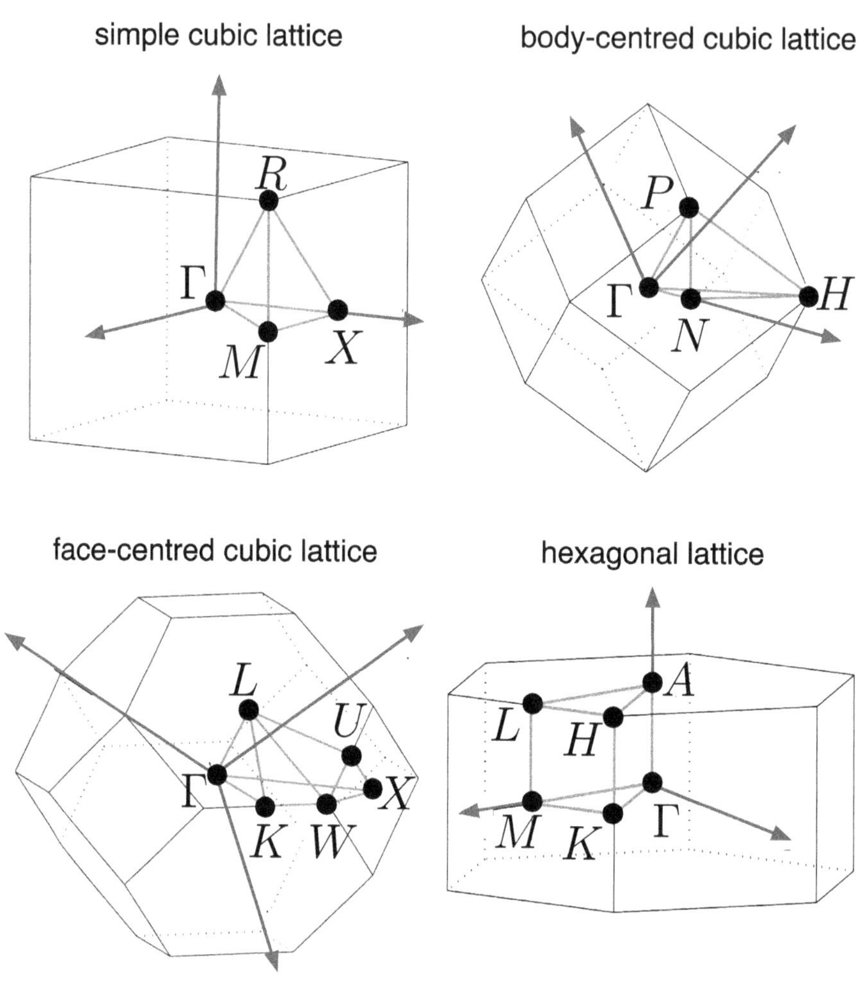

Figure 2.16. The first Brillouin zone of the sc, bcc, fcc, and hex lattices. Black dots and red lines represent the high symmetry points and the edges of the irreducible part, respectively. Blue arrows indicate the direction and the orientation (but not the modulus) of the reciprocal translation vectors $\{\mathbf{b}_1, \mathbf{b}_2, \mathbf{b}_3\}$ defined in table 2.2.

2.5 Lattice defects

The property of translational invariance extensively discussed in the previous sections generates ideally perfect crystalline structures. This is valid if we either consider an infinite lattice or apply Born–von Karman periodic boundary conditions. While useful in many circumstances to develop the constitutive ideas of solid state physics, this idealised situation is surely a strong approximation to reality: in fact, perfect crystals do not exist since *at any finite temperature a solid state system does contains defects*, i.e. lattice imperfections, that locally break the translational invariance. Their role is key in affecting many physical properties like, for instance, the transport of electric charge or thermal energy.

We will prove the unavoidable presence of defects in crystals by applying a simple thermodynamical argument to a monoatomic Bravais lattice containing N atoms and kept at constant non-vanishing temperature T and pressure P. Its energy content is provided by the Gibbs free energy $\mathcal{G} = \mathcal{U} - TS + PV = \mathcal{H} - TS$, where $\mathcal{U}$ and $\mathcal{H} = \mathcal{U} + PV$ are the internal energy and enthalpy, respectively (in appendix C, reference is made to the thermodynamic potentials used in this demonstration). The generation of a lattice defect (we mean: the local alteration of the crystal structure) requires a work

$$\Delta\mathcal{H}_f = \mathcal{H} - \mathcal{H}_0, \tag{2.15}$$

known as the *formation energy* of the defect. In this equation $\mathcal{H}_0$ represents the enthalpy of the pristine ideal crystal. In order to make physical concepts clear, we consider the actual case of a *lattice vacancy* and a *self-interstitial defect*: in the first case, a single atom is removed and taken far away from the crystal, while in the former case an extra atom of the same chemical nature is added to the crystal in a position not corresponding to any lattice point[12]. These defects are named *native*, since the chemistry of the crystal is unaffected by their existence. While these are specific (but realistic) situations, the reasoning developed below will lead to conclusions of general validity. We will further assume that the crystal is always in thermodynamical equilibrium, even after defects have been generated in it. A more thorough discussion on the formation of crystal defects is found elsewhere [16, 17].

The generation of a native defect will of course affect the crystal entropy and, therefore, we need to calculate its variation ΔS with respect to the pristine situation. Let us suppose to generate n similar defects at random and far away[13] positions. The number W of unalike configurations where the same number of native defects is differently distributed within the Bravais crystal is given by simple combinatorics

$$W = \frac{N!}{(N - n)!n!}, \tag{2.16}$$

so that the *entropy variation ΔS due to defects* is

$$\Delta S = S - S_0 = k_B \ln\left[\frac{N!}{(N - n)!n!}\right] - k_B \ln[1] = k_B \ln \frac{N!}{(N - n)!n!}, \tag{2.17}$$

where k_B is the Boltzmann constant and S_0 is the entropy of the perfect crystal for which $W = 1$ since there exists just one ideal lattice configuration. In order to process

---

[12] This means that there does not exist a lattice vector of the kind given in equation (2.1) giving the position of the self-interstitial defect.

[13] This assumption allows us to neglect the interactions among defects, by further considering that the vacancy and the self-interstitial are not electrically charged.

equation (2.17) we make use of another combinatorics result, namely the Stirling approximation, according to which for any sufficiently large number Ω it holds that

$$\ln \Omega! = \Omega \ln \Omega - \Omega. \tag{2.18}$$

By using the Stirling approximation in equation (2.17) we get

$$\begin{aligned}
\Delta S &= k_B[\ln N! - \ln(N - n)! - \ln n!] = \\
&= k_B[N \ln N - N - (N - n) \ln (N - n) + (N - n) - n \ln n + n] = \\
&= k_B[N \ln N - (N - n) \ln (N - n) - n \ln n].
\end{aligned} \tag{2.19}$$

By combining equations (2.15) and (2.19) we eventually obtain the *Gibbs free energy variation* $\Delta\mathcal{G}$ *caused by the generation of n native defects in the crystal*

$$\begin{aligned}
\Delta\mathcal{G} &= \mathcal{G} - \mathcal{G}_0 = n\Delta\mathcal{H}_f - T(S - S_0) = \\
&= n\Delta\mathcal{H}_f - k_BT[N \ln N - (N - n) \ln (N - n) - n \ln n].
\end{aligned} \tag{2.20}$$

Since the crystal is in equilibrium we have

$$\frac{d\Delta\mathcal{G}}{dn} = 0 \;\rightarrow\; \Delta\mathcal{H}_f - k_BT[\ln(N - n) - \ln n] = 0 \;\rightarrow\; \ln \frac{N - n}{n} = \frac{\Delta\mathcal{H}_f}{k_BT}, \tag{2.21}$$

which leads to[14]

$$n(T) = N \exp\left[-\frac{\Delta\mathcal{H}_f}{k_BT}\right], \tag{2.22}$$

which provides the *number $n(T)$ of native defects found in the crystal in equilibrium at temperature T*. This result holds in general[15], proving that at any finite temperature a crystal must necessarily contain any kind of defects, the number of which varies according to their formation energy $\Delta\mathcal{H}_f$.

In order to catalogue the various kinds of defects, in the next section we will follow a double criterion, according to which defects are distinguished both on the basis of their dimension (*point* and *extended* defects) and by considering their chemical nature (*native* and *non-native* character). Defects of any kind can combine within the same crystal, generating a structure that can markedly deviate from the ideal case of perfect lattice. Among the many possible configurations, two situations play an important role in solid state and materials physics, namely *alloys* and *polycrystals*; they are presented in appendix B. Here we also outline the intriguing case of *quasi-crystals*: atomic architectures made by regular repetitions of structural building blocks with no translational invariance.

---

[14] We remark that $\ln[(N - n)/n] = \ln(N/n - 1) \simeq \ln N/n$ since $N \gg n$.

[15] It is important to remark that we have implicitly assumed a *static lattice approximation*, that is: crystalline atoms have been considered as clamped at their lattice positions. Actually, as will be extensively discussed in the next chapters, atoms oscillate around their positions: therefore, we should also consider the vibrational energy contribution to $\mathcal{G}$. For the sake of simplicity, we have overlooked this feature, but this does not affect the general validity of the final conclusion.

2.5.1 Point defects

Lattice imperfections involving just a single atom are named *point defects*: their pictorial representation is reported in figure 2.17. They are more specifically referred to as *native* or *non-native*: the former ones correspond to the vacancy and self-interstitial case, previously introduced; the latter ones, instead, are generated whenever an atom is added of a chemical species not found in the pristine crystal. Of course, in lattices with a basis various kinds of vacancies and self-interstitials in fact exist. Furthermore, in two-atom compound crystals we can also find the *anti-site* defect, corresponding to a position exchange between two unalike nearest neighbouring atoms. Sometimes point defects gather to form small *defect aggregates*. The most common aggregates are *defect clusters* (where, for instance, a number of vacancies or self-interstitials precipitate, thus forming, respectively, a small void or an inclusion in the host crystal) and *Frenkel pairs* (where a bound pair of a vacancy and a self-interstitial is formed). Finally, an atom can replace a regular crystalline atom, belonging to a different chemical species: this configuration is known as a *substitutional impurity*.

An important feature missing so far in our discussion is that the *underlying lattice is deformed by the presence of defects*. We can say that this is *per se* another kind of defect which, in this case, is described in terms of an *induced lattice strain field*: interatomic distances are varied with respect to the ideal case and the bond network is accordingly distorted. In figure 2.18 we provide a rendering of this concept in the simple case of a two-dimensional square lattice which offers the possibility of a very intuitive graphics.

Impurity defects are generated in the host crystal by natural contamination or by some artificial process. In the first case, we refer to the very common *contamination from atmospheric environment* which injects impurities and, by affecting its chemical purity, alters many intrinsic physical properties of the crystal. Among the most

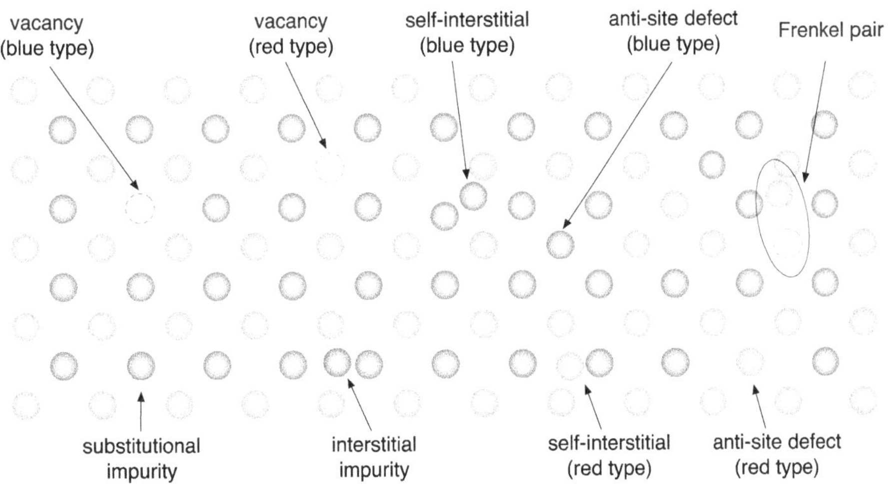

Figure 2.17. Point defects in a model two-dimensional compound crystal.

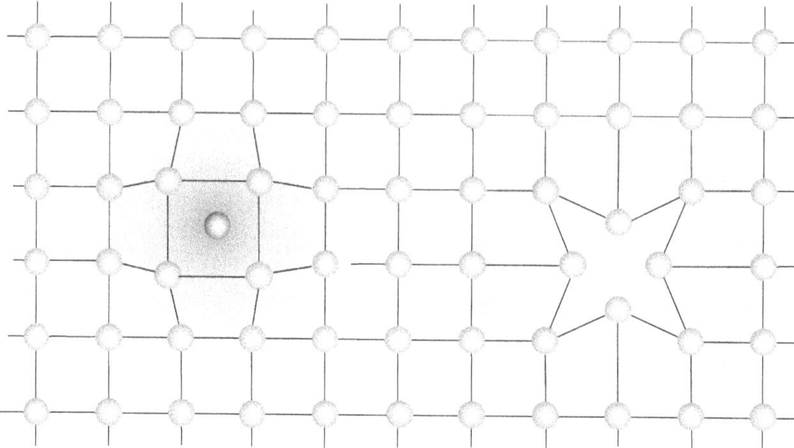

Figure 2.18. Pictorial representation of the lattice strain field (shaded regions) generated by an interstitial impurity (left) and a vacancy (right). The black lines represent the distorted network of the interatomic bonds.

common impurities we find hydrogen, nitrogen, carbon and oxygen. Another example of contamination is represented by the deposition on the surface of oxygen atoms, always present in the air humidity: this process, called oxidation, causes the formation of a surface oxide layer that often aggregates in a non-stoichiometric and structurally disordered form.

In the case of intentional contamination, impurities are implanted into the host lattice in arbitrary concentration. This process is normally called *doping*: it is widely used in technology whenever it is necessary to alter the intrinsic number of charge carriers of a material, so as to obtain the exact value useful for the application of interest. The most common doping technique is *ion implantation*: an ion beam is accelerated against a target sample; if the kinetic energy of the accelerated ions is sufficiently high (in practice this energy varies in the range 10^3–10^6 eV), they can penetrate the sample and, through a series of multiple inelastic collisions with the atoms of the target crystal, they lose their initial kinetic energy, eventually stopping inside the sample. This bombardment obviously generates quite a large lattice damage because many atoms of the implanted region are knocked out from their lattice position by collisions with the ions of the beam. Therefore, as a result of the implantation process, the target sample ends up having a very messy and chemically contaminated structure. By now applying suitable thermal annealing cycles, the lattice damage can be recovered: during the temperature-activated recrystallisation process the implanted impurities are incorporated into the lattice, mainly in the form of substitutional defects. The implantation process is sketched in figure 2.19 (top).

In order to understand how doping affects the electrical property of a material, we qualitatively discuss the prototypical case of a silicon sample which has been implanted by As atoms. At this stage it is useful to recall that Si and As belong to the group IV and group V, respectively (see appendix A), and therefore their chemical valence is 4 and 5: for this reason silicon crystallises as an elemental solid in the diamond structure with four-fold coordination. Whenever an As atom replaces a

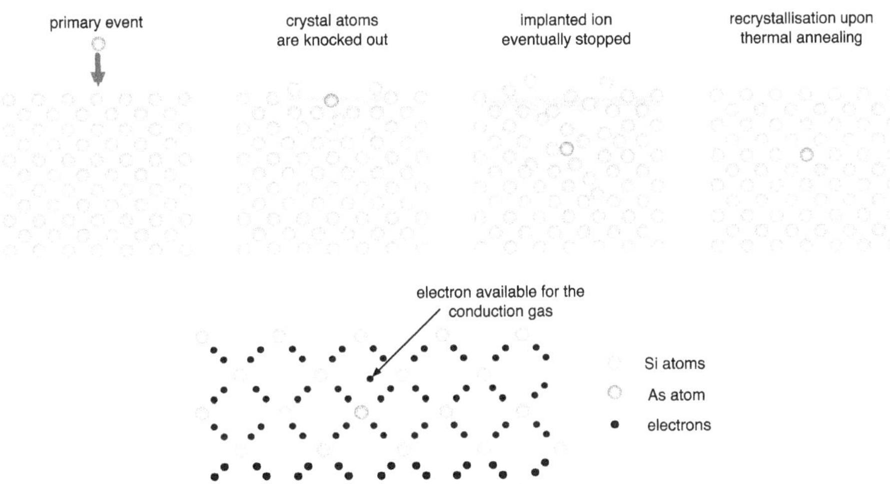

Figure 2.19. Top: pictorial representation of the ion implantation process. The grey shaded area indicates the region of maximum lattice damage. Bottom: graphical rendering of the doping process. For each As atom implanted into a silicon sample, an electron is made free to join a conduction gas of charge carriers.

Si one upon implantation, its first four electrons provide the same bonding as in the pristine sample; the fifth electron, instead, is available to join a conduction gas of free similar charge carriers generated by the other implanted dopant impurities. The number density of this gas is basically determined by the actual number of implanted As atoms per unit volume: its overall conduction properties can be therefore engineered as needed. The doping process is sketched in figure 2.19 (bottom).

2.5.2 Extended defects

Lattice defects involving multi-atomic configurations are called extended defects and represent lattice errors. The two most significant cases we limit our attention to are *dislocations* and *grain boundaries*.

Dislocations are line defects: a crystal lattice is 'dislocated' with respect to a line defined by an appropriate vector $\mathbf{L_d}$. The concept is illustrated in figure 2.20 (top) for the two different cases of *edge dislocation* and *screw dislocation*. Dislocations are described crystallographically by a set of two vectors: the first one is $\mathbf{L_d}$, while the second vector (indicated with the symbol $\mathbf{B_d}$) is called *Burgers vector* and it is graphically represented in figure 2.20 (bottom) in the case of an edge dislocation. Basically, $\mathbf{B_d}$ represents the difference in path when the dislocation core is short-circuited in the defective lattice or when the same path is followed in the perfect lattice[16]. By means of the pair $\{\mathbf{L_d}, \mathbf{B_d}\}$ we can distinguish the two kinds of extended line defects in that $\mathbf{L_d}$ and $\mathbf{B_d}$ are normal or parallel in edge or screw dislocations, respectively.

---

[16] It is understood that the individual steps of the path are represented by translation vectors.

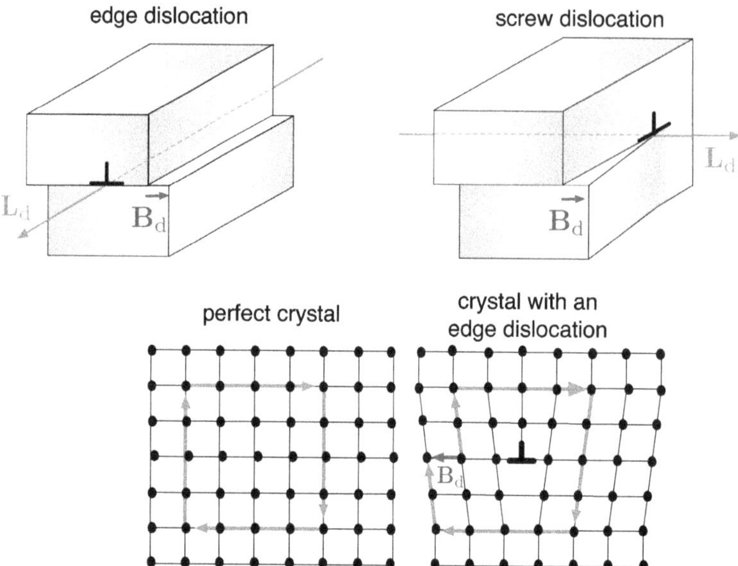

Figure 2.20. Top: an edge dislocation (left) and a screw dislocation (right) are shown together with their line vector $\mathbf{L}_d$ and Burgers vector $\mathbf{B}_d$. Bottom: the graphical definition of the Burgers vector in the case of an edge dislocation. The magenta arrows mark the path defining $\mathbf{B}_d$. The symbol $\perp$ indicate the position of the dislocation core.

Dislocations play a fundamental role in plasticity[17]. Qualitatively, a plastic deformation is due to the generation–migration–accumulation sequence of a dislocation forest. Another situation where dislocations are found is at the interface between two lattice-mismatched crystals. In this case, in an attempt to accommodate the unalike interatomic spacings on the planes parallel to the interface, the two crystals deform (by stretching or by compression provided that the lattice constant is smaller or larger, respectively) and thus they store elastic energy. When the lattice mismatch is sizeable, the accumulated elastic energy can be sufficient to be converted into formation work of dislocations, whose generation allows recovery of the pristine lattice spacings far away from the dislocation core.

Grain boundaries are planar defects: they form at the interface between two differently oriented crystal lattices. There are basically two different ways to generate a grain boundary (GB), both obtained by cutting a crystal along an imaginary plane: (i) a *twist* GB is formed by rotating one of the two semi-crystals around the normal direction of the imaginary plane; (ii) a *tilt* GB is instead generated by inclining one of the two semi-crystals with respect to the other one, through the imaginary plane. When the tilt or twist angle is small, a regular bond network is reconstructed at the interface between the two semi-crystals, although with a different topology with respect of the ideal crystal. On the other hand, when such angles are large enough, the accumulated elastic energy of lattice distortion is

---

[17] In solid mechanics plasticity is the tendency of a body under load to undergo permanent deformations.

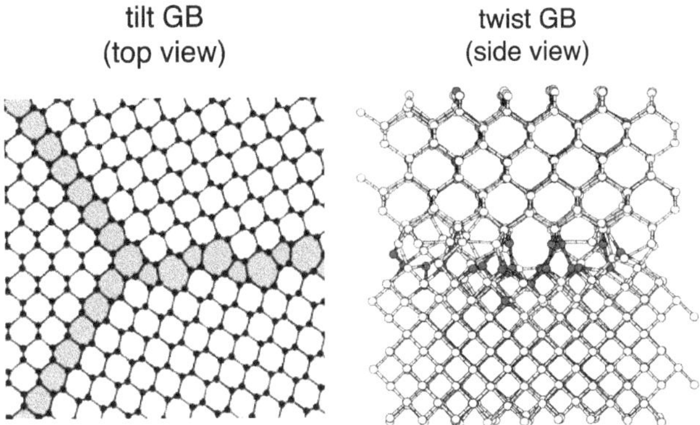

Figure 2.21. Left: a triple junction among three tilt GBs in silicon with small tilt angles; the shaded area represents the GB region. Right: a twist GB in silicon with a large twist angle; some structural disorder is found, both in atomic coordination and topology of the bond network. Pictures are obtained by atomistic simulations. Left: picture adapted from [18]. Right: picture adapted [19].

converted in formation work of an amorphous-like boundary layer; here atoms are found with coordination other than the pristine one, and bonds are randomly oriented. GBs of both kinds are shown in figure 2.21.

Grain boundaries represent accumulation regions for impurities or precipitation sites for dopant species. Qualitatively, we can explain this evidence by considering the possibly very disordered character of the bonding network in the boundary region: as a matter of fact, here a very high density of unsaturated bonds with high chemical reactivity is found (see for instance figure 2.21(right)): they work as gatherers of chemical species different from those of the host lattice.

2.6 Classification of solids

Symmetry has been largely used in the previous sections to rigorously classify crystalline solids. Other criteria can in fact be followed, based on physical properties of any kind. For instance, by looking at the mechanical response under load we can discriminate between elastic or plastic materials, as well as classify their fracture, failure, and yield behaviours; similarly, by investigating the interaction between electromagnetic waves and solids, we can separate them according to their optical properties. However, the most fruitful classification scheme based on a physical property is likely based on the *configuration assumed by the crystalline valence electrons*. This approach relies on the frozen-core approximation presented in section 1.3.2; since at this stage we have not yet developed any knowledge about the electronic structure of a solid state system, we will proceed at a more qualitative (but nevertheless useful in the next chapters[18]) level than allowed by symmetry classification.

---

[18] Also, it provides a justification for the previous reference to metallic or non-metallic solids we did in sections 2.3.2 and in appendix B.

The first step consists in making a difference between *metals* and *insulators*. While the most rigorous definition is only provided by the quantum band theory (see chapter 8), for the present discussion it is sufficient to rely just on phenomenological inputs: *metals are good electrical conductors*, while insulators are not. Whenever subjected to the action of an external electric field, metals are crossed by a current of charge carriers. This evidence is qualitatively attributed to the fact that valence electrons form a conduction gas of nearly-free charged particles: under the action of an external electric field, the conduction gas is accelerated, giving rise to current phenomena. On the other hand, in an insulator the density of such a gas is so small, that the resulting current density is comparatively negligible. From these considerations we draw a qualitative picture: in a metal system, ions sitting at lattice positions are embedded into a 'glue' of valence electrons which form an almost uniform charge distribution.

The case of insulators is definitely more complex, since quite different situations are found. Basically, they are broadly distinguished into four kinds:

- *molecular crystals*: solids made of noble elements Ne, Ar, Kr, Xe belonging to the group VIII A (see appendix A); their electronic structure is characterised by the fact that they completely lack valence electrons, since the configuration of the isolated atoms is only slightly affected in the crystalline state; their binding is due to weak van der Waals or electric dipole forces;
- *ionic crystals*: solids made of positive and negative ions, as consequence of electron transfer from cations to anions; their electronic structure is characterised by the fact the valence electron density is very highly localised nearby the cation and anion cores; consequently, they can be roughly treated as a periodic array of impenetrable positively or negatively charged spheres, respectively. Their physical properties are therefore largely dominated by electrostatic interactions; the most common ionic crystals are formed by atoms belonging to the I–VII A (also known as alkali-halides), or II–VI A, and III A–V A groups[19];
- *covalent crystals*: solids where the valence charge is mainly localised along chemical bonds; this configuration is explained by the so called octet rule[20], according to which elements tend to bond through electron sharing so that each atom reaches the same electronic configuration as a noble gas with eight electrons in the outer shell; covalent solids are made of atoms belonging to the IV A, or II–VI A, and III A–V A groups;
- *hydrogen-bonded crystals*: solids formed by electrostatic attraction between a hydrogen atom (working as the positive centre) and a negatively charged ion; since hydrogen is comparatively smaller than any other crystalline ion, it can easily fit into the crystal structure; typical hydrogen-bonded solids are organic ones or ice (solid water).

---

[19] In fact, II–VI A and III A–V A solids are better defined as partly ionic, partly covalent solids.

[20] The octet rule is used in chemistry as a rule of thumb for predicting the properties of the elements [20].

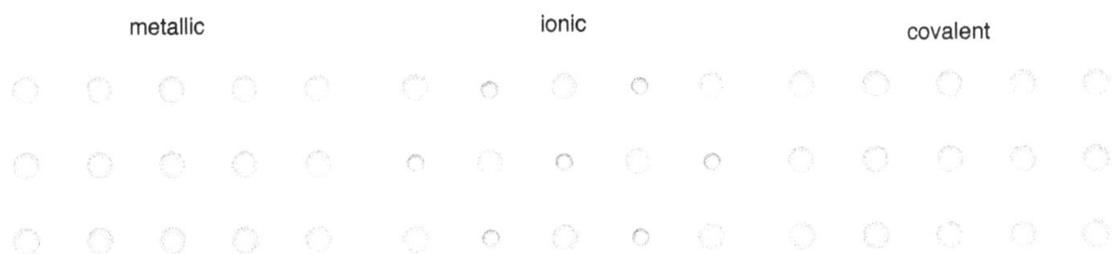

Figure 2.22. A pictorial representation of the valence charge distribution (grey) in metals and ionic or covalent insulators. Positive and negative ions are in red and blue colour, respectively.

The metallic and insulating bonding structure is schematically shown in figure 2.22, where a pictorial view of the valence charge distribution is drawn.

2.7 Cohesive energy

A final question still remains to be answered: *how much work E_{cohesive} is needed to assemble a set of atoms into a crystalline solid?* This quantity is defined as the difference in energy between a configuration where atoms lie at infinite distance and a configuration where they form a bound crystal. Let us name $E_T^{\text{free atoms}}$ the total energy of N free atoms (possibly of different chemical species) and E_T^{crystal} the energy of their crystalline configuration, then we calculate the *cohesive energy* per atom as

$$e_{\text{cohesive}} = \frac{1}{N}[E_T^{\text{free atoms}} - E_T^{\text{crystal}}], \tag{2.23}$$

while of course $E_{\text{cohesive}} = N e_{\text{cohesive}}$.

The most fundamental approach for calculating the cohesive energy is quantum mechanical [21], as formally shown in equation (1.13) (since E_{cohesive} does depend on *the ground-state energy of the solid*). This is the only way to proceed for metals and covalent crystals where valence electrons are totally or partially delocalised. On the other hand, for molecular and ionic crystal it is possible to follow a much simpler, although phenomenological, approach [10, 11, 15] where a number of simplifications are assumed, in that: atoms are treated classically (we place them at rest exactly in positions corresponding to crystalline lattice sites with no quantum zero-point energy effects included; the valence electron charge distribution is treated by classical electrostatics); the calculation is performed by imposing an arbitrary value of the lattice constant(s); and the temperature is set to zero (no free energy contributions are accounted for). By imposing that the derivative of the cohesive energy with respect to the lattice constant(s) is zero, we obtain their equilibrium values: those ones at which the real crystal is found in stable equilibrium at zero pressure.

In *molecular crystals* binding is due to the balance between interatomic attractive dipolar interactions (long-ranged sand weak) and Pauli repulsion between electron clouds (occurring at short distance and very strong). Usually the resulting pair interaction potential is written in the form known as the 6–12 Lennard-Jones potential [22]

$$V_{LJ}(R) = 4\epsilon\left[\left(\frac{\sigma}{R}\right)^{12} - \left(\frac{\sigma}{R}\right)^{6}\right], \tag{2.24}$$

where ϵ represents the attraction strength and σ the radius of the repulsive core (assumed spherical and centred on each lattice site). In equation (2.24) R is the interatomic distance. The cohesive energy E_{cohesive} of a molecular crystal crystal containing N atoms is

$$E_{\text{cohesive}} = \frac{N}{2} \sum_{\mathbf{R}_l \neq 0} V_{LJ}(R_l), \tag{2.25}$$

where the origin $\mathbf{R}_l = 0$ has been placed on an arbitrary lattice site. If we indicate by R_{nn} the first nearest-neighbour distance, then any translation vector is cast in the form $\mathbf{R}_l = \alpha_l R_{nn}$; in this way, we can more conveniently write the cohesive energy as

$$E_{\text{cohesive}}(R_{nn}) = 2N\epsilon\left[A_1\left(\frac{\sigma}{R_{nn}}\right)^{12} - A_2\left(\frac{\sigma}{R_{nn}}\right)^{6}\right], \tag{2.26}$$

where $A_1 = \sum_{\mathbf{R}_l \neq 0}\alpha_l^{-12}$ and $A_2 = \sum_{\mathbf{R}_l \neq 0}\alpha_l^{-6}$, both only depend on the crystal structure. Minimising $E_{\text{cohesive}}(R_{nn})$ leads to the theoretical prediction of equilibrium nearest-neighbour separation $R_{nn} = (2A_1/A_2)^{1/6}\sigma$. Finally, by a standard thermodynamics argument (see appendix C) and using the volume per atom $v = V/N$ (where V is the total volume of the N-atom crystal) and the cohesive energy per atom $e_{\text{cohesive}} = E_{\text{cohesive}}/N$ we easily get the bulk modulus B of molecular crystals as

$$B = v\,\frac{\partial^2 e_{\text{cohesive}}}{\partial v^2} = \frac{4\epsilon}{\sigma^3}\,A_1^{-3/2}\,A_2^{5/2} \tag{2.27}$$

The prediction of the Lennard-Jones potential, despite its phenomenological character, is remarkably in good agreement with experimental data, as shown in table 2.3.

In *ionic crystals* binding is due to the balance between electrostatic interactions (long-ranged and both attractive or repulsive, according to the signs of the charges for any selected ion pair) and Pauli repulsion between electron clouds (occurring at

Table 2.3. Lennard-Jones parameters (ϵ, σ), cohesive energy per atom (e_{cohesive}), nearest-neighbour separation R_{nn}, and bulk modulus B of some fcc molecular crystals made by noble elements. Values in parenthesis represent experimental data. For any crystal here represented we have $A_1 = 12.132$ and $A_2 = 14.454$.

	ϵ [10^{-4} eV]	σ [Å]	e_{cohesive} [eV/atom]	R_{nn} [Å]	B [atm]
Ne	32.11	2.74	0.02	2.99 (3.13)	1.79 (1.09)
Ar	107.25	3.40	0.08	3.71 (3.75)	3.15 (2.67)
Kr	144.49	3.65	0.12	3.98 (3.99)	3.42 (3.46)
Xe	205.50	3.98	0.17	4.34 (4.33)	3.77 (3.56)

short distance). The first term is by far the dominating one in these crystals and it is hard to evaluate correctly because of the long-range nature of electrostatic interactions: as a matter of fact, while there is no ambiguity in calculating the total electrostatic energy of a finite set of point-charges, the same quantity becomes ill defined for an infinite ionic crystal. The physical reason behind this is explained by a simple argument: let us suppose we reach the limit of an infinite crystal by assembling step-by-step increasingly large crystallites, each containing a finite number of anion/cation pairs; it is easy to figure out that each crystallite can be realised in many possible ways, each one with a different surface charge distribution. The infinite crystal limit is, therefore, reached by following different growth paths, resulting in a different total electrostatic potential energy because of the different surface electrostatics of its building blocks[21].

Methods mastering Coulomb lattice sums do nevertheless exist, among which the Edwald summation technique is the most widely used in solid state physics. Since it is anything but an elementary method, it is well beyond the level of our treatment (for a detailed description see [21]). Here we limit ourselves to outlining the main result of cohesion theory in ionic crystals. Let us consider a binary ionic crystal and let Q_β be the charge of the ion in position $\mathbf{R}_\beta$. The Coulomb potential $V_C(\mathbf{R}_\alpha)$ felt by the ion in position $\mathbf{R}_\alpha$ due to all remaining ions in the crystal is

$$V_C(\mathbf{R}_\alpha) = \frac{e}{4\pi\epsilon_0} \sum_{\beta \neq \alpha} \frac{Q_\beta}{R_{\alpha\beta}} \tag{2.28}$$

where $R_{\alpha\beta} = |\mathbf{R}_\alpha - \mathbf{R}_\beta|$ is the ion–ion distance. We now introduce the *Madelung constant* M_α for the α-type ions by writing

$$V_C(\mathbf{R}_\alpha) = \frac{e}{4\pi\epsilon_0 R_{nn}} M_\alpha, \tag{2.29}$$

where R_{nn} is the equilibrium first nearest-neighbour distance and

$$M_\alpha = \sum_{\beta \neq \alpha} Q_\beta \frac{R_{nn}}{R_{\alpha\beta}}, \tag{2.30}$$

a dimensionless constant which is calculated by some lattice summation method. The Coulomb potential energy $E_C(\mathbf{R}_\alpha)$ of the αth ion is eventually written as

$$E_C(\mathbf{R}_\alpha) = Q_\alpha e V_C(\mathbf{R}_\alpha) = \frac{e^2}{4\pi\epsilon_0 R_{nn}} Q_\alpha M_\alpha. \tag{2.31}$$

---

[21] More formally, it is proved that the sum of Coulomb interactions among positive and negative charges regularly distributed onto a translationally invariant lattice is only conditionally convergent: it provides any number, depending on the order in which the terms in the sum are added [21].

There are as many Madelung constants as ions occupying different lattice sites. For instance, for a typical binary ionic crystal we have two constants: one for each chemical species; however, since anions and cations occupy lattice sites with the very same symmetry, the two constants have opposite sign but same magnitude. Furthermore, M_α is specific of any crystal structure and, in particular, its magnitude is calculated to be 1.763 for the CsCl structure, 1.748 for the NaCl structure, and 1.638 for the zincblende structure.

In order to complete the model for the cohesion energy of ionic crystals, we must add a contribution describing the short-range Pauli repulsion. Its is usually written as an inverse power law of the distance between nearest-neighbour ions or, more accurately, as a rapidly decaying exponential. In summary, the cohesive energy is written as

$$E_{\text{cohesive}} = \sum_\alpha E_C(\mathbf{R}_\alpha) + \sum_{\text{ion pairs}} B_{\alpha\beta} e^{-R_{\text{nn}}/C_{\alpha\beta}}, \tag{2.32}$$

where the second term on the right-hand side of this equation is obtained by summing over all the ion pairs in the crystal and the $B_{\alpha\beta}$ and $C_{\alpha\beta}$ terms are phenomenological constants to be fitted on experimental information.

References

[1] Borchardt-Ott A 2012 *Crystallography–An Introduction* (Berlin: Springer)

[2] Hoffmann F 2020 *Introduction to Crystallography* 1 (Berlin: Springer)

[3] Glusker J P 2010 *Crystal Structure Analysis: A Primer* (Oxford: Oxford University Press)

[4] Hammond C 2015 *The Basics of Crystallography and Diffraction* (Oxford: Oxford Science Publications)

[5] Evarestov R A 1997 *Site Symmetry in Crystals: Theory and Applications* (Berlin: Springer)

[6] Müller U 2017 *Symmetry Relationships between Crystal Structures–Applications of Crystallographic Group Theory in Crystal Chemistry* (Oxford: Oxford Science Publications)

[7] Burns G and Glazer M 2013 *Space Groups for Solid State Scientists* 3rd edn (Waltham, MA: Academic)

[8] Wright J D 1995 *Molecular Crystals* (Cambridge: Cambridge University Press)

[9] Kitaigorodsky A I 1973 *Molecular Crystals and Molecules* (New York: Academic)

[10] Dove M T 2003 *Structure and Dynamics–An Atomic View of Materials* (Oxford: Oxford University Press)

[11] Kittel C 1996 *Introduction to Solid State Physics* 7th edn (Hoboken, NJ: Wiley)

[12] Ashcroft N W and Mermin N D 1976 *Solid State Physics* (London: Holt-Saunders)

[13] Kenyon I R 2008 *The Light Fantastic–A Modern Introduction to Classical and Quantum Optics* (Oxford: Oxford Science Publications)

[14] Bransden B H and Joachain C J 1983 *Physics of Atoms and Molecules* (Harlow: Addison Wesley)

[15] Hook J R and Hall H E 2010 *Solid State Physics* (Hoboken, NJ: Wiley)

[16] Tilley R J D 2008 *Defects in Solids* (New York: Wiley)

[17] Phillips R 2001 *Crystals, Defects and Microstructures* (Cambridge: Cambridge University Press)

[18] Costantini S, Alippi P, Colombo L and Cleri F 2000 *Phys. Rev.* B **63** 045302
[19] Cleri F, Keblinski P, Colombo L, Wolf D and Phillpot S R 1999 *Europhys. Lett.* **46** 671
[20] Pauling L 1970 *General chemistry* (New York: Dover Publications Inc)
[21] Grosso G and Pastori Parravicini G 2014 *Solid State Physics* 2nd edn (Oxford: Academic)
[22] Finnis M 2003 *Interatomic Forces* (Oxford: Oxford University Press)

Part II

Vibrational, thermal, and elastic properties

IOP Publishing

Solid State Physics
A primer
Luciano Colombo

Chapter 3

Lattice dynamics

Syllabus—*The dynamics of a crystal lattice is developed under the leading adiabatic and harmonic approximations. The main features are at first elaborated by studying the model cases of a monoatomic and a diatomic linear chain: here, the concepts of normal mode of vibration, dispersion relation, and acoustic or optical character of a vibration are extensively discussed. The dynamics of realistic three-dimensional crystals is formalised through the calculation of the dynamical matrix, which allows for the standard representation of the vibrational dispersion relations. Next, by a simple quantisation procedure of the ionic displacement field, we introduce the concept of phonon and interpret the vibrational spectrum of a crystal as the physics of a gas of such non-interacting pseudo-particles. Finally, we discuss the experimental determination of phonon dispersion relations by neutron spectroscopy and the vibrational density of states.*

3.1 Conceptual layout

The description of crystal structures developed in chapter 2 relies on an implicit (but really very strong) assumption, namely: ions are clamped at their lattice positions. This is, also, the situation assumed to define the eigenvalue problem for the total electron wavefunction given in equation (1.15) within the framework developed under the adiabatic approximation. While the assumption of *static lattice* is useful in the above contexts, it is either conceptually wrong and inadequate to describe many important solid state phenomena.

First of all, we recall that ions, although comparatively much more massive than electrons, have in fact a finite mass: therefore, according to fundamental quantum mechanics [1–4], they always (even at zero temperature) have a non-vanishing mean square momentum. We can reconcile crystallography with the quantum uncertainty principle by assuming that the *mean position of an ion* (obtained by averaging over its zero-point motion) corresponds to $\mathbf{R}_l$ in Bravais lattices or to $\mathbf{R}_l + \mathbf{R}_b$ for lattices with a basis (see equations (2.1) and (2.3), respectively).

This is not enough. Still considering ions at rest (although in some 'average' meaning) is inconsistent with a number of experimental evidences, including (but not limited to): thermal expansion, melting, thermal conductivity, sound propagation, inelastic scattering of electromagnetic waves or particles (electrons as well as neutrons). All together these phenomena provide a robust body of experimental evidence that *lattice ions do undergo some kind of motion*. The aim of this chapter is to fully characterise the corresponding *lattice dynamics*.

We will accomplish this task at first under the leading *adiabatic and classical approximations* (see sections 1.3.4 and 1.4.2, respectively). Non-classical dynamical features will appear later, by a suitable quantisation procedure operated on the ionic classical displacement field. In developing our classical phenomenological theory of lattice dynamics, we will assume that *there exists a many-body potential energy* $U = U(\mathbf{R})$ *governing the motion of the ions*[1]. Basically, U contains the ion–ion Coulomb interaction energy as well as their kinetic energy, as conceptualised in section 1.3.4.

The next step of simplification consists in assuming that *the displacement of each ion from its mean equilibrium position*[2] *is small as compared to the typical lattice spacing*. This assumption is proved by the very results of x-ray scattering: the crystal structure appears cleanly clear in spite of the fact that ions undergo movement during diffraction. In other words: ionic motion is neither diffusive nor so wide as to alter the underlying lattice structure.

In order to proceed rigorously, we must define a more accurate notion for ionic positions and displacements. To make it also compact, we agree to indicate the lattice position of the generic ion by $\mathbf{R}_l + \mathbf{R}_b$ and its displacement by $\mathbf{u}(lb)$, where it is understood that the indices (lb) run, respectively, over the lattice points and positions within the basis. Since electrons will not explicitly enter our discussion, Latin indices $i_1, i_2, \ldots, i_n$ will be used to label the Cartesian components.

The assumption of small displacements allows for the formal expansion of the many-body potential energy $U = U(\mathbf{R})$ into a power series of the ionic displacements. By setting to zero the constant term[3] U_0 for further convenience, we can write

$$
\begin{aligned}
U &= \sum_{lb,i} \left. \frac{\partial U}{\partial u_i(lb)} \right|_0 u_i(lb) + \frac{1}{2} \sum_{\substack{l_1b_1,i_1 \\ l_2b_2,i_2}} \left. \frac{\partial^2 U}{\partial u_{i_1}(l_1b_1)\partial u_{i_2}(l_2b_2)} \right|_0 u_{i_1}(lb)u_{i_2}(l'b') + \cdots \\
&= \sum_{n=1}^{+\infty} \frac{1}{n!} \sum_{\substack{i_1,i_2,\ldots,i_n \\ l_1b_1,l_2b_2,\ldots,l_nb_n}} U_{i_1i_2\cdots i_n}(l_1b_1, l_2b_2, \ldots, l_nb_n)\, u_{i_1}(l_1b_1)u_{i_2}(l_2b_2)\cdots u_{i_n}(l_nb_n),
\end{aligned}
\tag{3.1}
$$

---

[1] We recall that the symbol $\mathbf{R}$ indicates the full set of ionic positions.

[2] Hereafter equivalently referred to as its lattice position.

[3] This term represents the energy of a static lattice. Since it is constant, we can gauge the potential energy so that $U_0 = 0$ with no loss of generality.

where we have used the compact notation

$$\frac{\partial^n U}{\partial u_{i_1}(l_1 b_1)\partial u_{i_2}(l_2 b_2)\ldots\partial u_{i_n}(l_n b_n)}\bigg|_0 = U_{i_1 i_2\ldots i_n}(l_1 b_1,\ l_2 b_2,\ \ldots,\ l_n b_n). \qquad (3.2)$$

As indicated, all derivatives are calculated with all ions at their rest positions; in particular, this implies that

$$\frac{\partial U}{\partial u_i(lb)}\bigg|_0 = 0, \qquad (3.3)$$

which defines the rest condition for the crystal[4]. Since displacements are small, we can truncate the expansion to the first non-vanishing term, thus obtaining the many-body potential energy $U_{\text{harmonic}}^{\text{classical}}$ in *harmonic approximation*

$$U_{\text{harmonic}}^{\text{classical}} = \frac{1}{2}\sum_{\substack{l_1 b_1, i_1 \\ l_2 b_2, i_2}} U_{i_1 i_2}(l_1 b_1,\ l_2 b_2)\ u_{i_1}(l_1 b_1)u_{i_2}(l_2 b_2). \qquad (3.4)$$

This approximation is named 'harmonic' since the physics described by equation (3.4) is equivalent to that of a set of identical harmonic oscillators (as many as the ionic dynamical degrees of freedom) whose *force constants are given by the terms* $U_{i_1 i_2}(l_1 b_1,\ l_2 b_2)$. The analogy with classical mass-spring systems is further exploited by naming the ion motions as *vibrations*. In short, in our picture *ions harmonically vibrate around their lattice position*. This is the entry level for discussing lattice dynamics; other physical features will be added later.

Before addressing the physics of ionic vibrations, it is important to remark that the *force constants are subject to formal restrictions* [5, 6] which are imposed by some true physical features. The Taylor expansion given in equation (3.1) makes use of the full set of $3N$ vibrational degrees of freedom existing in a crystal made by N atoms. The expansion coefficients there appearing, i.e. the force constants, must nevertheless guarantee that the crystal energy U is invariant under a rigid body translation or rotation[5]. For instance, it is not difficult to prove that the invariance under a rigid translation dictates

$$\sum_{l_1 b_1, l_2 b_2, \ldots, l_n b_n} U_{i_1 i_2\ldots i_n}(l_1 b_1,\ l_2 b_2,\ \ldots,\ l_n b_n) = 0. \qquad (3.5)$$

Instead of using the $3N$ Cartesian components of the ionic displacement to expand U, we could rather use an alternative set of ($3N$-6) independent internal coordinates (such as interatomic distances or bond angles), carefully chosen to be invariant under rigid translations and rotations.

---

[4] The equilibrium condition of an infinite crystal is more complicated: ions must be at rest so that equation (3.3) is fulfilled and the corresponding configuration must be a stress-free one [5, 6]. The concept of stress will be introduced in the next chapters.

[5] These invariances reduce the number of degrees of freedom relevant for internal vibrations to ($3N - 6$), since translations and rotations each involve three of them.

The energy U must as well remain unchanged under any symmetry operation of the given crystal we are dealing with. Group theory [7] allows studying the action of any symmetry operation on each matrix whose elements are the 2nd-, 3rd-, 4th-, ⋯ nth-order force constants defined in equation (3.2). In this way we can calculate which are the independent non-zero elements of such matrices.

Finally, we remark that in the following *force constants will be considered as given parameters*, based on whose knowledge we will build the theory of lattice dynamics. This assumption is reasonable and adequate since, in general, force constants can be independently calculated by guessing some interatomic force field, whose parameters are typically fit on experimental vibrational frequencies (and, sometimes, using ionic displacement patterns as well). In appendix D some force constants models extensively used in literature are outlined. An alternative approach consists in developing a fully quantum mechanical theory of U, from which to derive the force constants; in general, they are much more accurate than their empirically determined counterparts, but they are obtained at the cost of a much higher computational burden. This more fundamental approach is based on the knowledge of the total energy of the crystalline ground state configuration: a task developed in chapter 10.

3.2 Dynamics of one-dimensional crystals

The most fundamental features of lattice dynamics can be brought out by investigating a one-dimensional crystal. The main advantage in studying such a simplified situation is that the heavy algebraic formalism indeed necessary to manage the full set of indices occurring in three-dimensional crystals is largely spared. A wire made by ions connected by effective harmonic springs is just enough to catch the essential physics: we will refer to this model system as the *harmonic linear chain*.

3.2.1 Monoatomic linear chain

Let us consider a linear chain where N identical ions of mass M are placed at distance a when they are at rest in equilibrium positions. This corresponds to a one-dimensional Bravais crystal with lattice spacing a; the primitive unit cell is obtained by the Wigner–Seitz construction as a segment of length a with the ion placed at its midpoint. By adopting Born–von Karman boundary conditions, the ionic positions are indicated as $R_l = la$ with $l = 0, 1, 2, \ldots, N - 1$. Finally, following the force constant approach discussed in the previous section, we represent the interactions between nearest neighbouring ions as harmonic springs. The situation is sketched in figure 3.1.

Let us consider a *longitudinal vibration* of the chain, that is a displacement pattern in which the ions move along the chain direction. The classical equation of motion for the lth ion is

$$M\ddot{u}_l = \gamma^{(L)}(u_{l+1} + u_{l-1} - 2u_l), \tag{3.6}$$

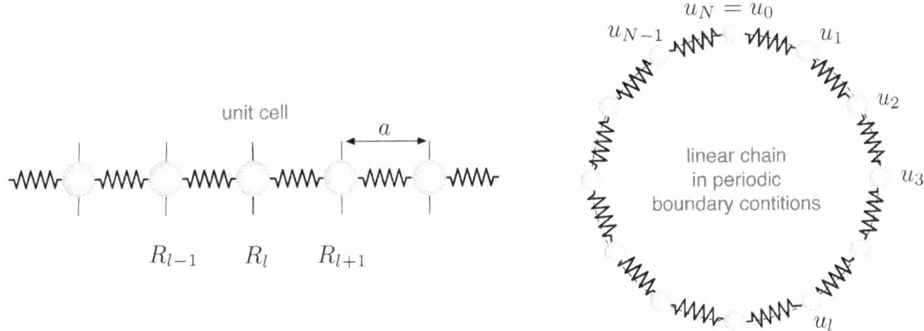

Figure 3.1. A monoatomic spring-mass linear chain (left) under the Born–Von Karman boundary condition (right); the unit cell is shaded in light-blue colour, with lattice constant a.

where $\gamma^{(L)}$ is the force constant of the effective spring. Suggested by the elementary mechanics of a vibrating wire, we seek a solution in the form

$$u_l = \frac{1}{\sqrt{NM}}\, |\mathcal{A}_q| \cos[qR_l - \omega(q)t + \varphi(q)], \tag{3.7}$$

where the normalising factor $(NM)^{-1/2}$ has been introduced for further convenience, while $|\mathcal{A}_q|$ and $\varphi(q)$ are the amplitude and the initial phase of the wave[6]. Of course, q and $\omega(q)$ are the wavenumber and the angular frequency of the travelling wave, respectively. Replacing equation (3.7) into equation (3.6) leads to

$$M\omega^2(q) = 2\gamma^{(L)}\,[1 - \cos(qa)] = 4\gamma^{(L)} \sin^2\!\left(\frac{1}{2}qa\right), \tag{3.8}$$

which is known as the *dispersion relation* and it is shown in figure 3.2(top). This representation is redundant since it ignores translational periodicity: it makes no difference in the displacement u_l by increasing $q \to q + G$ with $G = 2m\pi/a$ a reciprocal lattice vector of the linear chain crystal (m is any positive or negative integer number). It is therefore customary to adopt the *reduced zone scheme*: the dispersion relation is represented only for $q \in$ 1BZ or, equivalently, for $q \in [-\pi/a, +\pi/a]$ as shown in figure 3.2(bottom). The actual number of allowed q is determined by the imposed boundary conditions: since it must be $u_0 = u_N$ then

$$q = \frac{2\pi}{a}\frac{\xi}{N} \qquad \text{with} \quad \xi = 0, 1, 2, 3, \dots, N - 1 \tag{3.9}$$

This is a very important result: in an N-atom monoatomic chain there are only N independent values of the wavevector q associated with as many independent solutions of the equations of motion. In more physical terms: if we consider a one-dimensional chain containing N identical ions, there are only N different ways in which they can longitudinally oscillate around their equilibrium positions.

---

[6] Their relation is $\mathcal{A}_q = |\mathcal{A}_q| \exp[i\varphi(q)]$.

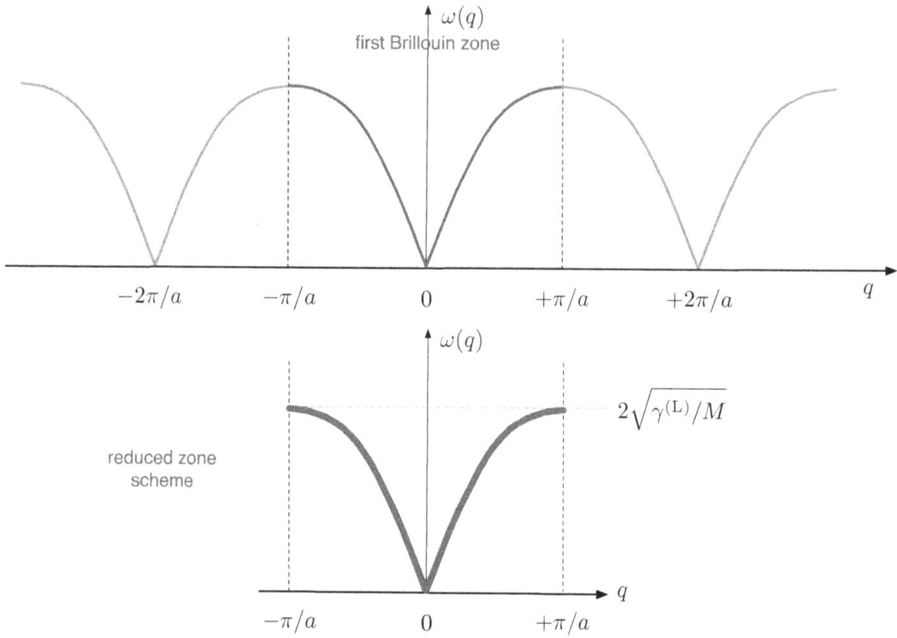

Figure 3.2. Top: dispersion relation $\omega = \omega(q)$ for the longitudinal vibrations of a monoatomic chain; the 1BZ is shaded in light-blue colour. Bottom: the same dispersion represented in the reduced zone scheme. The ion mass is M and the effective spring constant is $\gamma^{(L)}$.

The solutions of equation (3.8) are waves travelling along the chain with a group velocity $v_{\mathrm{g}} = d\omega(q)/dq$. In the limit $qa \ll 1$ the dispersion relation reduces to $\omega(q) = [\gamma^{(L)}/M]^{1/2}qa$, similarly to a wave propagating along a material wire with mass density M/a and subject to a mechanical tension $\gamma^{(L)}a$. In the long wavelength limit the dispersion relation is, therefore, that of a compressional wave propagating with group velocity $v_{\mathrm{g}} = \sqrt{\gamma^{(L)}/M}\,a$. In other words, in the limit $q \to 0$ (zone-centre wavevector) the equation (3.8) describes the propagation of a *longitudinal sound wave*. For this reason the $\omega = \omega(q)$ relation is referred to as the *acoustic dispersion*. In the opposite situation of $q = \pm\pi/a$ (zone-boundary wavevector), the vibrational frequency is found to reach the value $\omega(q = \pm\pi/a) = 2\sqrt{\gamma^{(L)}/M}$ which is understood as the *maximum frequency for longitudinal vibrations*.

We complete the above discussion by observing that *a linear chain can carry transverse oscillations* as well: in fact, ions can vibrate normally to the direction along which they are initially aligned to form the chain itself. The mathematical analysis of transverse oscillations is just the same as in the longitudinal case, with one obvious difference: the force constants describing the transverse effective springs may have a value $\gamma^{(T)} \neq \gamma^{(L)}$. So, at the mere cost of replacing in all the previous equations the appropriate force constant value, we can draw similar conclusions. In particular, we derive the *transverse dispersion relations* and get the speed of *transverse sound*. In general, the transverse force constants are smaller than longitudinal ones. Furthermore, since for a linear chain there are two equivalent transverse directions, the two force constants are equal $\gamma^{(T_1)} = \gamma^{(T_2)} < \gamma^{(L)}$ and two corresponding dispersion relations are degenerate.

Let us now move to consider displacements. In general, the oscillation paths of the chain are obtained from equation (3.7) by considering the N allowed discrete values of the wavenumber q, separated by the quantity $2\pi/Na$ within the interval $[-\pi/a, +\pi/a]$. In the presently discussed case of longitudinal oscillations the general solution for the displacement of the lth ion is

$$u_l = \frac{1}{\sqrt{NM}} \sum_q |\mathcal{A}_q| \cos[qR_l - \omega(q)t + \varphi(q)], \tag{3.10}$$

where the N possible amplitude-phase pairs[7] $[|\mathcal{A}_q|, \varphi(q)]$ are calculated by providing the N initial positions and the N initial velocities of the ions. The picture should at this point be clear: *each ion vibrates around its equilibrium position as a harmonic oscillator*, while *the chain as a whole is interested by N different normal vibrational modes*. The former describe single-particle motions, the latter provide information on collective movements.

In figure 3.3 the *displacement patterns* for longitudinal and transverse vibrations are shown, both at zone-centre and zone-boundary. The key point there illustrated is that if we consider ions belonging to nearest neighbour unit cells *they oscillate in phase for zone-centre vibrational modes*, while *they oscillate in phase opposition for zone-boundary vibrational modes*. The case of zone-centre vibrations deserves a further comment: figure 3.3 clearly shows that the travelling wave is a compressive wave in both the longitudinal and transverse polarisation, although the compression, respectively, occurs along or perpendicularly to the travelling direction of the wave. We can therefore distinguish between *longitudinal sound* and *transverse sound*. Experimental measurements of the sound speed in a real specimen provide evidence that the longitudinal sound (moving at speed $v_g^{(L)} = \sqrt{\gamma^{(L)}/M}\,a$) is faster than the transverse one (moving at speed $v_g^{(T)} = \sqrt{\gamma^{(T)}/M}\,a$): this is a direct confirmation that transverse force constants are smaller than longitudinal ones.

3.2.2 Diatomic linear chain

Let us now turn to consider the one-dimensional model of minimal complexity for a lattice with a basis, namely a diatomic linear chain. We need to define two ion masses M_1 and M_2 and two effective springs $\gamma^{(L)}$ and $\xi^{(L)}$, respectively, coupling ions within the same unit cell or belonging to nearest neighbouring unit cells. Ion positions are now indicated as $R_{l,1} = R_l + R_1$ and $R_{l,2} = R_l + R_2$, where R_l labels the lth unit cell, while R_1 and R_2 specify the ion within the basis. The situation is sketched in figure 3.4 and once again we start by considering *longitudinal oscillations*.

---

[7] One pair for each allowed wavenumber.

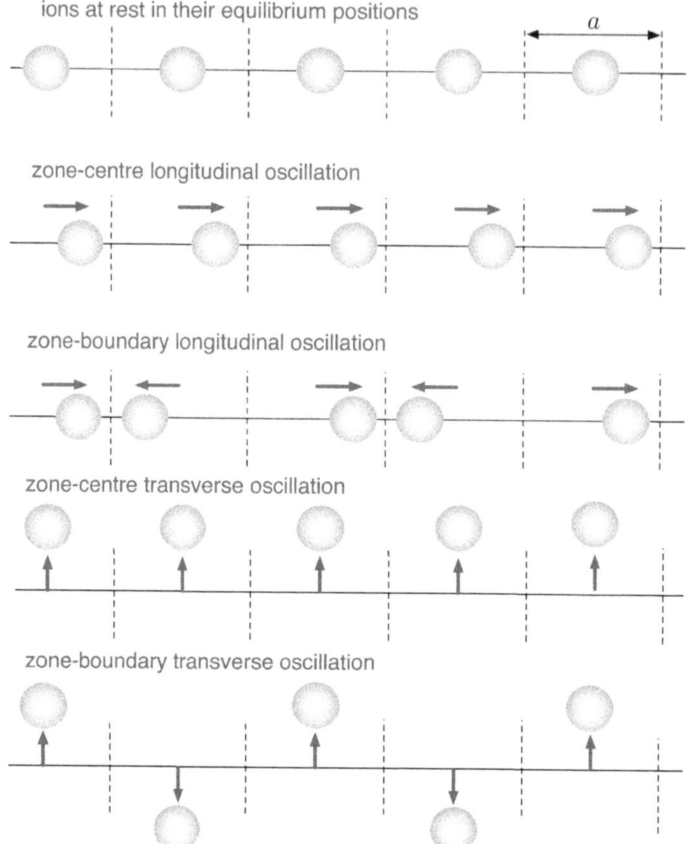

Figure 3.3. Zone-centre and zone-boundary displacement patterns of longitudinal and transverse vibrations in a monoatomic linear chain whose unit cell is shaded in light-blue colour in the top panel. Blue arrows indicate the direction and the orientation of the ionic displacement vector. The travelling direction of the wave is towards left or right provided that its wavevector is, respectively, positive or negative.

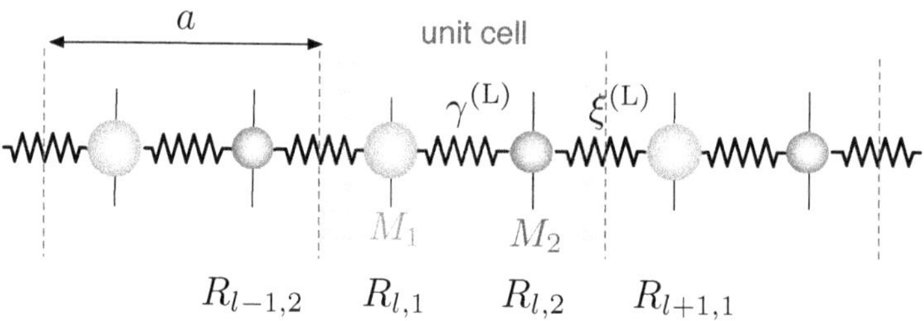

Figure 3.4. A diatomic spring-mass linear chain; the unit cell is shaded in light-blue colour, with lattice constant a. Ion masses M_1 and M_2 and force constants (for longitudinal oscillations) $\gamma^{(L)}$ and $\xi^{(L)}$ are indicated as in the main text. Ion positions entering in equation (3.12) are shown as well.

The equations of motion for the two ions in the lth unit cell form a system of two differential equations

$$\begin{cases} M_1\ddot{u}_{l,1} &= \gamma^{(L)}(u_{l,2} - u_{l,1}) + \xi^{(L)}(u_{l-1,2} - u_{l,1}) \\ M_2\ddot{u}_{l,2} &= \xi^{(L)}(u_{l+1,1} - u_{l,2}) + \gamma^{(L)}(u_{l,1} - u_{l,2}), \end{cases} \quad (3.11)$$

We seek solutions for this system of the same form given in equation (3.10). However, for further convenience, it is useful to rewrite the amplitude as $|\mathcal{A}_q| \to |\mathcal{A}_q|\,|a_i(q)|$ and the phase as $\varphi(q) \to \varphi(q) + \phi_i(q)$ (where $i = 1, 2$ labels the ion within the basis) since, as we will prove soon, the terms $|a_i(q)|$ and $\phi_i(q)$ are determined by the very equations of motion, rather than by the boundary conditions as instead $|\mathcal{A}_q|$ and $\varphi(q)$. By this choice, we write

$$u_{l,i} = \frac{1}{\sqrt{NM_i}} \sum_q |\mathcal{A}_q|\,|a_i(q)| \cos\left[qR_{l,i} - \omega(q)t + \varphi(q) + \phi_i(q) \right], \quad (3.12)$$

which, if inserted in equations (3.11), leads to the following matrix equation

$$\begin{pmatrix} \mathcal{D}_{11} & \mathcal{D}_{12} \\ \mathcal{D}_{21} & \mathcal{D}_{22} \end{pmatrix} \begin{pmatrix} a_1 \\ a_2 \end{pmatrix} = \omega^2 \begin{pmatrix} a_1 \\ a_2 \end{pmatrix}, \quad (3.13)$$

where the square 2×2 Hermitean[8] matrix $\{\mathcal{D}_{ij}\}_{i,j=1,2}$ has elements

$$\begin{aligned} \mathcal{D}_{11} &= \frac{\gamma^{(L)} + \xi^{(L)}}{M_1} \\ \mathcal{D}_{12} &= -\frac{\gamma^{(L)}}{\sqrt{M_1 M_2}} \exp[iq(R_{l,2} - R_{l,1})] - \frac{\xi^{(L)}}{\sqrt{M_1 M_2}} \exp[iq(R_{l-1,2} - R_{l,1})] \\ \mathcal{D}_{21} &= -\frac{\gamma^{(L)}}{\sqrt{M_1 M_2}} \exp[iq(R_{l,1} - R_{l,2})] - \frac{\xi^{(L)}}{\sqrt{M_1 M_2}} \exp[iq(R_{l+1,1} - R_{l,2})] \\ \mathcal{D}_{22} &= \frac{\gamma^{(L)} + \xi^{(L)}}{M_2}, \end{aligned} \quad (3.14)$$

and it is referred to as the *dynamical matrix* of the system: its eigenvalues $\omega = \omega_{\pm}(q)$ provide the *dispersion relations* for the diatomic chain, while its column eigenvector $\begin{pmatrix} a_1(q) \\ a_2(q) \end{pmatrix}$ provides information on the ionic displacements. It is now justified our choice of writing the oscillation amplitude as the product $|\mathcal{A}_q|\,|a_i(q)|$: the first term is determined by the initial conditions imposed to solve equation (3.13) (that is, by fixing the initial position and velocity of each ion), while the second term is determined by the equations of motion (3.11). In particular, the quantities $a_i(q)$ are named *polarisation vectors*.

---

[8] It is laborious but not difficult to prove that $\mathcal{D}_{12} = \mathcal{D}_{21}^*$.

By solving the matrix equation we obtain the eigenfrequencies

$$\omega_{\pm}^2(q) = \frac{1}{2}[\gamma^{(L)} + \xi^{(L)}]\frac{M_1 + M_2}{M_1 M_2}$$

$$\pm \frac{1}{2}\sqrt{[\gamma^{(L)} + \xi^{(L)}]^2\left(\frac{M_1 + M_2}{M_1 M_2}\right)^2 - 16\frac{\gamma^{(L)}\xi^{(L)}}{M_1 M_2}\sin^2\left(\frac{qa}{2}\right)}. \tag{3.15}$$

This result introduces a very important new feature: *for each q wavenumber two different vibrational frequencies $\omega_{\pm}(q)$ are found* or, equivalently, *in the diatomic linear chain we have two different longitudinal dispersion relations*, as shown in figure 3.5. The lower dispersion $\omega_-(q)$ has similar characteristics as found in the case of a monoatomic chain and for this reason is called the *acoustic branch*; once again, its slope in the limit $q \to 0$ corresponds to the speed of longitudinal sound. The upper dispersion $\omega_+(q)$ is instead referred to as the *optical branch* for reasons that can be easily understood by considering the displacement patterns in the unit cell. To this aim we must preliminarily discuss an important feature. By replacing the solutions $\omega_{\pm}(q)$ given by equation (3.15) into equation (3.13) we easily realise that we cannot determine the absolute magnitude of the eigenvector elements $\begin{pmatrix} a_1^{\pm}(q) \\ a_2^{\pm}(q) \end{pmatrix}$ but only their ratio. The calculation is easy by assuming for simplicity that $M_1 = M_2$ and also imposing the normalisation $|a_1^{\pm}(q)|^2 + |a_2^{\pm}(q)|^2 = 1$. Under these conditions, after some algebra we obtain

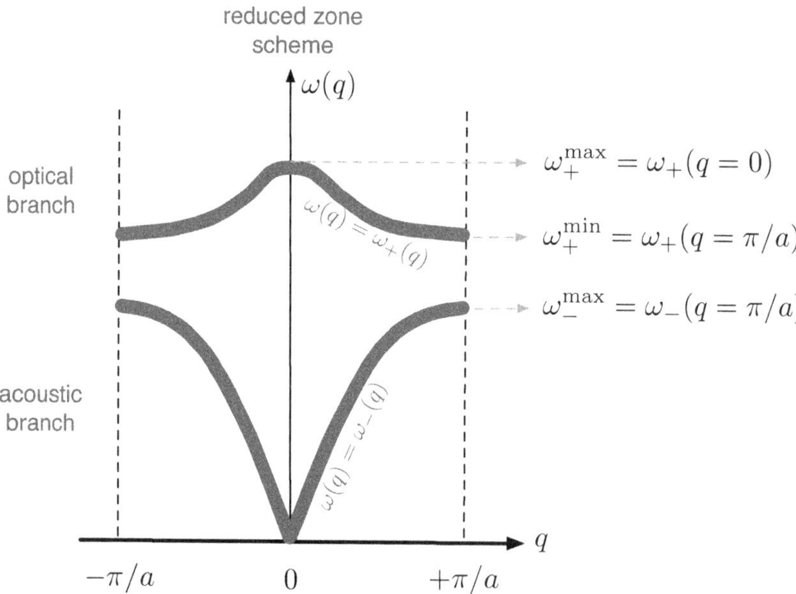

Figure 3.5. Dispersion relations $\omega = \omega_{\pm}(q)$ for the longitudinal vibrations of a diatomic chain in the reduced zone scheme. The ion masses are M_1 and M_2 and the effective spring constants are $\gamma^{(L)}$ and $\xi^{(L)}$. The $\omega_{\pm}^{min,max}$ values are calculated by equation (3.15).

$$\frac{a_1^+(q)}{a_2^+(q)} = -\frac{A(q)}{|A(q)|} = -\frac{a_1^-(q)}{a_2^-(q)},\qquad(3.16)$$

where $A(q) = [\gamma^{(L)} + \xi^{(L)}\exp(iqa)]\,\exp[iq(R_{l,1} - R_{l,2})]$. This result indicates that *in the limit of a vanishingly small wavevector the two ions of the unit cell always oscillate in phase or in phase opposition in an acoustic or an optic vibrational mode, respectively.*

More in general, the relative ionic motions are affected by the q-vector through a simple scheme: in vibrations of any kind (both acoustic and optical), nearest neighbouring unit cells move in-phase or in phase opposition for zone-centre and zone-boundary modes, respectively; on the other hand, in vibrations with any finite q-vector, ions within the same unit cell vibrate in-phase or in phase opposition for acoustic or optical modes, respectively. In figure 3.6 a graphical rendering of the displacement patters for zone-centre and zone-boundary modes is reported.

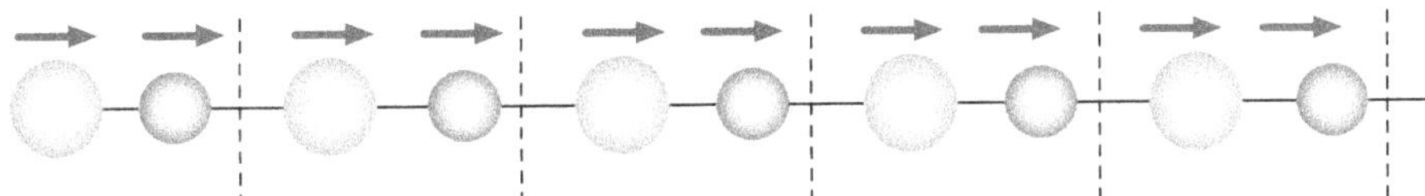

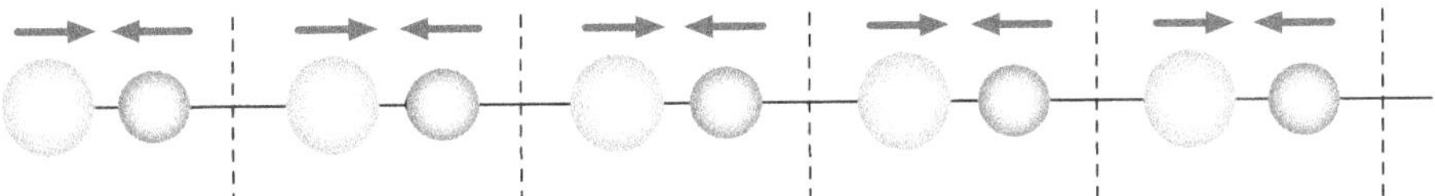

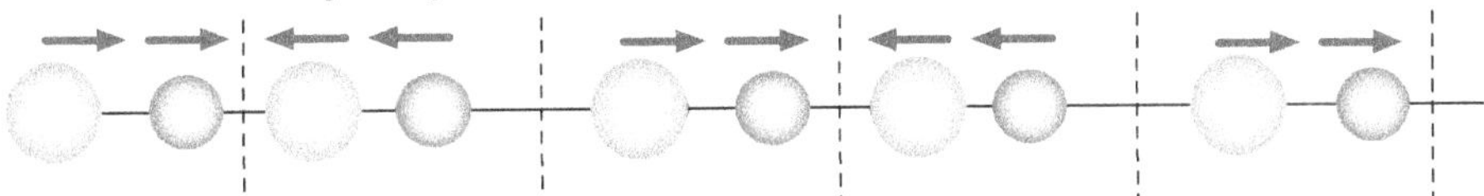

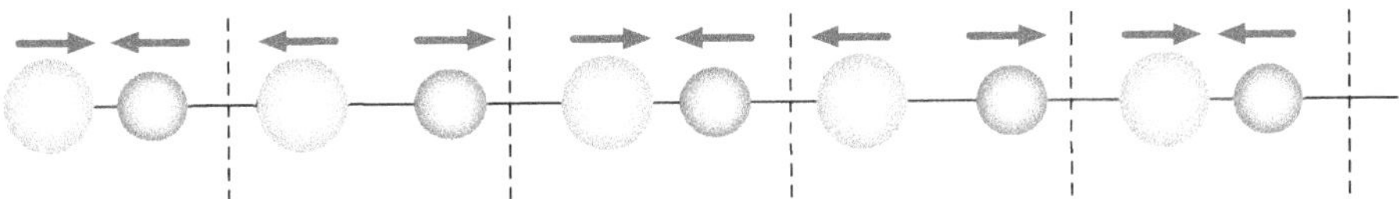

Figure 3.6. Zone-centre and zone-boundary displacement patters of longitudinal acoustic and optical vibrations in a diatomic linear chain. Blue arrows indicate the direction and the orientation of the ionic displacement vector (not its amplitude).

We can now understand why optical modes are so named: in ionic crystals the two ions have opposite charge and, therefore, long wavelength (zone-centre) optical modes generate fluctuating electric dipoles which can interact with electromagnetic radiation. Optical vibrational modes in fact dictate the overall optical behaviour of an ionic crystal.

We conclude the discussion of the vibrational modes in a diatomic chain by addressing its transverse oscillations. In analogy to the case of the monoatomic chain, we say that for each q-vector there are two degenerate transverse dispersions, both for the optical and the acoustic branch. This means that *in a diatomic linear chain, there are in total six dispersion relations*: two *transverse acoustic* (TA_1 and TA_2), one *longitudinal acoustic* (LA), two *transverse optical* (TO_1 and TO_2), and one *longitudinal optical* (LO).

3.3 Dynamics of three-dimensional crystals

The dynamical properties of a three-dimensional solid with arbitrary crystalline structure or basis configuration can only be described by means of a heavy formalism [5, 6, 8], which somewhat hides the underlying physics. This is the pedagogical reason why we have preliminarily treated the model system corresponding to a linear chain: we will extensively make use of concepts developed in that framework. As for the formal procedures, we will instead follow the same line of action adopted in section 3.1. In particular, we will assume that a suitable force field describing the interactions among ions is available (see appendix D); once for all, therefore, the force constants defined in equation (3.2) are given as known. More important, however, is the fact that we will rely on the harmonic approximation.

Before starting to develop our theory, we preliminarily remark that in a three-dimensional crystal containing N_{atom} atoms in its unit cell, we have $3N_{atom}$ ionic degrees of freedom (per unit cell) and, therefore, an equal number of branches in the vibrational dispersion relations; among them we will find 3 acoustic and $3(N_{atom} - 1)$ optical branches.

In the harmonic approximation, the equations of motion are written as[9]

$$M_b \ddot{u}_i(lb) = - \sum_{jl'b'} U_{ij}(lb, l'b')u_j(l'b'), \tag{3.17}$$

for which we guess solutions in the form

$$u_i(lb) = \frac{a_i(b|\mathbf{q})}{\sqrt{M_b}} \, e^{i\mathbf{q}\cdot\mathbf{R}_l} \, e^{-i\omega t}, \tag{3.18}$$

where M_b is the mass of the bth ion in the basis[10], $\mathbf{q}$ is the wavevector of the vibrational wave and $a_i(b|\mathbf{q})$ describes the amplitude of the corresponding ionic

---

[9] Since in harmonic approximation derivatives appear just up to the second order, we indicate Cartesian components by the two labels i, and j, instead of using the heavier notation of section 3.1.

[10] Of course, in a perfect crystal the ion mass does not depend on the cell index l.

oscillations. By inserting this guessed displacement into the equations of motion we get

$$\sum_{jb'} \mathcal{D}_{ij}(bb'|\mathbf{q})\mathsf{a}_j(b'|\mathbf{q}) = \omega^2 \mathsf{a}_i(b|\mathbf{q}), \tag{3.19}$$

where the quantities

$$\mathcal{D}_{ij}(bb'|\mathbf{q}) = \frac{1}{\sqrt{M_b M_{b'}}} \sum_{l'} U_{ij}(lb,\ l'b')e^{-i\mathbf{q}\cdot(\mathbf{R}_l - \mathbf{R}_{l'})}, \tag{3.20}$$

define the dynamical matrix of the crystal. It is important to remark that in this equation just a single summation of the cell index l' appears since the force constants $U_{ij}(lb,\ l'b')$ of an ideal crystal depend on the pair $(l,\ l')$ just through their difference[11]. This also reflects the choice of the guessed solution for the ionic displacements in the form of a Bloch wave.

Equation (3.20) is the three-dimensional counterpart of equation (3.13) and, similarly to the linear chain case, the $3N_{\text{atom}} \times 3N_{\text{atom}}$ dynamical matrix is once again found to be Hermitean: $\mathcal{D}_{ij}(bb'|\mathbf{q}) = \mathcal{D}^*_{ji}(b'b|\mathbf{q})$. Accordingly, *the lattice dynamics problem of a three-dimensional crystal is reduced to the diagonalisation of the dynamical matrix* written in equation (3.19). Nowadays this task is solved numerically, once the matrix has been built using either an empirical force field or by performing a full quantum mechanical calculation of $U(\mathbf{R})$ and its second-order derivatives.

Let us now discuss some model cases corresponding to metallic, ionic, and covalent crystals. We will follow the standard convention to represent the vibrational dispersion relations by choosing $\mathbf{q}$ along the borders of the irreducible part of the 1BZ.

In figure 3.7(left) we report the dispersion relations of copper, a prototypical fcc metal. Since Cu is a Bravais crystal (one atom in its unit cell), just three dispersion branches are found. As predicted by the elementary theory developed in section 3.2, in the limit of a vanishingly small wavevector the dispersion relations are linear: this allows one to predict the speed of longitudinal and transverse sound. LA vibrations have in general the higher frequency, although for some $\mathbf{q}$-values along the Σ direction the TA_2 branch overcomes the LA one. Interestingly enough, along the Δ and Λ directions the two transverse branches are degenerated, as we commented in the simple case of a linear chain. However, along the Σ direction such a degeneracy is removed because of a reduced symmetry. This is a general feature found in other crystals as well.

Another case of metallic system is shown in figure 3.7(right) where the dispersions of lead, another fcc crystal structure, are reported. While the main characteristics of the vibrational spectrum are similar to the previous case, we note a lowering of both the longitudinal and the transverse phonon frequency occurring at X point of the

---

[11] This is related to the fact that the total classical energy of a harmonic ideal crystal is invariant upon rigid body translations and rotations.

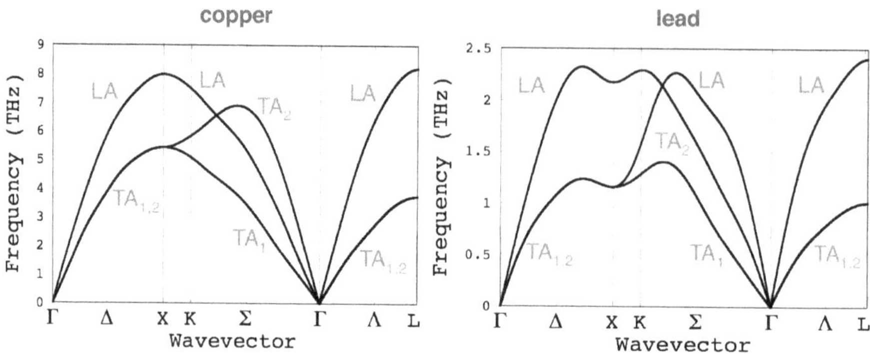

Figure 3.7. Vibrational dispersion relations of copper (left) and lead (right). The calculations have been performed by using the QUANTUM ESPRESSO integrated suite of Open-Source computer codes for materials modelling; see website https://www.quantum-espresso.org (by courtesy of Aleandro Antidormi).

Brillouin zone. This feature is named *Kohn anomaly* and it is the signature of a strong electron–phonon interaction effect: in fact, its full explanation requires one to develop a theory beyond the adiabatic approximation. A Kohn anomaly occurs whenever it is observed that the dielectric screening[12] of the Coulomb interactions among bare nuclei abruptly changes: this affects the ionic motion or, equivalently, their vibrational frequencies. The change is theoretically explained by the Lindhard theory for the dielectric function of a free electron gas [9–11] which is found to diverge for certain wavevectors: exactly those ones at which the anomaly is found in the dispersion relations. The modes with lowered frequency are named *soft modes*. Sometimes the Kohn anomaly can be so huge that the vibrational frequency is lowered to zero: in this case a static distortion of the crystal lattice is generated.

We now move to consider the case of two covalent solids, namely silicon and gallium arsenide, shown in figure 3.8. They both are fcc crystals with a two-atom basis and a diamond-like or zincblende-like structure, respectively. Because of that, both acoustic and optical dispersions are found, summing up to six branches, as shown along the low-symmetry Σ direction (while along the Δ and Λ ones vibrations with transverse polarisation are degenerate). Once again, the acoustic branches are correctly linear as $\mathbf{q} \to 0$. A very meaningful difference between Si and GaAs is found in the zone-centre optical modes: while transverse and longitudinal vibrations are degenerate in Si, they are not in GaAs. In this latter case one observes the so called LO–TO *zone-centre splitting* which is basically due to the fact that the two atoms in the unit cell are not the same and, therefore, the crystal is partly ionic. In this situation, an electric dipole moment occurs when the two ions are displaced from their equilibrium positions; in other words, the ionic motion causes an electric polarisation, giving rise to a Coulomb field. On its part, the field affects the motion of the ions in two ways: (i) it acts as an additional external force (thus modifying the force constants entering in the dynamical matrix); and (ii) it polarises the ions, thus enhancing their dipole moments. The microscopic theory needed to deal with these

---

[12] The screening is provided by the system of the valence electrons.

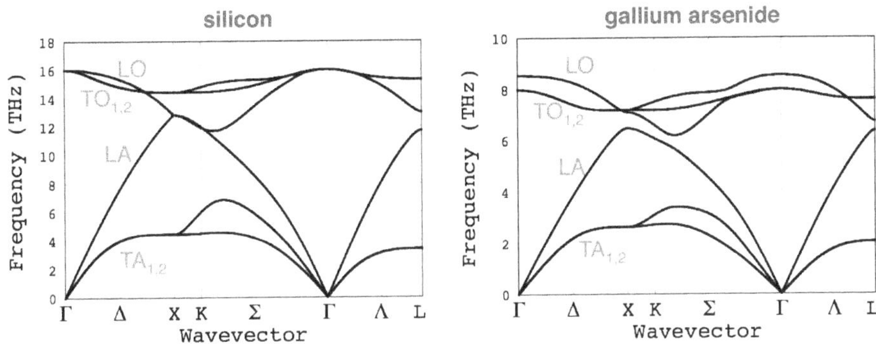

Figure 3.8. Vibrational dispersion relations of silicon (left) and gallium arsenide (right). The calculations have been performed by using the QUANTUM ESPRESSO integrated suite of open-source computer codes for materials modelling; see website https://www.quantum-espresso.org (by courtesy of Aleandro Antidormi).

subtle phenomena [6, 8] goes far beyond the level of treatment we are developing. However, by following a more phenomenological approach, it possible to prove that

$$\omega_{LO}(\mathbf{q} = 0) = \sqrt{\frac{\epsilon_0}{\epsilon_\infty}} \; \omega_{TO}(\mathbf{q} = 0), \tag{3.21}$$

a very general result referred to as the *Lyddane–Sachs–Teller relation* (LST relation), which is derived in the next section. In this equation ϵ_0 and ϵ_∞ are the static and high-frequency[13] dielectric constant, respectively: they rule over the polarisation phenomena mentioned above. Since $\epsilon_0 > \epsilon_\infty$, we have $\omega_{LO}(\mathbf{q} = 0) > \omega_{TO}(\mathbf{q} = 0)$. In silicon this effect is not observed since it is an elemental solid and, therefore, no first-order dipole moments are generated by optical modes.

For sake of completeness, we finally consider the case of a pure ionic crystal like KBr whose dispersions are shown in figure 3.9. The main features previously discussed are found even in this case, including the expected large LO–TO zone-centre splitting: the fingerprint of a strong ionic character.

In concluding this section we observe that if we consider the whole crystal, instead of just its unit cell, the total number of vibrational degrees of freedom must be appropriately counted. If the crystal is made of N_{cell} unit cells, each containing N_{atom} atoms, in total we have $N_{cell} \times 3N_{atom}$ vibrational modes. We describe the overall dynamics as the superposition of $N_{cell} \times 3N_{atom}$ non-interacting harmonic vibrational modes labelled by a *branch index s* and a **q** wavevector. The branch index runs over the full set of TA, TO, LA and LO modes, while **q** takes those values within the 1BZ which are allowed by the periodic boundary conditions.

3.4 The physical origin of the LO–TO splitting

The derivation of the LST relation anticipated in equation (3.21) is rigorously framed only within the theory of the dielectric properties of crystalline solids [9, 10], indeed an

---

[13] High-frequency here means a frequency of the order of optical vibrational frequencies or beyond.

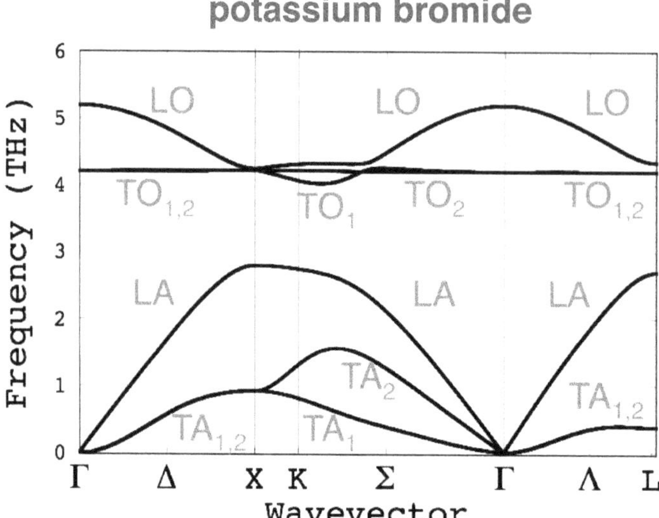

Figure 3.9. Vibrational dispersion relations of KBr. The calculations have been performed by using the QUANTUM ESPRESSO integrated suite of open-source computer codes for materials modelling; see website https://www.quantum-espresso.org (by courtesy of Aleandro Antidormi).

advanced topic of solid state physics. Good for us, it is possible to elaborate a phenomenological model which, although derived under some important simplifying assumptions, nevertheless leads to a result of general validity. More specifically, we are going to consider a *dielectric ionic crystal* containing just two atoms in its unit cell.

Let the dielectric crystal be subject to the action of an external macroscopic electric field **E**. Because of polarisation phenomena, the *local electric field* $\mathbf{E}_{\text{loc}}$ found at any position **r** within the crystal differs from the applied one: the theory of the dielectric properties of crystalline solids displays exactly here. We are not developing this calculation; rather, we assume that the local field is known. The electrostatic action on the two ions within the unit cell causes their displacements, but since such a perturbation occurs on a length scale much longer that the typical interatomic distances, we can assume that *equally charged ions move as a whole*. Accordingly, in harmonic approximation we can guess the ionic equations of motion in the form

$$\begin{cases} m_+\ddot{\mathbf{u}}_+ = -K(\mathbf{u}_+ - \mathbf{u}_-) + e\mathbf{E}_{\text{loc}} \\ m_-\ddot{\mathbf{u}}_- = +K(\mathbf{u}_+ - \mathbf{u}_-) - e\mathbf{E}_{\text{loc}}, \end{cases} \tag{3.22}$$

where for sake of simplicity we have assumed just one force constant K for any interaction and indicated by $m_{\pm}$ and $\mathbf{u}_{\pm}$ respectively the mass and the displacement of the positive (+) and negative (−) ion. By further setting $\mathbf{w} = (\mathbf{u}_+ - \mathbf{u}_-)$ and $1/m = 1/m_+ + 1/m_-$ we derive a forced oscillator equation

$$\ddot{\mathbf{w}} = \frac{e}{m}\mathbf{E}_{\text{loc}} - \frac{K}{m}\mathbf{w}, \tag{3.23}$$

which describes the relative motion within the unit cell. If we assume a harmonic time dependence of the electric field $\mathbf{E}_{\text{loc}} = \mathbf{E}_{\text{loc}}^{(0)} \exp(-i\omega t)$ we calculate $\mathbf{w}(t) = \mathbf{w}_0 \exp(-i\omega t)$ with

$$\mathbf{w}_0 = \frac{e}{m} \frac{1}{\omega_0^2 - \omega^2} \mathbf{E}_{\text{loc}}^{(0)}, \tag{3.24}$$

with $\omega_0 = \sqrt{K/m}$. Since the oppositely charged ions are unequally displaced, an *induced displacement electric dipole moment* $\mathbf{p}(t) = e\mathbf{w}(t) = \mathbf{p}_0 \exp(-i\omega t)$ is generated within the unit cell with

$$\mathbf{p}_0 = e\mathbf{w}_0 = \frac{e^2}{m(\omega_0^2 - \omega^2)} \mathbf{E}_{\text{loc}}^{(0)}, \tag{3.25}$$

and we can accordingly define

$$\alpha_{\text{displ}} = \frac{e^2}{m(\omega_0^2 - \omega^2)}, \tag{3.26}$$

as its *displacement polarisability*.

So far we have implicitly assumed a sort of *rigid ion approximation*, that is, we have neglected atomic polarisation phenomena. Atomic physics in fact provides us with the fundamental notion that the atomic electronic clouds can be distorted [12], that is, atoms are polarisable. By assuming that ionic polarisabilities $\alpha_{\pm}$ are known, we can refine the above model by introducing the ultimate expression for the unit cell polarisability α_{tot} as

$$\alpha_{\text{tot}} = \alpha_{\text{displ}} + \alpha_+ + \alpha_-. \tag{3.27}$$

This expression can be safely used in the Clausius–Mossotti relation of macroscopic electrostatics [13, 14]

$$\frac{\epsilon_r(\omega) - 1}{\epsilon_r(\omega) + 2} = \frac{1}{3\epsilon_0} \alpha_{\text{tot}} = \frac{1}{3\epsilon_0}\left[\frac{e^2}{m(\omega_0^2 - \omega^2)} + \alpha_+ + \alpha_- \right], \tag{3.28}$$

and is valid for ionic crystals. In this equation $\epsilon_r(\omega)$ is the frequency-dependent relative dielectric constant (or permittivity) of the material. It is easy to calculate the limiting behaviour of equation (3.28) at very low frequencies $\omega \ll \omega_0$, a condition where we can replace $\epsilon_r(\omega)$ with the static dielectric constant $\epsilon_r(0)$ and write

$$\frac{\epsilon_r(0) - 1}{\epsilon_r(0) + 2} = \frac{1}{3\epsilon_0}\left[\frac{e^2}{m(\omega_0^2 - \omega^2)} + \alpha_+ + \alpha_- \right]. \tag{3.29}$$

Similarly, we can evaluate the high-frequency $\omega \to \infty$ behaviour[14] for which we can replace $\epsilon_r(\omega)$ with the $\epsilon_r(\infty)$ and write

---

[14] In this discussion an 'infinite' frequency means a frequency much larger than the lattice vibrational ones, but still lower that the typical frequencies associated with electronic transitions.

$$\frac{\epsilon_r(\infty) - 1}{\epsilon_r(\infty) + 2} = \frac{1}{3\epsilon_0}(\alpha_+ + \alpha_-),\tag{3.30}$$

where $\epsilon_r(\infty)$ is of course the dielectric constant of the material at optical frequencies. By combining equations (3.29) and (3.30) we get

$$\epsilon_r(\omega) = \epsilon_r(\infty) + \frac{\epsilon_r(\infty) - \epsilon_r(0)}{\omega^2/\bar{\omega}^2 - 1},\tag{3.31}$$

where $\bar{\omega}$ is at present just a shortcut for

$$\bar{\omega}^2 = \frac{\epsilon_r(\infty) + 2}{\epsilon_r(0) + 2}\omega_0^2.\tag{3.32}$$

Let us now introduce the lattice dynamics by considering *a long-wavelength optical mode*: ions in the same unit cell move in phase opposition, thus giving rise to an electric dipole moment. This in turn implies the onset of a non-zero polarisation **P** within the crystal. By adding the assumption that the material is homogeneous and isotropic, the resulting electrostatics can be summarised in the equation

$$\mathbf{D} = \epsilon_0\mathbf{E} + \mathbf{P} = \epsilon_0\mathbf{E} + \epsilon_0\chi\mathbf{E} = \epsilon_0(1 + \chi)\mathbf{E} = \epsilon_0\epsilon_r\mathbf{E},\tag{3.33}$$

where **D** is the electric displacement field and χ the electric susceptibility of the material. We now add three more simplifying assumptions, namely: (i) there are no free charges within the crystal and, therefore, $\nabla \cdot \mathbf{D} = 0$; (ii) we neglect retardation effects or, equivalently we set the electrostatic approximation $\nabla \times \mathbf{E} = 0$; and (iii) we impose the same space variation $\mathbf{C} = \mathbf{C}_0 \exp(i\mathbf{k} \cdot \mathbf{r})$, for the three relevant vectors $\mathbf{C} = \mathbf{E}$, **D** or **P**. These assumptions lead to the noteworthy conclusions that: whenever $\mathbf{k} \cdot \mathbf{D}_0 = 0$ then either $\mathbf{D}_0 = 0$ or $\mathbf{D}_0, \mathbf{E}_0, \mathbf{P}_0 \perp \mathbf{k}$ (condition 1); on the other hand, whenever $\mathbf{k} \times \mathbf{E}_0 = 0$ then either $\mathbf{E}_0 = 0$ or $\mathbf{D}_0, \mathbf{E}_0, \mathbf{P}_0 \| \mathbf{k}$ (condition 2). These constraints have different, but equally meaningful, consequences for longitudinal and transverse modes. In a *longitudinal mode* the polarisation vector is parallel to **k** and, therefore, condition 1 imposes that the electric displacement field must vanish. Since $\mathbf{E} \neq 0$, equation (3.33) dictates that $\epsilon_r = 0$. On the other hand, in a *transverse mode* the polarisation vector is normal to **k** and, therefore, condition 2 imposes that the electric field must vanish: this time equation (3.33) imposes that $\epsilon_r = \infty$. This situation is only found if $\omega = \bar{\omega}$ in equation (3.31). Therefore, *we can identify $\bar{\omega}$ with the zone-center frequency of a transverse optical mode* and accordingly set $\bar{\omega} = \omega_{TO}$. Finally, the condition $\epsilon_r = 0$ valid for a longitudinal optical mode leads, once again through equation (3.31), to the LST relation anticipated in equation (3.21).

Interestingly enough, this analysis also provides a valuable model for *the frequency dependent dielectric constant of a diatomic ionic crystal*, as reported in figure 3.10, left. It is interesting to observe that for $\omega_{TO} < \omega < \omega_{LO}$ the dielectric constant is negative and, therefore, *no electromagnetic mode can propagate within the crystal in this frequency range*, as indeed found experimentally. The dispersion relation $\omega = kc/\sqrt{\epsilon}$ of electromagnetic modes outside such a forbidden gap is

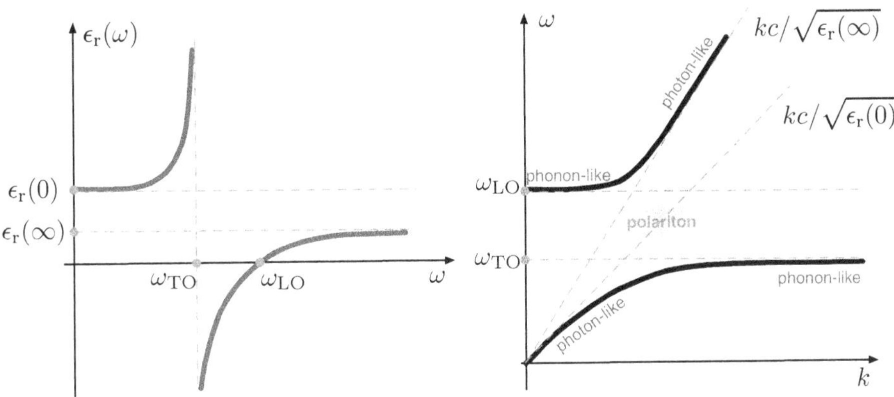

Figure 3.10. Left: frequency-dependent dielectric constant of a diatomic ionic crystal. Right: dispersion relation for transverse electromagnetic modes in a diatomic ionic crystal.

reported in figure 3.10, right. It must be understood that the plot only refers to the long wavelength region (that is, the range of photon k-values reported here is very small as compared with the typical phonon wavevectors). The picture suggests that the coupling between the transverse electromagnetic wave and the vibrational optical modes of the crystal has determined a dramatic variation of the linear ω versus k dependence observed in free space: the character smoothly changes from phonon-like to photon-like or vice versa according to the actual branch considered. In the region where the character changeover occurs, the excitation is better defined as a *polariton*, a sort of combination of a photon and a phonon.

3.5 Quantum theory of harmonic crystals

Moving to a quantum description, as simple as it may appear, represents a major conceptual step forward in our search for a truly fundamental description of lattice dynamics. To appreciate its relevance, we anticipate a result more extensively discussed in the next chapter. The specific heat of a crystal described as an assembly of classical harmonic oscillators is calculated to be independent of temperature (Dulong–Petit law). Contrary to this prediction, experimental measurements provide evidence that the specific heat becomes vanishingly small as $T \to 0$, thus proving that it is in fact temperature-dependent. Only a full quantum treatment is able to reconcile theoretical predictions to measurements.

Based on the theory developed in the previous section, *we will agree to describe each classical ($s\mathbf{q}$) vibrational mode as a quantum one-dimensional harmonic oscillator* [1–3] whose energy is restricted to the values $\left(n_{s\mathbf{q}} + 1/2\right)\hbar\omega_s(\mathbf{q})$ where $n_{s\mathbf{q}} = 0, 1, 2, \ldots$ is the *vibrational quantum number* and $\omega_s(\mathbf{q})$ is obtained by diagonalising the dynamical matrix. Since the vibrational energy levels are equally spaced, we can look at the state with energy $\left(n_{s\mathbf{q}} + 1/2\right)\hbar\omega_s(\mathbf{q})$ as a single $n_{s\mathbf{q}}$th excited state or, equivalently, as *the state obtained by adding $n_{s\mathbf{q}}$ identical energy quanta $\hbar\omega_s(\mathbf{q})$.* We will adopt this second approach since it is especially effective in describing the dynamical and thermal characteristics of a crystal lattice through the properties of *a gas of pseudo-particles*, hereafter named *phonons*. This choice introduces a corpuscular description of lattice dynamics, where phonons are the energy quanta of the ionic displacement field[15].

Let us now consider in some detail the physics of phonons. First of all, we clarify that phonons, similarly to photons, are named *pseudo*-particles since they do not have a mass. Furthermore, in addition to an energy $\hbar\omega_s(\mathbf{q})$, they also carry a momentum $\hbar\mathbf{q}$. Since such a momentum is exact, the uncertainty principle imposes that the phonon position is totally undetermined and, therefore, they must be understood as *delocalised pseudo-particles*. This is consistent with the fact that their corresponding non-interacting classical vibrational modes extend throughout the system[16].

The phonon momentum must be treated with some care. We know that, because of translational invariance, each classical vibrational mode remains unaffected by replacing $\mathbf{q}$ with $\mathbf{q} + \mathbf{G}$, where $\mathbf{G}$ is a reciprocal lattice vector. This implies that $\hbar\mathbf{q}$ does not actually represent the kinematic momentum of the phonon since $\hbar\mathbf{q}$ and $\hbar(\mathbf{q} + \mathbf{G})$ are equivalently good choices for it. For this reason the phonon momentum is better referred to as its *crystal momentum*. An elastic phonon scattering event will conserve the crystal momentum unless an $\hbar\mathbf{G}$ term, as discussed in the next section.

Similarly to the photon case, *the total number of phonons is not conserved*: they can be created or annihilated by increasing/decreasing temperature or by phonon–phonon scattering events. This urges us to understand their statistics. Let us consider a gas of phonons with energies $\left(n_{sq} + \frac{1}{2}\right)\hbar\omega_s(\mathbf{q})$ in thermal equilibrium at temperature T. The average energy $\langle U \rangle$ of the gas is calculated according to statistical physics [15] as

$$\langle U \rangle = \frac{1}{Z} \sum_{s\mathbf{q}} (n_{sq} + 1/2)\hbar\omega_s(\mathbf{q}) \exp\left[-\frac{(n_{sq} + 1/2)\hbar\omega_s(\mathbf{q})}{k_B T}\right], \qquad (3.34)$$

where

$$Z = \sum_{s\mathbf{q}} \exp\left[-\frac{(n_{sq} + 1/2)\hbar\omega_s(\mathbf{q})}{k_B T}\right], \qquad (3.35)$$

is the partition function[17] for the canonical (i.e. constant-temperature) ensemble. This energy is the quantum counterpart of equation (3.4). After some algebra[18] the series are summed obtaining the *quantum vibrational energy of a harmonic crystal*

---

[15] It could be helpful to consider the analogy with the case of classical electromagnetic waves described by the Maxwell equations: the first step toward the quantisation of the electromagnetic field is to describe it as a flux of radiation quanta, referred to as photons.

[16] We remark that we are considering an ideal and harmonic crystal. The situation looks very different when considering the role of anharmonicity and crystal defects.

[17] We adopt a widely used abuse of notation. For quantum systems, as the phonon gas indeed is, the partition function is more rigorously defined as the trace of the Boltzmann factor $\exp(-\hat{H}/k_B T)$, where $\hat{H}$ is the Hamiltonian operator (in the present case: the total energy operator of an assembly of quantum harmonic oscillators). By assuming that we are working with a basis set for which the matrix representation of $\hat{H}$ is diagonal, the above notation for Z is correct.

[18] We remember that $\sum_n \exp(-nx) = [1 - \exp(-x)]^{-1}$ and $\sum_n n \exp(-nx) = \exp(x)[\exp(x) - 1]^{-2}$.

$$\langle U \rangle = U_{\text{harmonic}}^{\text{quantum}} = \sum_{s\mathbf{q}} [n_{\text{BE}}(s\mathbf{q},\, T) + 1/2]\, \hbar\omega_s(\mathbf{q}), \tag{3.36}$$

where

$$n_{\text{BE}}(s\mathbf{q},\, T) = \cfrac{1}{\exp\left[\cfrac{\hbar\omega_s(\mathbf{q})}{k_{\text{B}}T}\right] - 1}, \tag{3.37}$$

represents the *average number of phonons with energy* $\hbar\omega_s(\mathbf{q})$ (also referred to as the *phonon population*) found in the crystal at temperature T. As indicated, this average number is provided by the Bose–Einstein statistics, proving that *phonons are bosons* as consistent with the fact that they are spinless pseudo-particles (see appendix E).

In summary, in the harmonic approximation a crystal is quantum mechanically described as a *gas of non-interacting phonons* of different character (s = TA, LA, TO, LO) and wavevector $\mathbf{q}$. Their number varies upon temperature according to the Bose–Einstein statistics given in equation (3.37). Interesting enough, in the limit $T \to 0$ we find $U_{\text{harmonic}}^{\text{quantum}} = 1/2 \sum_{s\mathbf{q}} \hbar\omega_s(\mathbf{q})$ representing the *crystalline zero-point energy*: in the quantum description, therefore, the vibrational energy of a crystal is never zero. Finally, we remark that the dispersion relations $\omega = \omega_s(\mathbf{q})$ discussed in section 3.3 are usually referred to as *the phonon dispersion relations*.

3.6 Experimental measurement of phonon dispersion relations

The experimental determination of the $\omega = \omega_s(\mathbf{q})$ dispersion relations over the entire 1BZ needs a probe fulfilling two conditions: (i) its wavelength must be comparable with the typical interatomic distances in the crystal structure and (ii) its energy must be of the same order of typical phonon quanta $\hbar\omega_s(\mathbf{q})$, which range mostly in the interval $[1, 10^2]$ meV. Optical probes are unsuitable: x-rays have the right wavelength, but a too high energy of the order $\mathcal{O}(10^4\text{eV})$; other kinds of photons, instead, can only explore the $\mathbf{q} \sim 0$ region of the Brillouin zone, i.e. they can only detect (some) zone-centre phonons. A probe consisting in a flux of electrons is also impractical for a twofold reason: (i) their surface scattering is very strong and, therefore, they are unable to probe the bulk region of the crystal; (ii) multiple scattering is likely to occur in the case of electrons and this makes the analysis of the experiment a very challenging task. In contrast, both requirements of suitable wavelength and energy are guaranteed by a flux of *thermal neutrons* which have typical wavelengths of the order of just a few Å and energies in between a few and a few tens of meV. Accordingly, *neutron spectroscopy* is the most powerful technique for measuring the phonon dispersion relations [16, 17].

The typical experimental setup for neutron spectroscopy is shown in figure 3.11. A continuous flux of neutrons is emitted by a source, typically a fission reactor; the beam is thermalised by collisions within the reactor moderator and, therefore, neutrons are produced with a Maxwell–Boltzmann distribution of velocities corresponding to the room temperature (for this reason they are referred to as 'thermal neutrons'). The thermalised beam emerging from the source is Bragg

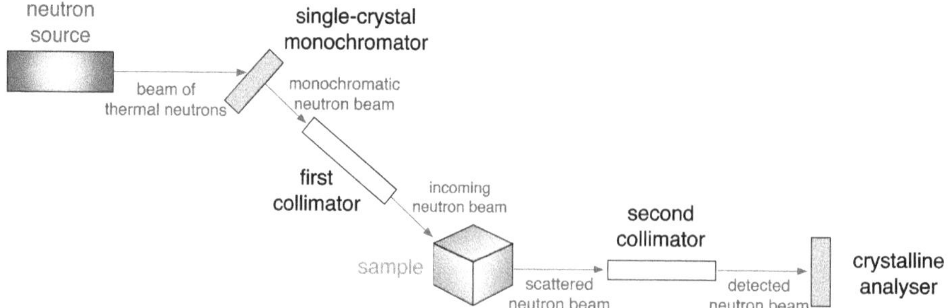

Figure 3.11. Schematic representation of a triple-axis neutron spectrometer.

reflected by a single-crystal monochromator which selects neutrons of same momentum $\hbar\mathbf{p}_{\text{in}}$ and energy $E_{\text{in}} = \hbar^2 p_{\text{in}}^2/2m_{\text{n}}$ where m_{n} is the neutron mass. The resulting monochromatic beam is now focussed by a collimator and then it falls on a sample with known crystalline orientation. The momentum vector $\hbar\mathbf{p}_{\text{out}}$ of the scattered neutrons is defined by letting the beam pass through a second collimator and eventually arrive at the analyser. Here the magnitude p_{out} and the corresponding energy $E_{\text{out}} = \hbar^2 p_{\text{out}}^2/2m_{\text{n}}$ of the detected neutrons are determined by measuring the Bragg reflection angles from the planes of the crystalline analyser.

Two possible physical mechanisms occur in a scattering event. First of all, neutrons are scattered by very strong nuclear forces: basically, the nuclei of the atoms in the sample scatter neutrons as they were hard spheres. In addition, provided that such atoms have a non-zero magnetic moment[19], neutrons are scattered by magnetic interactions between their intrinsic magnetic dipole moment and the atomic magnetic moments. While this latter interaction is helpful to investigate magnetic ordering in solids, we will take into consideration only the first mechanism which is active in any (even non-magnetic) material.

To the aim of investigating a scattering event, we will keep things easy by considering just a single vibrational mode of wavevector $\mathbf{q}$ and frequency ω in the sample hit by the monochromatic beam of neutrons. We will also use a simplified version of the notation introduced in section 3.2 according to which the position $\mathbf{R}(t)$ of an atom sitting at rest in the equilibrium site $\mathbf{R}_{\text{eq}}$ is written as

$$\mathbf{R}(t) = \mathbf{R}_{\text{eq}} + \mathbf{u} \cos(\mathbf{q} \cdot \mathbf{R}_{\text{eq}} - \omega t), \qquad (3.38)$$

where $\mathbf{u}$ contains information on both the amplitude and the polarisation of the selected vibrational mode. In addition, to take profit from our previous knowledge it is useful to develop our arguments in analogy with the discussion of x-ray scattering (see section 2.4.1). Therefore, we will at first consider the neutron beam as an incoming wave whose quanta (neutrons) have energy $\hbar\Omega_{\text{in}}$; next, we will interpret the scattering outcome using the phonon corpuscular language.

Similarly to equation (2.9) the amplitude of the scattered wave is written as

---

[19] This situation is found in atoms with an electronic configuration giving a net magnetic moment or with an uncompensated electron spin condition [12].

$$\mathcal{A}_{\text{out}} = \sum_{\mathbf{R}} b_{\mathbf{R}} \exp[-i(\mathbf{P} \cdot \mathbf{R} + \Omega_{\text{in}}t)], \tag{3.39}$$

where $b_{\mathbf{R}}$ is the so called scattering length[20]. In this case the scattering vector is defined as $\mathbf{P} = (\mathbf{p}_{\text{out}} - \mathbf{p}_{\text{in}})$. By inserting equation (3.38) into equation (3.39), after some algebra we get

$$\begin{aligned}
\mathcal{A}_{\text{out}} = &\sum_{\mathbf{R}_{\text{eq}}} b_{\mathbf{R}} \exp[-i(\mathbf{P} \cdot \mathbf{R}_{\text{eq}} + \Omega_{\text{in}}t] \\
&- \frac{i}{2}(\mathbf{P} \cdot \mathbf{R}_{\text{eq}}) \sum_{\mathbf{R}_{\text{eq}}} b_{\mathbf{R}} \exp\{-i[(\mathbf{P} + \mathbf{q}) \cdot \mathbf{R}_{\text{eq}} + (\Omega_{\text{in}} - \omega)t]\} \\
&- \frac{i}{2}(\mathbf{P} \cdot \mathbf{R}_{\text{eq}}) \sum_{\mathbf{R}_{\text{eq}}} b_{\mathbf{R}} \exp\{-i[(\mathbf{P} - \mathbf{q}) \cdot \mathbf{R}_{\text{eq}} + (\Omega_{\text{in}} + \omega)t]\} \\
&+ \mathcal{O}(u^2),
\end{aligned} \tag{3.40}$$

where, by virtue of the fact that we are in harmonic approximation, the underlying assumption is that atomic displacements are small and, therefore, the development has been safely truncated at the linear terms in $\mathbf{u}$. The maximum conditions for $\mathcal{A}_{\text{out}}$ are found in three occurrences: the first term imposes that the scattering vector is equal to a translational vector of the reciprocal lattice of the sample, that is: $\mathbf{P} = \mathbf{G}$; in turn, the second and third terms provide the conditions $\mathbf{P} \pm \mathbf{q} = \mathbf{G}$. In the first case the maximum peak of the scattered wave is found at the frequency $\Omega_{\text{out}} = \Omega_{\text{in}}$ and, therefore, the scattered neutrons have the same energy of the incident ones: this is the condition of *elastic scattering*. In the two remaining cases the scattered neutrons have instead a different energy since the maximum peak is, respectively, found at frequency $\Omega_{\text{out}} = \Omega_{\text{in}} \mp \omega$, corresponding to the condition of *inelastic scattering*. The analysis is completed by also considering the momentum exchange described in equation (3.40): by passing to the corpuscular picture this leads us to identify two unalike scattering conditions

$$\begin{cases} \hbar\Omega_{\text{out}} = & \hbar\Omega_{\text{in}} - \hbar\omega \\ \hbar\mathbf{p}_{\text{out}} = & \hbar\mathbf{p}_{\text{in}} - \hbar\mathbf{q} + \hbar\mathbf{G} \end{cases} \quad \text{or} \quad \begin{cases} \hbar\Omega_{\text{out}} = & \hbar\Omega_{\text{in}} + \hbar\omega \\ \hbar\mathbf{p}_{\text{out}} = & \hbar\mathbf{p}_{\text{in}} + \hbar\mathbf{q} + \hbar\mathbf{G}, \end{cases} \tag{3.41}$$

which are easily interpreted as *conservation laws*. A neutron with energy $\hbar\Omega_{\text{in}}$ can be inelastically scattered by *emitting or absorbing a phonon* of energy $\hbar\omega$. The corresponding conservation of the momentum is more subtle: in fact, inelastic scattering events conserve momentum to within a reciprocal lattice vector[21]. This

---

[20] In the theory of neutron scattering it is customary to write the total cross section for a scattering event as $4\pi b_{\mathbf{R}}^2$. It plays a role similar to the atomic form factor.

[21] In our treatment this important result has been obtained phenomenologically. It is, however, possible to formally prove it by invoking the general result of quantum theory which makes a correspondence between the symmetry properties of the Hamiltonian operator and some conservation laws of physical observables. More specifically, the crystal Hamiltonian is invariant upon discrete translations and this implies that momenta are conserved to within a translational vector of the reciprocal lattice [9, 10].

result experimentally confirms what we anticipated in section 3.5 when the phonon momentum was better referred to as 'crystal momentum': while in an inelastic scattering event the crystalline sample does absorb a momentum $\hbar\mathbf{P}$, it is just a matter of convention to assign the part $\hbar\mathbf{q}$ to a phonon and the part $\hbar\mathbf{G}$ to the crystal as a whole; equivalently, we could totally assign it to the phonon with momentum $\hbar(\mathbf{q} + \mathbf{G})$. There is no difference in physics, since the two phonons with wavevector $\mathbf{q}$ or $(\mathbf{q} + \mathbf{G})$ represent exactly the same vibrational mode.

This state of affairs suggests that the neutron scattering spectrum consists in peaks, each corresponding to a possible scattering event. In a real experiment it is possible to vary $\mathbf{p}_{\mathrm{in}}$ and/or $\mathbf{p}_{\mathrm{out}}$ still keeping the scattering vector $\mathbf{P} = \mathbf{p}_{\mathrm{out}} - \mathbf{p}_{\mathrm{in}}$ at the exact value of the phonon wavevector $\mathbf{q}$ whose energy we want to measure. This situation fulfils the conservation laws (3.41) and, therefore, each peak appearing in the spectrum corresponds to a different existing phonon of wavevector $\mathbf{q}$. By repeating the measurement for a different scattering vector we can move along the edges of the irreducible part of the 1BZ (see section 2.4.4) and, therefore, we obtain the complete phonon dispersion relations.

3.7 The vibrational density of states

In many frameworks it is necessary to explicitly take into account *the distribution of phonon frequencies over the full vibrational spectrum*. To this aim it is useful to define the quantity $G(\omega)$, hereafter referred to as the *vibrational density of states* (vDOS), defined such that $G(\omega)d\omega$ represents *the number of phonons* (that is: vibrational modes) *with frequency in the interval* $[\omega, \omega + d\omega]$. In order to calculate a general expression for $G(\omega)$ we will proceed by two steps of increasing generality.

Let us preliminarily consider the model case of a monoatomic sc crystal with first nearest neighbours distance a and subject to Born–von Karman periodic boundary conditions. For convenience we choose the crystal in the form of a cube with edge length L and assume that there is only one dispersion relation: basically, this is the three-dimensional counterpart of the monoatomic linear chain. Accordingly, by generalising equation (3.9) we understand that the allowed phonon wavevectors $\mathbf{q}$ have Cartesian components $q_i = 2\pi\xi_i/L$ with $L = Na$, while $i = x, y, z$ and $\xi_i = 0, 1, 2, \ldots(N - 1)$ if N is the total number of atoms in the crystal. Their number density in the reciprocal space is straightforwardly calculated as $L^3/(2\pi)^3 = V/(2\pi)^3$ where $V = L^3$ is the volume of the sample. The number of allowed wavevectors corresponding to vibrational modes with a frequency in between ω and $\omega + d\omega$ is given by the product between their number density and the infinitesimal volume $4\pi q^2 dq$ of the reciprocal space. Since we have just one phonon branch, this number equals the number of vibrational frequencies in the selected interval. We accordingly write

$$G(\omega)d\omega = \frac{V}{(2\pi)^3}\, 4\pi q^2 dq. \tag{3.42}$$

We remark that $4\pi q^2 dq$ is the volume of the spherical shell precisely bounded by the two constant-frequency surfaces corresponding to ω and $\omega + d\omega$, respectively.

Therefore, equation (3.42) is strictly true only if we assume that the phonon properties depend on their magnitude but not their direction, indeed a simplification we have implicitly used.

This constraint must be removed if we seek for a general theory. To this aim we consider the more general case of a three-dimensional crystal and we focus at first just on its sth dispersion $\omega = \omega_s(\mathbf{q})$: in this case, the vibrational frequencies are not only determined by the magnitude of the wavevector $\mathbf{q}$, but also by its orientation within the 1BZ. This calls for some care since we are forced to use vector formalism. Similarly to the simple model previously discussed, the number of allowed wave-vectors in the sth branch for which the frequency is in between ω_s and $\omega_s + d\omega$ is once again given by the product between their number density and an appropriate infinitesimal volume $d\mathbf{q}$ of the reciprocal space. Since we are considering just the sth phonon branch, this number equals the number of its allowed modes in the selected frequency interval. The vDOS we are looking for is

$$G_s(\omega)d\omega = \frac{V}{(2\pi)^3}d\mathbf{q}, \tag{3.43}$$

where we made use of the same number density $V/(2\pi)^3$ as before; the subscript s reminds us that we are only considering this dispersion branch. The difficult task is to calculate $d\mathbf{q}$ which now corresponds to the generic volume of the reciprocal space between the two surfaces with $\omega_s(\mathbf{q}) = $ constant and $\omega_s(\mathbf{q}) + d\omega = $ constant. We remark that their shape is not known and it can be in principle very complex; in other words, the calculation of $d\mathbf{q}$ cannot be traced back to that of a simple known volume: we need a more general formalism.

If we consider an element dS_{ω_s} on the inner surface $\omega_s(\mathbf{q}) = $ constant, we can name $dq_\perp(\mathbf{q})$ its distance to the outer surface $\omega_s(\mathbf{q}) + d\omega = $ constant; as shown in figure 3.12, such a distance is $\mathbf{q}$-dependent since it varies according to the selected surface element dS_{ω_s}. This procedure allows us to write the volume element $d\mathbf{q}$ as

$$d\mathbf{q} = \int_{S_{\omega_s}} dS_{\omega_s}\, dq_\perp(\mathbf{q}), \tag{3.44}$$

where it is understood that the integral is extended to all points of the surface S_{ω_s} where we know that $\omega_s(\mathbf{q}) = $ constant. The infinitesimal difference in frequency $d\omega$ between the two surfaces can be written as

$$d\omega = \left| \nabla_{\mathbf{q}}\omega(\mathbf{q}) \right| dq_\perp(\mathbf{q}), \tag{3.45}$$

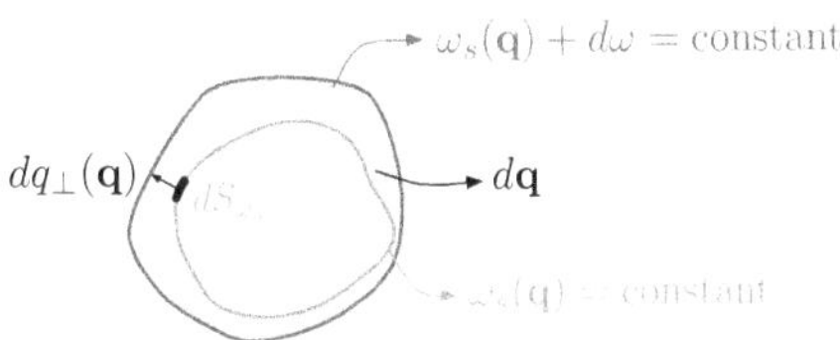

Figure 3.12. The red and blue lines represent two constant-frequency surfaces embedding the reciprocal space volume $d\mathbf{q}$. The two surfaces are separated by the distance $dq_\perp(\mathbf{q})$ which is shown for the element dS_{ω_s} on the inner surface.

where $\nabla_{\mathbf{q}}\omega_s(\mathbf{q})$ is the gradient of $\omega_s(\mathbf{q})$. By using this result in equation (3.44) we get for the volume element in the reciprocal space the following general expression

$$d\mathbf{q} = d\omega \int_{S_{\omega_s}} \frac{dS_{\omega_s}}{\left| \nabla_{\mathbf{q}}\omega(\mathbf{q}) \right|}, \qquad (3.46)$$

which is used in equation (3.43) to obtain the vDOS of the sth vibrational branch

$$G_s(\omega) = \frac{V}{(2\pi)^3} \int_{S_{\omega_s}} \frac{dS_{\omega_s}}{\left| \nabla_{\mathbf{q}}\omega_s(\mathbf{q}) \right|}, \qquad (3.47)$$

where it is understood that the integral is extended to all points of the surface S_{ω_s} defined by $\omega_s(\mathbf{q}) = \omega$. This result is used to calculate the *total vibrational density of states* simply by summing over all branches

$$G(\omega) = \sum_s G_s(\omega) = \frac{V}{(2\pi)^3} \sum_s \int_{S_{\omega_s}} \frac{dS_{\omega_s}}{\left| \nabla_{\mathbf{q}}\omega_s(\mathbf{q}) \right|}, \qquad (3.48)$$

where $\mathbf{q} \in 1BZ$.

It is important to remark an intriguing feature of the vDOS as formalised in equation (3.48): the denominator appearing in the integrated function is nothing else than the magnitude of the group velocity $v_{g,s}(\mathbf{q}) = \left| \nabla_{\mathbf{q}}\omega_s(\mathbf{q}) \right|$ of the $(s, \mathbf{q})$ phonon. This allows us to predict that whenever such velocity is zero, a singularity must appear in the vDOS: they are referred to as *van Hove singularities*. This situation occurs whenever a phonon branch flattens, as typically found for zone-centre optical modes and zone-boundary acoustic ones. The van Hove singularities are clearly shown in figure 3.13, where we report the calculated vDOS for metallic Cu and covalent Si. By comparing this figure with the corresponding phonon dispersion relations reported in figures 3.7 and 3.8, respectively, it is easy to spot the singularities.

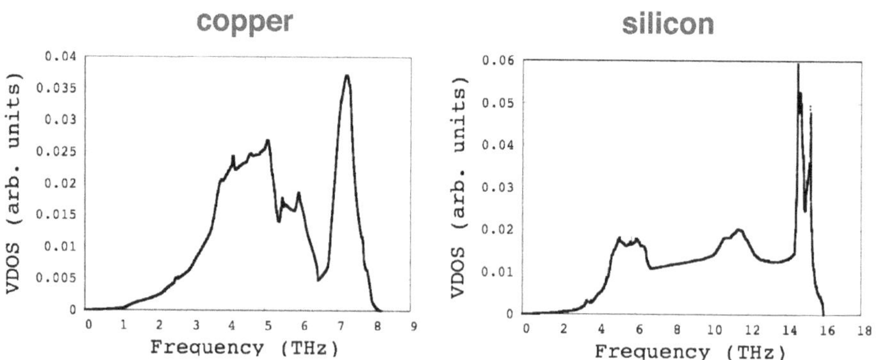

Figure 3.13. The vibrational density of states (vDOS) of Cu (left) and Si (right). The calculations have been performed by using the QUANTUM ESPRESSO integrated suite of open-source computer codes for materials modelling; see website https://www.quantum-espresso.org (by courtesy of Aleandro Antidormi).

Another interesting characteristics emerging from figure 3.13 (but not easy to see in equation (3.48)) is that $G(\omega) \sim \omega^2$ in the small frequency range. This behaviour is explicitly predicted in the Debye theory of the specific heat, as extensively discussed in section 4.1.2.

We finally remark that equation (3.48), although providing the most general theoretical formulation for the vibrational density of states, is not that easy to handle in practice. However, nowadays solid state theory makes extensive use of computational methods and, therefore, the $\omega_s(\mathbf{q})$ frequencies are in most cases determined numerically. This means that a more practical formulation for the vDOS is the following

$$G(\omega) = \frac{V}{(2\pi)^3} \sum_s \int \delta(\omega - \omega_s(\mathbf{q}))d\mathbf{q}, \tag{3.49}$$

where the integral is over the full 1BZ.

The vDOS is also referred to as the *phonon density of states*, as allowed by the direct correspondence between the classical and quantum pictures, respectively, formulated in terms of vibrational modes and phonons.

References

[1] Miller D A B 2008 *Quantum Mechanics for Scientists and Engineers* (New York: Cambridge University Press)

[2] Bransden B H and Joachain C J 2000 *Quantum Mechanics* (Upper Saddle River, NJ: Prentice Hall)

[3] Sakurai J J and Napolitano J 2011 *Modern Quantum Mechanics* 2nd edn (Reading, MA: Addison-Wesley)

[4] Griffiths D J and Schroeter D F 2018 *Introduction to Quantum Mechanics* 3rd edn (Cambridge: Cambridge University Press)

[5] Born M and Huang K 1954 *Dynamical Theory of Crystal Lattices* (Oxford: Oxford University Press)

[6] Böttger H 1983 *Principles of the Theory of Lattice Dynamics* (Berlin: Akademie)

[7] Burns G and Glazer M 2013 *Space Groups for Solid State Scientists* 3rd edn (Waltham, MA: Academic)

[8] Maradudin A A, Montroll E W, Weiss G H and Ipatova I P 1971 *Theory of Lattice Dynamics in the Harmonic approximation* (New York: Academic)

[9] Ashcroft N W and Mermin N D 1976 *Solid State Physics* (London: Holt-Saunders)

[10] Grosso G and Pastori Parravicini G 2014 *Solid State Physics* 2nd edn (Oxford: Academic)

[11] Martin R M 2012 *Electronic Structure–Basic Theory and Practical Methods* (Cambridge: Cambridge University Press)

[12] Colombo L 2019 *Atomic and Molecular Physics: A Primer* (Bristol: IOP Publishing)

[13] Feynman R P, Leighton R B and Sands M 1963 *The Feynman Lectures on Physics* (Reading: Addison-Wesley)

[14] Panofsky W K H and Phillips M 1990 *Classical Electricity and Magnetism* (New York: Dover)

[15] Glazer M and Wark J 2001 *Statistical Mechanics–A Survival Guide* (Oxford: Oxford University Press)
[16] Hook J R and Hall H E 2010 *Solid State Physics* (Hoboken, NJ: Wiley)
[17] Dove M T 2003 *Structure and Dynamics–An Atomic View of Materials* (Oxford: Oxford University Press)

IOP Publishing

Solid State Physics
A primer
Luciano Colombo

Chapter 4

Thermal properties

Syllabus—The need of a more fundamental theory of lattice dynamics going beyond the approximations adopted in the previous chapter is discussed by investigating the thermal properties of a crystalline solid. The calculation of the heat needed to increase the temperature of insulating materials is developed at different levels of increasing complexity, eventually providing the general quantum theory of the heat capacity. Next, the anharmonic features affecting the ionic oscillations are extensively discussed by considering at first the thermal expansion phenomenon. A full quantum picture of anharmonicity is then worked out, based on the phonon language which allows one to quantitatively describe their scattering mechanisms. A rigorous treatment of thermal transport in non-metallic crystals is eventually presented, based on the fundamental Boltzmann transport equation.

4.1 The lattice heat capacity

4.1.1 Historical background

The energy content of a physical system is thermodynamically accounted for by its internal energy $\mathcal{U}$ (see appendix C) whose derivative with respect to temperature

$$C_V = \frac{d\mathcal{U}}{dT}\bigg|_V, \tag{4.1}$$

is known as the *heat capacity* at constant volume V: it represents the amount of heat we need to quasi-statically provide in order to increase the system temperature by one degree.

The first attempts to derive a microscopic theory for C_V in crystalline solids were developed at the dawn of the XXth century. In many respects, we can consider these investigations as the beginning of quantum solid state physics [1]. Developing a microscopic theory was certainly worthy of effort since the classical theory of C_V is contradicted by the experimental evidence. In order to outline this theory, outdated

doi:10.1088/978-0-7503-2265-2ch4

but still valuable for our pedagogical approach to the thermal properties, we preliminarily remark that there are three main contributions to the heat capacity of a crystal, respectively, deriving from lattice vibrations, conduction electrons, and magnetic ordering. In non-magnetic insulators the first one is by far the leading one and in this chapter we focus just on it[1].

Classically the internal energy $\mathcal{U}$ of a crystal containing N atoms corresponds to the vibrational energy of $3N$ one-dimensional harmonic oscillators, as calculated by means of the equipartition theorem: if the crystal is in equilibrium at temperature T, an average energy $k_B T/2$ is attributed to each energy contribution which is quadratic either in general coordinates or momenta. Therefore, the average energy of each atomic oscillator is estimated to be $\langle u(T) \rangle = k_B T$ so that

$$\mathcal{U} = 3N\langle u(T) \rangle = 3Nk_B T = 3RT, \tag{4.2}$$

where we have hereafter assumed that $N = \mathcal{N}_A$ (i.e. we have an Avogadro number of atoms in the crystal), while $R = 8.314$ J K^{-1} mol^{-1} is the universal gas constant. The corresponding classical prediction for the heat capacity $C_V = 3R$ is known as the *Dulong–Petit law*. Contrary to this, experimental measurements provide evidence that $C_V \to 0$ for $T \to 0$. More specifically, it is found that $C_V \sim T^3$ in the range of vanishingly small temperatures. The measured C_V approaches the predicted value only at very high temperature. In conclusion, the classical theory is unable to justify the observed $C_V = C_V(T)$ trend over the full range of temperatures.

A major step forward to a more fundamental theory was taken in 1907 by A Einstein who replaced the classical harmonic oscillators by quantum ones further assuming that *all oscillators vibrate at the same frequency* ω_E, known as the Einstein frequency. Following the arguments developed in section 3.5, we write the average energy of the single one-dimensional Einstein oscillator at temperature T as

$$\langle u_E(T) \rangle = \left[n_{BE}(\omega_E) + \frac{1}{2} \right] \hbar\omega_E, \tag{4.3}$$

where $n_{BE}(\omega)$ is the Bose–Einstein distribution introduced in equation (3.37) and more formally discussed in appendix E. In figure 4.1 a comparison is shown between the energy of a classical and a quantum Einstein oscillator as a function of temperature. We accordingly calculate the heat capacity as

$$C_V^{\text{Einstein}}(T) = 3R \left(\frac{\hbar\omega_E}{k_B T} \right)^2 \frac{\exp(\hbar\omega_E/k_B T)}{[\exp(\hbar\omega_E/k_B T) - 1]^2}, \tag{4.4}$$

which fulfils the Dulong–Petit law at high temperature, while becoming vanishingly small at zero temperature. However, it is easy to prove that that $C_V^{\text{Einstein}}(T) \sim \exp(-1/T)$ as $T \to 0$; in other words, *the Einstein model is qualitatively correct, but it provides a wrong behaviour for the heat capacity at small temperatures.*

---

[1] The other contributions to C_V will be discussed in the following chapters.

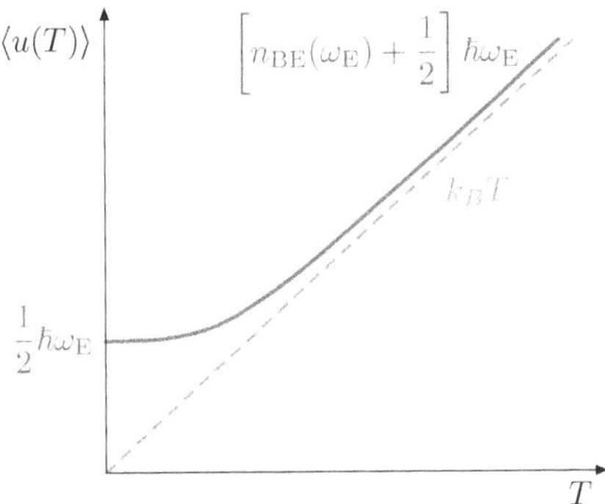

Figure 4.1. The average energy $\langle u(T)\rangle$ of a classical (red dashed line) and a quantum Einstein (blue full line) oscillator as a function of the temperature T.

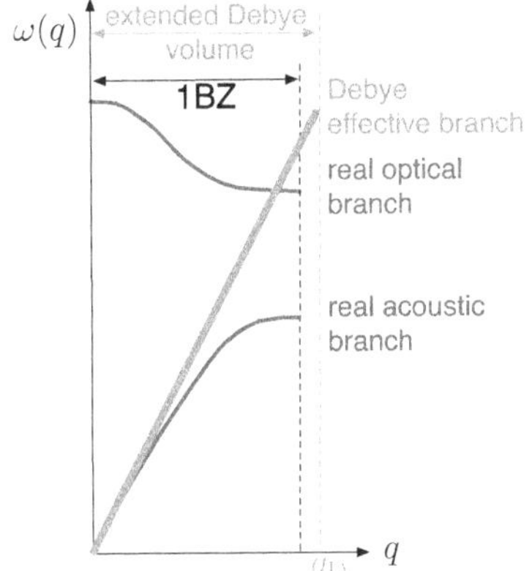

Figure 4.2. Definition of the effective acoustic Debye branch (red) and corresponding Debye wavevector q_D.

A more refined model is actually needed; since the theory must necessarily be quantum, we will refer to vibrational modes as phonons, as explained in section 3.5.

4.1.2 The Debye model for the heat capacity

The Einstein model is correct in treating atomic vibrations as quantum oscillators, but it fails in attributing the same frequency to all of them: simply, this is inconsistent with the knowledge of the dispersion relations we developed in chapter 3. We must therefore introduce in the theory the fundamental notion that *atomic oscillators can vibrate at different frequencies*. Within the *Debye model* this notion is developed in a simplified way which allows us to carry on a clean analytical calculation of the heat capacity.

According to Debye, *all phonon dispersion relations are effectively described by only three effective acoustic branches* whose extension in wavevector, however, exceeds the boundary of the 1BZ. This is shown in figure 4.2: the low and high q-values of the effective branch, respectively, describe an acoustic and an optical vibration of the real crystal. Furthermore, since for any direction there are in fact three possible phonon polarisations, the linearisation of their dispersions must properly take care to distinguish between one effective longitudinal and two effective transverse branches with slope $v_g^{(L)}$ and $v_g^{(T)}$, respectively (see section 3.2.1). To this aim it is useful to introduce the *effective speed of sound* v_{eff} defined as

$$\frac{3}{v_{\mathrm{eff}}^3} = \frac{1}{\left[v_g^{(L)}\right]^3} + \frac{2}{\left[v_g^{(T)}\right]^3}. \tag{4.5}$$

We can now calculate the density of vibrational states in the Debye model $G_{\mathrm{D}}(\omega)$ by making use of equation (3.42) elaborated for a single branch so that

$$G_{\mathrm{D}}(\omega) = 3\,\frac{V}{2\pi^2}q^2\frac{1}{d\omega/dq} = 3\frac{V}{2\pi^2}\frac{\omega^2}{v_{\mathrm{eff}}^3} = \frac{V}{2\pi^2}\left\{\frac{1}{\left[v_g^{(L)}\right]^3} + \frac{2}{\left[v_g^{(T)}\right]^3}\right\}\omega^2, \tag{4.6}$$

where we set $\omega = v_{\mathrm{eff}}\,q$ consistently with the linearisation procedure; the factor 3 takes into account the three possible polarisations. Interesting enough, *we find a quadratic dependence of the VDOS upon the frequency*: this feature is indeed found in real materials in the acoustic region of the vibrational spectrum, that is, exactly where the phonon dispersion relations are linear as supposed in the Debye model (see section 3.7). Since $G_{\mathrm{D}}(\omega)d\omega$ is the number of vibrational modes with frequency in the range $[\omega, \omega + d\omega]$, it is straightforward to impose the following normalisation condition to a crystal containing N atoms

$$\int_0^{\omega_{\mathrm{D}}} G_{\mathrm{D}}(\omega)\,d\omega = 3N, \tag{4.7}$$

where ω_{D}, known as *the Debye frequency*, dictates that no phonons with higher frequency are found in the crystal. By using the expression given in equation (4.6) for the vibrational density of states we easily obtain

$$\omega_{\mathrm{D}} = \sqrt[3]{\frac{6\pi^2 N}{V}}\,v_{\mathrm{eff}} = \sqrt[3]{\frac{18\pi^2 N}{V}}\left\{\frac{1}{\left[v_g^{(L)}\right]^3} + \frac{2}{\left[v_g^{(T)}\right]^3}\right\}^{-3}, \tag{4.8}$$

which in turn defines the maximum Debye wavevector introduced above since the two quantities are related as $\omega_{\mathrm{D}} = v_{\mathrm{eff}}\,q_{\mathrm{D}}$.

The full quantum expression for the internal energy[2] is eventually written as

$$\begin{aligned}
\mathcal{U} &= \int_0^{\omega_{\mathrm{D}}} [n_{\mathrm{BE}}(\hbar\omega, T) + 1/2]\,\hbar\omega\,G_{\mathrm{D}}(\omega)\,d\omega \\
&= \frac{9N\hbar}{\omega_{\mathrm{D}}^3}\int_0^{\omega_{\mathrm{D}}} [n_{\mathrm{BE}}(\hbar\omega, T) + 1/2]\omega^3 d\omega \\
&= \frac{9}{8}N\hbar\omega_{\mathrm{D}} + \frac{9N\hbar}{\omega_{\mathrm{D}}^3}\int_0^{\omega_{\mathrm{D}}} n_{\mathrm{BE}}(\hbar\omega, T)\,\omega^3 d\omega,
\end{aligned} \tag{4.9}$$

---

[2] We recall once more that this is just the lattice contribution, the leading one for non-magnetic insulators.

where $n_{\rm BE}(\hbar\omega, T)$ is the Bose–Einstein distribution law (see appendix E). This result is the Debye counterpart of the more general expression given in equation (3.36), where the sum over the index s labelling the discrete vibrational modes has been replaced by the integral over a continuum of frequencies: this is certainly valid for a large enough N. The first term on the right-hand side is the Debye estimate for the quantum zero-point energy: it is independent of temperature and, therefore, gives no contribution to the heat capacity; accordingly we calculate

$$C_V^{\rm Debye}(T) = 9R \left(\frac{T}{T_{\rm D}}\right)^3 \int_0^{T_{\rm D}/T} \frac{x^4 \exp(x)}{[\exp(x) - 1]^2} \, dx \ \text{ with } x = \frac{\hbar\omega}{k_{\rm B}T}, \tag{4.10}$$

where has been introduced the *Debye temperature* $T_{\rm D} = \hbar\omega_{\rm D}/k_{\rm B}$ and, as before, it has been assumed $N = \mathcal{N}_{\rm A}$. It should be clear from its very definition that *the Debye temperature is an empirical parameter* as much as the Debye frequency or wavevector: it can be determined by fitting experimental data with equation (4.10). In this respect, the Debye model we worked out should be more rigorously looked at as a simple interpolation scheme. Nevertheless, we can attribute to $T_{\rm D}$ a very intuitive physical meaning by calculating $C_V^{\rm Debye}(T)$ in the limit of a vanishingly small or of a very high temperature, respectively corresponding to $T \ll T_{\rm D}$ or $T \gg T_{\rm D}$. It is easy to prove that[3]

$$\begin{cases} \lim_{T \to 0} C_V^{\rm Debye}(T) = \dfrac{12R\pi^4}{5T_{\rm D}^3} \, T^3 \\[2ex] \lim_{T \to +\infty} C_V^{\rm Debye}(T) = 3R, \end{cases} \tag{4.11}$$

while the trend for intermediate temperatures is reported in figure 4.3. The predicted behaviour exactly corresponds to the experimental observation that $C_V^{\rm expt}(T) \sim T^3$ at low temperature, thus making the Debye interpolation scheme superior to the simpler Einstein model which, as already commented, failed in this temperature regime predicting $C_V^{\rm Einstein}(T) \cdot \exp(-1/T)$. On the other hand, in the high-temperature limit the Debye formula recovers the classical model and, once again, it is in perfect agreement with measurements.

An important remark must be added in commenting on figure 4.3 where the $C_V^{\rm Einstein}(T)$ is also reported: *the Einstein and the Debye models use a different temperature scale* which we have so far defined only in the latter case by introducing $T_{\rm D}$. In contrast, the only way to define the notion of temperature in the Einstein model is setting $\hbar\omega_{\rm E} = k_{\rm B}T_{\rm E}$, where $T_{\rm E}$ will be hereafter addressed as the *Einstein temperature*: the problem is finding its relationship with $T_{\rm D}$. The sole vibrational frequency defined in the Einstein model should necessarily be linked to the speed of sound, if we imagine the sound propagation within the crystal as a sequence of hits between oscillating ions. Accordingly, the relationship between the effective speed $v_{\rm eff}$ defined in equation (4.5) and the Einstein frequency is

---

[3] We remark that if $T \ll T_{\rm D}$ then $T_{\rm D}/T \to +\infty$ and the integral appearing in equation (4.10) is tabulated to be $\int_0^{+\infty} x^4 \exp(x)/[\exp(x) - 1]^2 \, dx = 4\pi^4/15$.

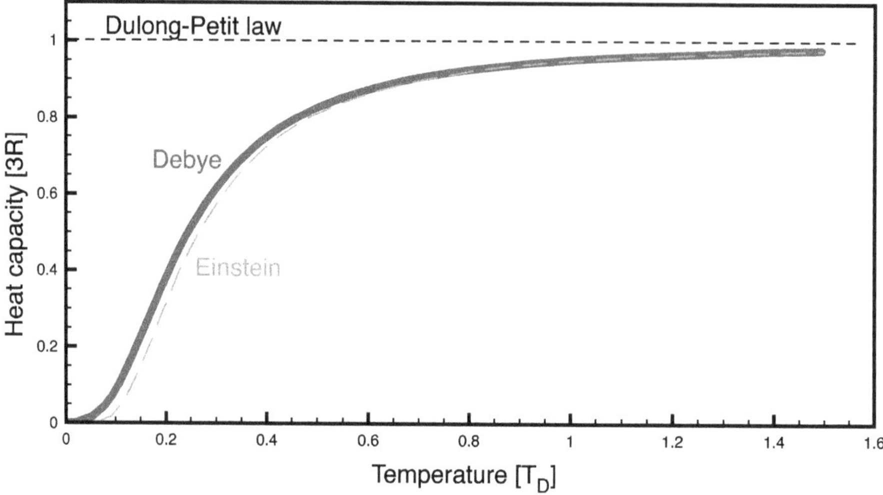

Figure 4.3. The lattice heat capacity calculated according to the Debye model (full blue line) for a crystalline non-magnetic insulator containing $N = \mathcal{N}_A$ atoms. The same quantity calculated according to the Einstein model (red dashed line) is shown for comparison. Temperatures are measured in units of the Debye temperature T_D and heat capacities in units of $3R$.

$$v_{\text{eff}} = \frac{\omega_E}{2\pi} \lambda_E, \tag{4.12}$$

where λ_E corresponds to the minimum wavelength of the ionic oscillations sustained by the crystal. For a system in the form of a cube with edge L and containing N atoms we have $\lambda_E/2 = L/\sqrt[3]{N}$ and therefore

$$T_E = \frac{h}{2k_B} \sqrt[3]{\frac{N}{V}}\, v_{\text{eff}}, \tag{4.13}$$

where $V = L^3$ for the adopted hypotheses. It easy to prove that

$$\frac{T_E}{T_D} = \sqrt[3]{\frac{\pi}{6}}, \tag{4.14}$$

which allows for a direct comparison between the Einstein and the Debye model, as shown in figure 4.3.

In conclusion, the performances of the Debye model allow us to interpret T_D as *the threshold temperature below which the crystal behaves quantum mechanically, while above it a classical picture is reasonably adequate.* In other words, we can say that if $T > T_D$ then all vibrational modes have been excited, while if $T < T_D$ some of them (just a few or very many, according to the Bose–Einstein distribution) are still not populated. The Debye temperature for some crystalline solids is reported in table 4.1.

Table 4.1. The calculated Debye temperature T_D (in Kelvin units) for some crystalline solids with metallic, covalent, or ionic bonding.

	Metallic					Covalent			Ionic		
	Al	Cu	Au	Fe	Pb	C[a]	Si	Ge	NaCl	LiF	KBr
T_D	428	343	170	470	105	2230	645	374	321	730	173

[a] Diamond crystalline structure.

4.1.3 The general quantum theory for the heat capacity

Although the Debye expression for the lattice heat capacity is rather accurate over a wide range of temperatures for most materials, deviations from laboratory measurements are nevertheless found. The most effective way to develop the comparison is to fit experimental data taken at different temperatures by means of equation (4.10), while keeping T_D as the only calibration parameter for the fitting. For many systems this procedure returns a Debye temperature varying within few tens of Kelvin degrees: this is the fingerprint of some failure of the interpolation scheme, which is conceptually based on the existence of a unique Debye temperature. Good for us, these deviations are small for many practical applications and, therefore, the Debye model can be used as a very good approximation.

If, however, a high degree of accuracy is needed, then there is no better solution than using a full quantum theory where the lattice contribution to the internal energy $\mathcal{U}$ is calculated according to equation (3.36) so that

$$\mathcal{U} = \mathcal{U}_0 + \sum_{s\mathbf{q}}[n_{BE}(s\mathbf{q}, T) + 1/2]\,\hbar\omega_s(\mathbf{q}), \tag{4.15}$$

where $\mathcal{U}_0$ is the total energy content of the static lattice. We accordingly calculate[4]

$$C_V^{\text{quantum}}(T) = \frac{\partial}{\partial T}\sum_{s\mathbf{q}}\frac{\hbar\omega_s(\mathbf{q})}{\exp[\hbar\omega_s(\mathbf{q})/k_B T] - 1}$$

$$= \sum_{s\mathbf{q}}\hbar\omega_s(\mathbf{q})\frac{\partial n_{BE}(s\mathbf{q}, T)}{\partial T} = \sum_{s\mathbf{q}}C_{V,s\mathbf{q}}(T), \tag{4.16}$$

where we used equation (3.37) for the phonon population $n_{BE}(s\mathbf{q}, T)$ and we introduced the specific contributions $C_{V,s\mathbf{q}}(T)$ of each $(s, \mathbf{q})$ mode to the heat capacity.

This expression is more easily handled *in the limit of a very large crystal*: the density of allowed $\mathbf{q}$ wavevectors in the reciprocal space is so high that we can treat them as a continuum

$$C_V^{\text{quantum}}(T) = \frac{V}{(2\pi)^3}\frac{\partial}{\partial T}\sum_s\int\frac{\hbar\omega_s(\mathbf{q})}{\exp[\hbar\omega_s(\mathbf{q})/k_B T] - 1}\,d\mathbf{q}, \tag{4.17}$$

where the integral is taken for $\mathbf{q} \in 1BZ$.

---

[4] The zero-point energy contribution has been ignored since it is independent of temperature.

The quantum expression $C_V^{\text{quantum}}(T)$ (both in discrete or continuum form) approaches the Dulong–Petit law at high temperature, while going to zero as $\sim T^3$ for a vanishingly small temperature. In the intermediate-temperature regime it provides the correct behaviour for the lattice heat capacity without any simplifying assumption. The price to pay is a complete (numerical) calculation of all phonon frequencies.

4.2 Anharmonic effects

The crystal lattice dynamics has been so far described under the harmonic approximation which allowed us to understand many fundamental intrinsic properties of solids. It is, however, just an approximation, as emerged from the discussion developed in section 3.1 where it has been presented as a convenient truncation of a Taylor expansion of the total ionic potential energy $U = U(\mathbf{R})$ (see equation (3.1) and relative discussion). Beyond this formal argument, *robust experimental evidences suggest that a real system is in fact not purely harmonic*; they are mostly related to thermal properties like:

- if $k_\mathrm{B} T/\hbar$ is much larger than typical phonon frequencies, deviations of the predicted heat capacity from the experimental data are actually observed: they are the onset of anharmonic effects, not yet explicitly included in the theory leading to equation (4.17);
- a crystalline solid differently resists to positive or negative strains of identical magnitude; since any volume variation reflects a change in the lattice spacing, this suggests that ions are confined nearby their equilibrium positions by a non-parabolic (that is, non harmonic) potential;
- real solids undergo thermal expansion; this would not be possible if the ions thermally oscillate under the action of a perfectly parabolic potential since the average ion–ion distance would not increase upon temperature;
- finally, a beam of phonons travelling along a given direction within an infinite defect-free crystal would propagate with no damping if anharmonic effects were not included (harmonic vibrational modes overlap without interference); this would imply an infinite lattice thermal conductivity.

In the following we are going to treat separately thermal expansion and thermal conduction in the next subsections.

4.2.1 Thermal expansion

Thermal expansion is due to *the dependence of vibrational frequencies on the crystal volume*. To exploit this notion, we will make use of some fundamental thermodynamic definitions reported in appendix C.

Our first goal is to work out an equation of state $P = P(V, T)$ relating the pressure P acting on the system to its volume V and temperature T. To this aim, we use the Helmholtz free energy $\mathcal{F}$, since we assume that our solid is coupled to a heat reservoir, that is $T =$ constant. We also understand that no matter is added to or removed from the system and, therefore, the numbers of moles of any chemical species are also constant. Under these assumptions, we can write (see equations (C.4) and (C.8))

$$dF = d(U - TS) = -PdV - SdT, \qquad (4.18)$$

so that the equation of state for the pressure is cast in the form

$$P = -\left.\frac{\partial F}{\partial V}\right|_T, \qquad (4.19)$$

which is conveniently developed as follows

$$
\begin{aligned}
P &= -\left.\frac{\partial(U - TS)}{\partial V}\right|_T \\
&= -\frac{\partial}{\partial V}\left(U - T\int_0^T \left.\frac{\partial S}{\partial T'}\right|_V dT'\right)_T \qquad (4.20)\\
&= -\frac{\partial}{\partial V}\left(U - T\int_0^T \frac{1}{T'}\left.\frac{\partial U}{\partial T'}\right|_V dT'\right)_T,
\end{aligned}
$$

where we used the identity $T(\partial S/\partial T)_V = (\partial U/\partial T)_V$. In order to proceed we need an explicit expression for the internal energy: by using equation (4.15) in the equation of state, after some non trivial algebra we obtain

$$
P(V, T) = \underbrace{-\left.\frac{\partial U_0}{\partial V}\right|_T - \frac{1}{2}\sum_{s\mathbf{q}}\left.\frac{\partial\hbar\omega_s(\mathbf{q})}{\partial V}\right|_T}_{T=0 \text{ contribution}} \\
+ \underbrace{\sum_{s\mathbf{q}}\left[-\left.\frac{\partial\hbar\omega_s(\mathbf{q})}{\partial V}\right|_T\right]n_{\mathrm{BE}}(s\mathbf{q}, T)}_{T>0 \text{ contribution}}, \qquad (4.21)
$$

indicating that *the pressure depends on V and T through the variations of the phonon frequencies upon volume changes and through the Bose–Einstein phonon population,* respectively.

In a purely harmonic crystal we have $(\partial\hbar\omega_s(\mathbf{q})/\partial V)_T = 0$ for any $(s, \mathbf{q})$ mode, since the vibrational frequencies do not depend on the crystal volume[5]: therefore, the pressure turns out to be independent of temperature and we accordingly conclude that *in the harmonic approximation the equation of state reduces to a simple law* $P = P(V)$. Consistently, the thermal expansion coefficient β of the harmonic crystal (see appendix C) is calculated to be

$$\beta = \frac{1}{B}\left.\frac{\partial P}{\partial T}\right|_V = 0, \qquad (4.22)$$

---

[5] To make it simple: the frequencies of an array of harmonic atomic oscillators just depend on the force constants of the springs connecting them and on their masses, while they are independent of the interatomic separations, as explicitly shown in section 3.2.

where B is the bulk modulus of the system. In short: *a harmonic crystal does not undergo thermal expansion*, contrary to experience. Another unphysical consequence of the harmonic approximation is that by combining equation (C.14) with equation (4.22) we immediately obtain that the constant-pressure and constant-volume heat capacities should be equal, contrary to experimental evidence. We need to do better.

The most obvious option to improve our physical picture would be to explicitly introduce the anharmonic terms in the evaluation of the vibrational crystal energy, as explained in equation (3.1). This can be done either classically or quantum-mechanically, but it requires some non-trivial formal developments [2]. We will rather follow a more phenomenological approach, nevertheless leading to meaningful conclusions.

Let us assume that $(\partial \hbar \omega_s(\mathbf{q})/\partial V)_T \neq 0$ for any $(s, \mathbf{q})$ mode: basically, *we are assuming that the vibrational frequencies are no longer purely harmonic* or, equivalently, that ions oscillate in non-parabolic potential wells: their frequencies now depend on the crystal volume. For further convenience, it is useful to define the mode-specific *Grüneisen parameter* γ_{sq} as

$$\gamma_{sq} = -\frac{\partial \ln \omega_s(\mathbf{q})}{\partial \ln V} = -\frac{V}{\omega_s(\mathbf{q})}\frac{\partial \omega_s(\mathbf{q})}{\partial V}, \tag{4.23}$$

so that it is possible to recast the equation of state (4.21) in the new form

$$P(V, T) = -\frac{\partial \mathcal{U}_0}{\partial V}\bigg|_T + \frac{1}{V}\sum_{sq}\gamma_{sq}\,\hbar\omega_s(\mathbf{q})\left[\frac{1}{2} + n_{\mathrm{BE}}(s\mathbf{q}, T)\right], \tag{4.24}$$

from which we calculate the *thermal expansion coefficient of an anharmonic crystal* as

$$\begin{aligned}
\beta &= \frac{1}{B}\frac{\partial P}{\partial T}\bigg|_V = \frac{1}{VB}\sum_{sq}\gamma_{sq}\,\hbar\omega_s(\mathbf{q})\frac{\partial n_{\mathrm{BE}}(s\mathbf{q}, T)}{\partial T} \\
&= \frac{1}{VB}\sum_{sq}\gamma_{sq}\,\mathcal{C}_{V,sq}(T) \\
&= \frac{\gamma}{VB}\mathcal{C}_V(T),
\end{aligned} \tag{4.25}$$

where we have introduced the weighted Grüneisen parameter

$$\gamma = \frac{\sum_{s,q}\mathcal{C}_{V,sq}(T)\gamma_{sq}}{\mathcal{C}_V(T)}. \tag{4.26}$$

If we consider γ as an empirical material-specific parameter[6], we can use equation (4.25) as a meaningful phenomenological model, usually referred to as the *Grüneisen model*, for the thermal expansion coefficient, properly taking into account the volume-dependence of the vibrational modes: it effectively explains the experimental

---

[6] For instance, we could calculate γ within the Debye model by setting $\gamma = -\partial \ln \omega_{\mathrm{D}}/\partial \ln V$.

evidence that real crystals expand upon heating. In table 4.2 we report the thermal expansion coefficient and the bulk modulus of some selected crystals.

The Grüneisen model predicts that the temperature dependence of β is dominated by the heat capacity term, since the temperature dependence of the volume and bulk modulus is much weaker in most materials. This implies that the ratio $BV\beta/C_V = \gamma$ should be almost constant in temperature. Both features[7] are roughly found in many materials, although deviations have been predicted in some cases [8]. This is a first evidence that the Grüneisen model, although very useful, can be improved: as a matter of fact, the most accurate way to proceed is to calculate β as reported in the first row of equation (4.25). Another limitation of the model is highlighted by the direct evidence that *anharmonicity really affects each mode in a different way* [2], as shown in in table 4.3: there we report some calculated mode-specific Grüneisen parameter in covalent materials. Not only do we observe large variations in the absolute value of γ_{sq}, but this parameter is even found either positive and negative. Finally, the complex role played by anharmonicity is especially evident in ferro-electric crystals which undergo *displacive transitions* [8, 9]: at a particular temperature, the frequency of some vibrational modes (known as *soft modes*) is lowered down to zero by anharmonicity. Accordingly, the ion displacements assume a

Table 4.2. Room temperature bulk modulus B (in units of 10^2 GPa) and the thermal expansion coefficient β (units of 10^{-6} K^{-1}) for some crystalline solids with metallic, covalent, or ionic bonding.

	Metallic					Covalent			Ionic		
	Al	Cu	Au	Fe	Pb	C[a]	Si	Ge	NaCl	LiF	KBr
B^b	0.72	1.37	1.73	1.68	0.43	4.43	0.99	0.77	0.24	0.67	0.15
β^c	70.8	51.0	41.7	35.1	86.4	2.4	6.9	18.3	118.5	98.7	115.5

[a] Diamond bulk modulus is measured at $T = 4$ K.
[b] Data taken from: [3, 4].
[c] Data taken from: [5, 6, 7].

Table 4.3. Some calculated Grüneisen parameter for selected zone-centre (Γ) and zone-boundary (X) vibrational modes in covalent crystals.

	TO(Γ)	LO(Γ)	LA(X)	LO(X)	TO(X)	TA(X)
Si[a]	0.90	0.90	1.30	1.30	0.90	-1.50
Ge[a]	0.90	0.90	1.40	1.40	1.00	-1.50
GaAs[b]	1.42		1.11	0.91	1.56	-3.48

[a] Data taken from: [10].
[b] Data taken from: [11].

---

[7] Namely: $\lim_{T \to 0} \beta(T) = \lim_{T \to 0} C_V(T)$ and $\gamma \sim$ constant.

permanent character which causes a variation of the lattice parameters and the crystal is faced with a structural phase transition[8].

4.2.2 Phonon–phonon interactions

The Grüneisen model has phenomenologically treated anharmonic effects in lattice dynamics through volume-dependent vibrational frequencies. The most fundamental description of anharmonicity is, however, based on quantum theory and exploits the phonon language: while in the harmonic approximation phonons are described as a gas of free pseudo-particles, in a most realistic anharmonic crystal they actually undergo mutual interactions.

Phonon–phonon interactions are not so strong to fully invalidate the harmonic picture: this is proved by the true existence of well-resolved peaks in neutron scattering spectra (see section 3.6), each peak being the fingerprint of a specific harmonic phonon mode. Therefore, *anharmonicity can be treated as a perturbation on the quantum states of the harmonic crystal*: while its energy spectrum remains basically unaffected by phonon–phonon interactions (that is, we can still speak about phonon frequencies and vibrational modes with different character s and wavevector $\mathbf{q}$), *anharmonicity causes transitions between different states of quantum harmonic oscillator*.

The formal treatment of such a perturbation is non trivial and falls beyond the present level of discussion [2, 12, 13], but we can assimilate the underlying physical concept by means of an analogy with atomic physics: the energy spectrum of, say, an isolated hydrogen atom remains unaffected by a low-intensity electromagnetic field[9], whose perturbative effect is only to promote electronic transitions between the discrete stationary-state levels of the atom [14]. We can say that, for both absorption or emission transitions, the occupation of the initial and final stationary state has been varied by -1 and $+1$, respectively, while a photon has been annihilated (absorption) or created (emission). This is a three-particle event involving two-electron and one-photon populations.

Similarly, any anharmonic term of the vibrational Hamiltonian appearing in equation (3.1) causes transitions among harmonic eigenstates, correspondingly affecting their phonon populations. The physical picture is simple: we can say that the nth order term (with $n \geqslant 3$) in the Taylor expansion of the lattice potential energy activates interactions among n phonons, which we will refer to as *n-phonon scattering events*. Since the phonon number is not conserved, during a scattering event phonons of some harmonic mode are annihilated (their population is decreased), while other phonons of different modes are created (their population is increased).

Let us consider the specific case of a three-phonon scattering event: the phonon populations of any vibrational mode remain unchanged except for the three modes

---

[8] Ferroelectric crystals are especially prone to this kind of displacive transitions since the soft modes generate a permanent electric dipole moment.

[9] This is strictly true only in the semi-classical picture, where the atomic energy spectrum is treated by quantum mechanics, while the electromagnetic field is described by Maxwell equations.

$(s_1, \mathbf{q}_1)$, $(s_2, \mathbf{q}_2)$, and $(s_3, \mathbf{q})_3$ which are coupled by a cubic term of the Hamiltonian. This may occur in two different ways, namely

$$
\begin{cases}
n_{s_1\mathbf{q}_1} \rightarrow n_{s_1\mathbf{q}_1} - 1 \\
n_{s_2\mathbf{q}_2} \rightarrow n_{s_2\mathbf{q}_2} + 1 \\
n_{s_3\mathbf{q}_3} \rightarrow n_{s_3\mathbf{q}_3} + 1
\end{cases}
\quad \text{and} \quad
\begin{cases}
n_{s_1\mathbf{q}_1} \rightarrow n_{s_1\mathbf{q}_1} - 1 \\
n_{s_2\mathbf{q}_2} \rightarrow n_{s_2\mathbf{q}_2} - 1 \\
n_{s_3\mathbf{q}_3} \rightarrow n_{s_3\mathbf{q}_3} + 1,
\end{cases}
\tag{4.27}
$$

which correspond to a *phonon generation event* (left) or to a *phonon annihilation event* (right). The two events are shown in figure 4.4 (left).

The rate of occurrence per unit time $P^{(3)}_{i \rightarrow f}$ of a three-phonon event can be calculated according to first-order quantum perturbation theory by means of the Fermi golden rule [15–17] as

$$
P^{(3)}_{i \rightarrow f} = \frac{2\pi}{\hbar} \, |\langle f | \hat{V}_3 | i \rangle|^2 \, \delta(\hbar\omega_{s_1\mathbf{q}_1} \mp \hbar\omega_{s_2\mathbf{q}_2} - \hbar\omega_{s_3\mathbf{q}_3}),
\tag{4.28}
$$

where $\hat{V}_3$ is the quantum operator describing the cubic anharmonic perturbation [2, 12, 13]. For a phonon creation event, the initial $|i\rangle$ and final $|f\rangle$ states correspond to the quantum states with phonon populations $(n_{s_1\mathbf{q}_1}, n_{s_2\mathbf{q}_2}, n_{s_3\mathbf{q}_3})$ and $(n_{s_1\mathbf{q}_1} - 1, n_{s_2\mathbf{q}_2} + 1, n_{s_3\mathbf{q}_3} + 1)$, respectively. On the other hand, for a phonon annihilation event, the final state is rather corresponding to $(n_{s_1\mathbf{q}_1} - 1, n_{s_2\mathbf{q}_2} - 1, n_{s_3\mathbf{q}_3} + 1)$. The quantum calculation appearing in equation (4.28) is rather complicated; we skip it and just focus on the δ-term which imposes that *the total energy of the phonon system is conserved during the scattering event*; more specifically, we get

$$
\begin{aligned}
\text{phonon creation:} \qquad & \hbar\omega_{s_1\mathbf{q}_1} = \hbar\omega_{s_2\mathbf{q}_2} + \hbar\omega_{s_3\mathbf{q}_3} \\
\text{phonon annihilation:} \quad & \hbar\omega_{s_1\mathbf{q}_1} + \hbar\omega_{s_2\mathbf{q}_2} = \hbar\omega_{s_3\mathbf{q}_3},
\end{aligned}
\tag{4.29}
$$

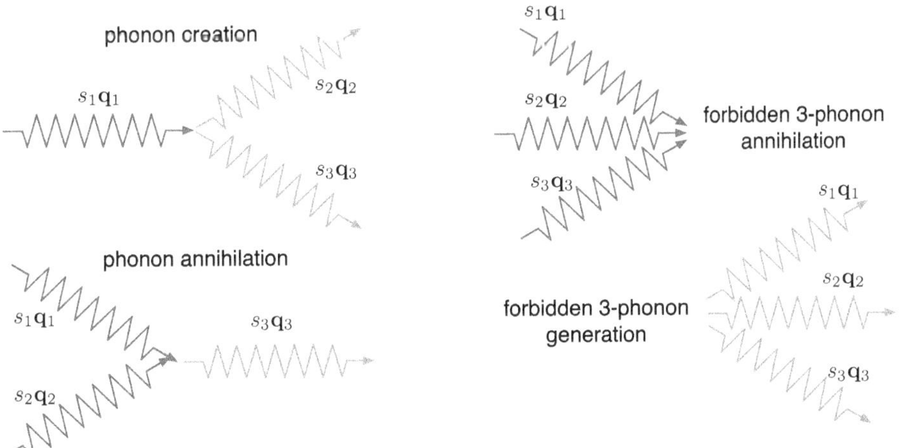

Figure 4.4. Pictorial representation of allowed (left) and forbidden (right) three-phonon scattering events. As usual $s_n\mathbf{q}_n$ are the branch index and wavevector of the three $n = 1, 2, 3$ harmonic vibrational modes undergoing mutual interaction.

where it is understood that the energy contribution of all remaining modes not of interest to the scattering event remains unaffected. This explains why the three-phonon events shown in figure 4.4 (right) are never observed: simply, they are forbidden since they do not conserve energy. If we still retain the approximation that ionic oscillations have small amplitudes, the dominant terms ruling over anharmonicity are expected to be the cubic and quartic ones. These latter are treated similarly to the discussion above, provided that a suitable expression $\hat{V}_4$ for the quantum operator describing the quartic anharmonic perturbation is worked out. However, in the attempt to keep our development simple, we will only consider three-phonon events in the following.

Another important anharmonic feature is that *a phonon scattering also conserves the crystal momentum*. In particular, with reference to figure 4.4 a three-phonon event must obey the following rules

$$\text{phonon creation:} \qquad \hbar\mathbf{q}_1 = \hbar\mathbf{q}_2 + \hbar\mathbf{q}_3 \pm \hbar\mathbf{G}$$
$$\text{phonon annihilation:} \quad \hbar\mathbf{q}_1 + \hbar\mathbf{q}_2 = \hbar\mathbf{q}_3 \pm \hbar\mathbf{G}, \qquad (4.30)$$

where $\mathbf{G}$ is a reciprocal lattice vector. We can read these conservation laws by stating that $\mathbf{q}_1$, $\mathbf{q}_2$, and $\mathbf{q}_3$ directly involve three specific vibrational modes, while $\mathbf{G}$ is absorbed or provided by the crystal as a whole. In studying three-phonon scattering events it is customary to distinguish between *normal processes* (or N-processes) with $\mathbf{G} = 0$ and *Umklapp processes*[10] (or U-processes) with $\mathbf{G} \neq 0$. They are represented in figure 4.5 in the case of a phonon annihilation process occurring in a two-dimensional crystal for which the graphics is easier to draw and to understand (but, of course, the basic notion is also valid in one- and three-dimensions). For further convenience, it is important to remark here that the true phonon momentum is not conserved in a U-process, because of the $\pm\mathbf{G}$ term appearing in equations (4.30). For this reason, it is often said that Umklapp processes destroy (or generate) momentum.

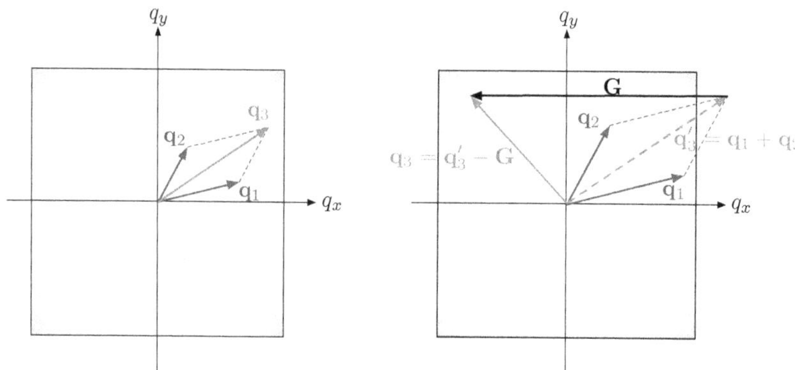

Figure 4.5. Pictorial representation of a normal (left) and an Umklapp (right) three-phonon scattering event in a two-dimensional square lattice. The grey-shaded area represents the 1BZ.

---

[10] The German word 'umklapp' means 'flipped over': this wording will be explained soon.

We conclude this section by remarking that the rigorous perturbative treatment of phonon–phonon scattering [2, 12] proves that harmonic vibrations are slightly *shifted in frequency* $\omega_s(\mathbf{q}) \rightarrow \omega_s(\mathbf{q}) + \Delta_{s\mathbf{q}}$ and *damped in amplitude*. This second feature is translated in the corpuscular phonon language by saying that *because of anharmonic effects each phonon mode has a finite lifetime* $\tau_{s\mathbf{q}}$. Our simplified description of phonon–phonon interactions is consistent with this notion: for instance, if we consider a three-phonon annihilation process, then the inverse of its scattering rate per unit time given in equation (4.28) is related to the lifetime of the decaying phonon. This concept will be fully exploited in the theory of thermal transport developed in the next section where *phonons will be treated as the microscopic heat carriers*. Each carrier will be treated as a $\hbar\omega_s(\mathbf{q})$ quantum of energy travelling with velocity $\mathbf{v}_g(s\mathbf{q}) = d\omega_s(\mathbf{q})/d\mathbf{q} = \lambda_{s\mathbf{q}}\tau_{s\mathbf{q}}$, where $\lambda_{s\mathbf{q}}$ is its *average mean free path* (that is, the average distance covered by the phonon between two successive scattering events).

4.3 Thermal transport

In order to set up the investigation on thermal transport phenomena in a solid state system, let us consider the situation represented in figure 4.6, where (i) a small temperature gradient is imposed along the x direction of (ii) a homogeneous insulating crystal. This is the minimal complexity framework containing all the most relevant physical features which rule over the thermal energy transport. Under these constitutive hypotheses we can assume that *the microscopic heat carriers are the lattice vibrations*[11] and, therefore, the most appropriate approach is based on the phonon language. Furthermore, under a small imposed thermal gradient dT/dx experiments provide evidence that, after a transient time, *a steady state thermal*

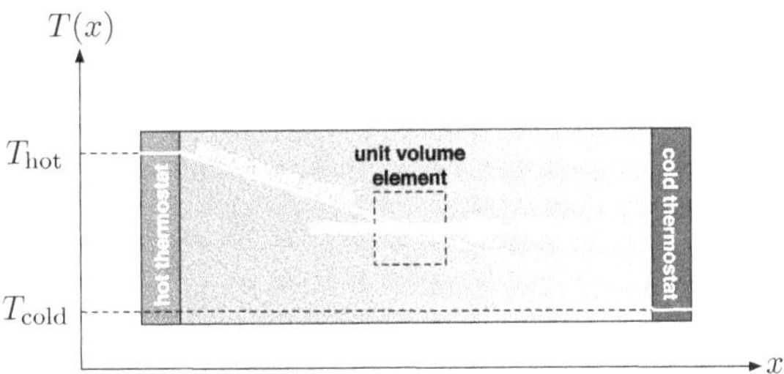

Figure 4.6. A homogeneous crystalline specimen is coupled to a hot (left) and cold (right) thermostat, respectively, at temperature T_{hot} and $T_{\text{cold}} < T_{\text{hot}}$. The white full line represents the temperature profile across the sample in the steady-state condition. The indicated volume element is used in equation (4.34) for the continuity equation of the thermal current described by the heat flux $J_{\text{h},x}$.

---

[11] In metals there is an additional important contribution due to conduction electrons. It will be investigated in section 7.2. For this reason the parameter κ_l introduced just below will be referred to as the *lattice* thermal conductivity.

conduction regime is established, which is pretty well described by the phenomeno-logical linear Fourier law [18, 19]

$$J_{h,x} = -\kappa_l \frac{dT}{dx}, \tag{4.31}$$

where $J_{h,x}$ is the heat flux along x, namely the amount of thermal energy crossing a unit area normal to the x direction per unit time (it is measured in units J m^{-2} s^{-1}). The key physical parameter is the *lattice thermal conductivity* κ_l: it is a material-specific quantity making the difference between thermal insulators (low κ_l values) and good thermal conductors (large κ_l vaules). While the non-equilibrium thermo-dynamics fundamentals of thermal transport can be found elsewhere [20], here *we aim at developing a microscopic theory of* κ_l based on the phonon language.

The heat flux $J_{h,x}$ can be calculated by summing over all phonons $(s, \mathbf{q})$ the product [energy carried] $\times$ [number of phonons] $\times$ [speed of propagation] $\times$ [1/V] (where V is the volume crossed by the thermal energy flux) or equivalently[12]

$$J_{h,x} = \frac{1}{V} \sum_{s,\mathbf{q}} \hbar\omega_s(\mathbf{q}) \, \bar{n}(s\mathbf{q}, T) \, v_{g,x}(s\mathbf{q}), \tag{4.32}$$

where it must be noted that *the phonon population $\bar{n}(s\mathbf{q}, T)$ is not given by the Bose–Einstein distribution* since the condition represented in figure 4.6 does not correspond to an equilibrium situation[13]. Our next task is, therefore, to calculate the non-equilibrium population $\bar{n}(s\mathbf{q}, T)$ for the steady state condition we are interested in. To this aim we resume the full phonon picture including their mutual interactions developed in the previous section and follow a simple argument: if we assume that the temperature gradient is removed, then we expect the system to recover the equilibrium situation after a suitable time. This implies that during such a transient regime each single-mode phonon population $\bar{n}(s\mathbf{q}, T)$ must relax back to its equilibrium value $n_{BE}(s\mathbf{q}, T)$. The simplest situation we can guess is that such relaxation processes are independent or, equivalently, we can assume that *the evolution of each single phonon population is unaffected by the relaxation of all the other populations*. This is the essence of the so-called *single-mode relaxation time approximation* (SM-RTA) which allows us to write

$$\frac{\partial \bar{n}(s\mathbf{q}, T)}{\partial t} = -\frac{\bar{n}(s\mathbf{q}, T) - n_{BE}(s\mathbf{q}, T)}{\tau_{s\mathbf{q}}}, \tag{4.33}$$

where the single-mode relaxation time $\tau_{s\mathbf{q}}$ is identified with the lifetime of the corresponding vibrational mode, as discussed in the previous section. We remark

---

[12] We incidentally remark that equation (4.32) explains why in equilibrium conditions no thermal current occurs: if the temperature is uniform throughout the whole system, then $J_{h,x} = 0$ since the phonon group velocities are isotropically distributed.

[13] We will nevertheless rely on the *local equilibrium hypothesis* [20] which allows us to define anywhere within the system a local temperature and to use standard thermodynamics equations (although extensive quantities must be replaced by their volume densities).

that at this stage only anharmonicity (i.e. phonon–phonon interactions) has been included in our analysis. This will be soon critically readdressed in the following.

Let us now consider the unit volume element V shown in figure 4.6 by a dashed line. In steady-state conditions we can write a continuity equation for the thermal current

$$\frac{1}{V}\frac{\partial \mathcal{U}}{\partial t} = \frac{\partial J_{h,x}}{\partial x},\tag{4.34}$$

where the internal energy $\mathcal{U}$ is calculated by means of equation (4.15) with the non-equilibrium population $\bar{n}(s\mathbf{q},\, T)$; in equation (4.34) no term describing the rate of energy generation/removal appears since, consistently with the assumed homogeneity of the system, the selected unit volume does not contain any energy source or sink. This equation basically proclaims energy conservation within the selected volume element, properly taking into account that the incoming and the outcoming thermal energy fluxes are not equal because of the space variation of $\bar{n}(s\mathbf{q},\, T)$ due to the temperature profile established across the system. By inserting equations (4.15) and (4.34) into equation (4.34) we easily obtain

$$\frac{\partial \bar{n}(s\mathbf{q},\, T)}{\partial t} - v_{g,x}(s\mathbf{q})\,\frac{\partial \bar{n}(s\mathbf{q},\, T)}{\partial T}\,\frac{\partial T}{\partial x} = 0.\tag{4.35}$$

In order to proceed we will further assume that *the temperature-dependence in the non-equilibrium population is the same as in the Bose–Einstein distribution*, that is

$$\frac{\partial \bar{n}(s\mathbf{q},\, T)}{\partial T} = \frac{\partial n_{\mathrm{BE}}(s\mathbf{q},\, T)}{\partial T}.\tag{4.36}$$

This procedure eventually leads to the *linearised Boltzmann transport equation* (BTE)

$$\bar{n}(s\mathbf{q},\, T) = n_{\mathrm{BE}}(s\mathbf{q},\, T) - \tau_{s\mathbf{q}}\,v_{g,x}(s\mathbf{q})\,\frac{\partial n_{\mathrm{BE}}(s\mathbf{q},\, T)}{\partial T}\,\frac{\partial T}{\partial x},\tag{4.37}$$

where we explicitly made use of the SM-RTA expression given in equation (4.33). The Boltzmann transport equation allows for a straightforward calculation of the non-equilibrium phonon population $\bar{n}(s\mathbf{q},\, T)$ once:
- the temperature gradient $\partial T/\partial x$ is assigned;
- the phonon dispersion relations are known, so that the phonon group velocities $v_{g,x}(s\mathbf{q})$ are easily calculated;
- the relaxation times $\tau_{s\mathbf{q}}$ are calculated by quantum perturbation theory, as outlined in the previous section.

We duly remark that so far we have considered just one single physical mechanism for the determination of the phonon lifetimes, namely phonon–phonon interactions. As a matter of fact, they are many more: phonons are also scattered by lattice point defects (like isotope impurities, native defects, contaminant or doping impurities), by extended surface (grain boundaries) or line (dislocations) defects, as well as by the boundaries of the crystal[14]. If each scattering mechanism were to act alone, we could define a different scattering rate for each mechanism or, equivalently, a different relaxation time. If we assume that the scattering rates are additive, we can calculate the total *relaxation time* τ_{sq}^{tot} for the $(s\mathbf{q})$ mode according to the *Matthiessen rule* [2]:

$$\frac{1}{\tau_{sq}^{tot}} = \sum_i \frac{1}{\tau_{sq,i}}, \qquad (4.38)$$

where the label i runs over all possible phonon scattering mechanisms. The Matthiessen rule is similarly applied to the physics of electron scattering; while violations of this rule are in fact observed [13], it is in practice rather well satisfied in many circumstances and, therefore, we will fully rely on it.

We can now calculate the heat flux

$$J_{h,x} = \frac{1}{V} \sum_{sq} \hbar\omega_s(\mathbf{q}) \, n_{BE}(s\mathbf{q}, T) \, v_{g,x}(s\mathbf{q})$$
$$- \frac{1}{V} \sum_{sq} \hbar\omega_s(\mathbf{q})\tau_{sq}^{tot} \, v_{g,x}^2(s\mathbf{q}) \frac{\partial n_{BE}(s\mathbf{q}, T)}{\partial T} \frac{\partial T}{\partial x}, \qquad (4.39)$$

where we have inserted the non-equilibrium population provided by the Boltzmann transport equation into equation (4.32). The first term on the right-hand side obviously does not contribute to the heat current since it only depends on equilibrium quantities. Therefore, by comparing equations (4.31) and (4.39) we eventually obtain a *microscopic equation for the lattice thermal conductivity*[15]

$$\kappa_l(T) = \frac{1}{V} \sum_{sq} \hbar\omega_s(\mathbf{q}) \, \tau_{sq}^{tot} \, v_{g,x}^2(s\mathbf{q}) \frac{\partial n_{BE}(s\mathbf{q}, T)}{\partial T}$$
$$= \frac{1}{3V} \sum_{sq} \tau_{sq}^{tot} \, \langle v_{g,sq}^2 \rangle \, C_{V,sq}(T) \qquad (4.40)$$
$$= \frac{1}{3} \langle \tau \rangle \langle v_g^2 \rangle \, c_V(T),$$

which we have written in the three more commonly used forms. More specifically, the second form is obtained by using equation (4.16) and by defining a mode-specific mean square group velocity $\langle v_{g,sq}^2 \rangle/3 = v_{g,x}^2(s\mathbf{q}) = v_{g,y}^2(s\mathbf{q}) = v_{g,z}^2(s\mathbf{q})$, while the third form is found by further attributing the same mean value of relaxation time $\langle \tau \rangle$ and square group velocity $\langle v_g^2 \rangle$ to all phonon modes, as well as by using the constant-volume specific heat $c_V(T) = (1/V)\sum_{sq} C_{V,sq}(T) = C_V(T)/V$. Interestingly enough,

---

[14] This latter scattering mechanism plays an important role whenever the phonon average mean free paths are longer than or comparable to the system dimension.

[15] This value of κ is referred to as 'lattice thermal conductivity' to highlight the fact that the contribution from the valence electron system is not accounted for.

this third form for $\kappa_1(T)$ is the same provided by elementary kinetic theory, where the heat carriers are treated as an interacting phonon gas [13].

In figure 4.7 we report the SM-RTA-BTE calculated thermal conductivity of crystalline silicon. In this case the system has been considered isotopically pure, totally defect-free, and large enough to exclude boundary scattering. Therefore, the resulting picture can be considered as representative of the general $\kappa = \kappa_1(T)$ trend observed in most non-metallic crystals when only phonon–phonon scattering events (that is, anharmonic effects) are at work. The agreement with experimental data (red dots) is impressive, proving that the adopted hierarchy of approximations is rather physically sound. We can distinguish three regimes:

- *very low temperature*: the lattice thermal conductivity increases with temperature simply because the higher the temperature the larger are the phonon populations; in this temperature range, however, phonon scattering plays a minor role, since vibrational modes are still poorly populated; overall the temperature-dependence is dictated by the specific heat and, therefore, we have $\kappa_1(T) \sim T^3$, reflecting into a linear trend in the reported bi-logarithmic plot;

- *intermediate temperatures*: a larger and larger number of phonons is generated and, therefore, scattering plays an increasingly important role in this temperature range; the maximum lattice thermal conductivity value is found at the temperature where the rate of generation of new phonons per unit temperature increase equals the rate of annihilation because of scattering events; above this temperature, the thermal conductivity begins to decrease as $\kappa_1(T) \sim \exp[(1/b)T_D/T]$ where it is typically found that $2 < b < 3$ according to the selected material;

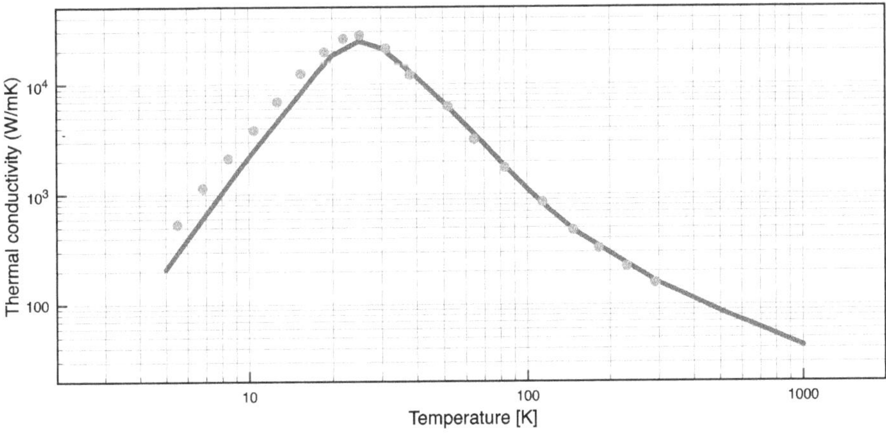

Figure 4.7. Full blue line: the lattice thermal conductivity of crystalline silicon calculated by the SM-RTA-BTE approach. The calculations have been performed by using the QUANTUM ESPRESSO integrated suite of open-source computer codes for materials modelling; see website https://http://www.quantum-espresso.org (by courtesy of Giorgia Fugallo). Red dots: experimental data taken from [21].

- *high temperatures*: phonon annihilation mechanisms are more efficient than generation ones and the lattice thermal conductivity is found to decrease according to a power law $\kappa_l(T) \sim T^{-d}$, where for most materials it is found that $1 < d < 2$.

Finally, it is interesting to observe that in the harmonic approximation there are no phonon–phonon interactions and, therefore, the lifetime $\tau_{s\mathbf{q}}$ of any vibrational mode is infinite. According to equation (4.40), this would suggest an infinite lattice thermal conductivity: another indeed striking failure of the purely harmonic crystal picture.

We now address more in detail the different role played by normal and Umklapp processes (see section 4.2.2) in affecting lattice thermal conductivity. We begin by considering a general feature which we elaborate by a simple order-of-magnitude estimate of the energies and momenta of phonons which undergo three-phonon scattering events. Let us assume a system temperature below the Debye temperature: then, the frequency of the three involved phonons must necessarily be much smaller than the Debye frequency. This is in fact a direct consequence of the Bose–Einsten distribution which phonons must obey: at any temperature T the only populated phonon modes are those with vibrational energy $\hbar\omega \lesssim k_B T$. Accordingly, at $T < T_D$ the only incoming phonons that can undergo scattering events have energy $\hbar\omega_{in} \ll \hbar\omega_D$. Because of energy conservation, the outcoming phonons emerging from the scattering event must similarly have $\hbar\omega_{out} \ll \hbar\omega_D$ or, equivalently, the magnitude of their wavevector must be $|\mathbf{q}_{out}| \ll q_D$. This is possible only if $\mathbf{G} = 0$ in equation (4.30). In conclusion, *at low temperature mainly N-processes occur*; on the other hand, U-processes must be duly considered at moderate or high temperature where scattering events are compatible with the $\pm\mathbf{G}$ exchange of a reciprocal lattice vector with the crystal as a whole.

Next, we address the different effect due to U- and N-processes on the heat current and phonon populations; to make things easy, let us refer to figure 4.5 and assume a situation such that the thermal transport occurs along the positive orientation of the q_x-axis (that is, we assume a rightward thermal current). It is evident that the three-phonon Umklapp process shown on the right flips over the momentum of the outcoming scattered phonon with respect to the direction of the heat current: this deprives the net thermal energy current of its energy contribution. Because of this, *U-processes are said to be resistive*, meaning that their main contribution is to decrease lattice thermal conductivity or, equivalently, to increase thermal resistivity. On the other hand, *N-processes have the main role to repopulate each single phonon mode*, which is continuously affected by opposite annihilation events.

By combining all the results discussed so far, we classify different thermal conduction regimes. When the size of the crystal is comparable with the phonon mean free path, the dominant scattering mechanisms are due to dislocations, grain boundaries and surfaces; this is referred to as the *ballistic regime* since phonons freely propagate, until they meet some extrinsic[16] scattering source. This typically occurs at very low temperatures (no intrinsic[17] resistive scattering processes are active) and, as commented above, in the ballistic regime we have $\kappa(T) \sim T^3$. By increasing temperature we enter the *Poiseuille regime* where the phonon gas flows under the action of the temperature gradient within the system, whose boundaries work as a friction source; this situation is similar to that found in fluid hydrodynamics. Resistive processes of any kind (that is both intrinsic U- or extrinsic ones) occur at higher temperatures, giving rise to the *Ziman regime*; here N-processes are still dominant, but they are increasingly contrasted by resistive ones (although boundary mechanisms are no longer important). Finally, in the high-temperature *kinetic regime* thermal transport is dominated by anharmonic resistive processes. The Ziman and kinetic regimes are together referred to as *diffusive thermal transport*; here it is found that $\kappa(T) \sim T^{-d}$ (with $1 < d < 2$).

We conclude the investigation on thermal transport by considering a very special situation referred to as *second sound*. Ordinary sound is the propagation of a compressive wave of frequency ω_s within a gas made of massive particles [22]. This phenomenon is observed provided that $\omega_s \ll 1/\tau$, where τ is the average relaxation time for the collisions among gas particles (in other words, $1/\tau$ is the rate of particle–particle collisions). This condition is necessary to guarantee that local thermodynamical equilibrium is always reached by the gas at each instant of the sound wave oscillation. It is important to stress that collisions among gas particles conserve energy and momentum. We now turn to look at phonons as a gas: we must preliminarily admit that *there are two average relaxation times*, namely τ_N and τ_U for normal and Umklapp processes, respectively. In a U-process the true phonon momentum is actually not conserved (see equations (4.30) and related discussion) and, therefore, their occurrence should be very rare to observe a sound-like phenomenon in the phonon gas. On the other hand, normal processes should obey the condition valid for ordinary sound. Accordingly, provided that the condition $1/\tau_U \ll \omega_{ss} \ll 1/\tau_N$ is fulfilled, the analogue of mechanical sound is found in the phonon gas: the 'second sound' is a wavelike oscillation in temperature[18] occurring at frequency ω_{ss}. Since the necessary condition is that $\tau_N \gg \tau_U$, second sound is typically observed at very low temperature.

References

[1] Eisberg R and Resnick R 1985 *Quantum Physics of Atoms, Molecules, Solids, Nuclei, and Particles* 2nd edn (Hoboken, NJ: Wiley)

[2] Srivastava G P 1990 *The Physics of Phonons* (Bristol: Adam Higler)

---

[16] In this framework 'extrinsic' means not related to anharmonicity. Extrinsic processes are due to isotope, point-defect, grain boundary, dislocation, and surface scattering.

[17] In this framework 'intrinsic' means related to anharmonicity. Intrinsic processes are only due to direct phonon–phonon interactions.

[18] An oscillating density of phonons reflects in an oscillation of their energy density; since their energy is thermal, this corresponds to a temperature oscillation.

[3] Birch F 1966 *Handbook of Physical ConstantsGSA Memoirs* vol 97 (McLean, VA: The Geological Society of America) p 97

[4] Tosi M P 1964 *Solid State Physics*vol 16 (New York: Pergamon) p 44

[5] Pearson W B 1958 *A Handbook of Lattice Spacings and Structures of Metals and Alloys* (New York: Pergamon)

[6] White G K 1965 *Proc. R. Soc. London* **A286** 204

[7] Wolfe C M, Holonyak N and Stillman G E 1989 *Physical Properties of Semiconductors* (Englewood Cliffs, NJ: Prentice-Hall)

[8] Dove M T 2003 *Structure and Dynamics–An Atomic View of Materials* (Oxford: Oxford University Press)

[9] Kittel C 1996 *Introduction to Solid State Physics* 7th edn (Hoboken, NJ: Wiley)

[10] Yin M T and Cohen M L 1982 *Phys. Rev.* B **26** 3259

[11] Kunc K and Martin R M 1981 *Phys. Rev.* B **24** 2311

[12] Böttger H 1983 *Principles of the Theory of Lattice Dynamics* (Berlin: Akademie)

[13] Ashcroft N W and Mermin N D 1976 *Solid State Physics* (London: Holt-Saunders)

[14] Colombo L 2019 *Atomic and Molecular Physics: A Primer* (Bristol: IOP Publishing)

[15] Sakurai J J and Napolitano J 2011 *Modern Quantum Mechanics* 2th edn (Reading, MA: Addison-Wesley)

[16] Miller D A B 2008 *Quantum Mechanics for Scientists and Engineers* (New York: Cambridge University Press)

[17] Griffiths D J and Schroeter D F 2018 *Introduction to Quantum Mechanics* 3rd edn (Cambridge: Cambridge University Press)

[18] Incoprera F P and Dewitt D P 2011 *Fundamentals of Heat and Mass Transfer* (New York: Wiley)

[19] Lienhard H H IV and Lienhard J H V 2006 *A Heat Transfer Textbook* (Cambridge, MA: Phlogiston)

[20] Kjelstrup S and Bedeaux D 2008 *Non-equilibrium Thermodynamics of Heterogeneous Systems* (Singapore: World Scientific)

[21] Inyushin A V, Taldenkov A N, Gibin A M, Gusev A V and Pohl H J 2004 *Phys. Status Solidi* C **1** 2995

[22] Feynman R P, Leighton R B and Sands M 1963 *The Feynman Lectures on Physics* (Reading, MA: Addison-Wesley)

IOP Publishing

Solid State Physics
A primer
Luciano Colombo

Chapter 5

Elastic properties

Syllabus—We model the discrete atomistic structure of a solid by a continuous distribution of matter and describe its deformations occurring on a length scale much larger than the interatomic spacings by means of the strain and stress tensors. Within this continuum picture, the constitutive equation for the elastic behaviour of a solid body is derived under the assumption of small deformations, defining the core of linear elasticity. The constitutive stress–strain relation is derived in the paradigmatic case of a homogeneous and isotropic solid, whose thermoelastic behaviour is thoroughly discussed.

5.1 Basic definitions

5.1.1 The continuum picture

We have so far developed our theory of the crystalline state on the basis of the atomistic picture (see section 1.1). We are now about to consider two situations that are more effectively described by looking at a solid in a rather different way.

Let us start by considering ionic vibrations in the very long wavelength limit: in this case, the deformation of the crystal lattice with respect to its ideal crystallo-graphic configuration occurs on a length scale which is much larger than the typical interatomic spacings. This corresponds to the propagation of elastic waves (sound) in the material. The situation is even more marked if we suppose that the solid is subject to some external action of mechanical, electric or magnetic nature. For instance, this happens, respectively, when a compressive/tensile or bending load is applied to the system, when the onset of a coupling between a state of charge and a state of deformation is observed in the material response to some mechanical stress (a phenomenon known as piezoelectricity), or when during the magnetisation process of a material its shape or dimension are changed (a phenomenon known as magnetostriction). Once again, in all cases the resulting deformation typically unfolds on a macroscopic scale that, although it may result much shorter than the specimen dimensions, is nevertheless definitely much longer than interatomic distances.

In order to formally describe this kind of situations on a general ground, *it is convenient to switch to a new conceptual paradigm*: the discrete atomistic structure of the crystal is now replaced by *a continuous distribution of matter*, and its deformations of any origin are better described by *a continuous displacement field*. We will refer to this approach as the *continuum picture* which lies at the foundation of solid mechanics, a huge and independent scientific discipline with many applications in physics and engineering [1–3]. In this chapter we will just outline the fundamentals of elasticity theory which represents the sub-field of solid mechanics which mostly overlaps to solid state physics.

In the continuum picture we no longer deal with atomic positions onto a lattice but, rather, with a *continuous position variable* $\mathbf{r}$ representing the location of each material point within the solid. Its components are calculated with respect to a Cartesian frame of reference and, contrary to the notation adopted elsewhere in this volume, they will be labelled by Latin indices[1] $\mathbf{r} = \{r_i\}_{i=1,2,3}$. For consistency, the Cartesian axes will be indicated by the set (x_1, x_2, x_3). The *continuous displacement vector* $\mathbf{u}(\mathbf{r})$ is introduced by the following relation

$$\mathbf{r} \rightarrow \mathbf{r}' = \mathbf{r} + \mathbf{u}(\mathbf{r}), \tag{5.1}$$

where the change $\mathbf{r} \rightarrow \mathbf{r}'$ is caused by some external action. When the solid is described by the $\mathbf{r}$ positions is said to be in its *reference configuration*, while the set of $\mathbf{r}'$ positions is usually referred to as the *deformed configuration*. The *Jacobian matrix* $\mathbb{J} = \{J_{ij}\}_{i,j=1,2,3}$ describing the change from the reference configuration $\mathbf{r}$ to the deformed one $\mathbf{r}'$ can be written as

$$\begin{aligned} J_{ij} &= \frac{\partial u_i}{\partial x_j} \\ &= \underbrace{\frac{1}{2}\left(\frac{\partial u_i}{\partial x_j} + \frac{\partial u_j}{\partial x_i}\right)}_{\text{symmetric part}} + \underbrace{\frac{1}{2}\left(\frac{\partial u_i}{\partial x_j} - \frac{\partial u_j}{\partial x_i}\right)}_{\text{antisymmetric part}}, \end{aligned} \tag{5.2}$$

where the symmetric and antisymmetric part of $\mathbb{J}$ describe deformations of rather different nature. In order to provide a simple explanation of this statement[2] we consider a *pure infinitesimal local rotation* described by the matrix $\mathbb{R}$. In this case, the Jacobian matrix for any point $\mathbf{r}$ in the volume element under rotation is easily calculated:

$$\mathbf{r}' = \mathbb{R}\mathbf{r} = \mathbf{r} + \mathbf{u}(\mathbf{r}) \quad \rightarrow \quad \mathbf{u}(\mathbf{r}) = (\mathbb{R} - \mathbb{I})\mathbf{r} \quad \rightarrow \quad \mathbb{J} = \mathbb{R} - \mathbb{I}, \tag{5.3}$$

---

[1] This choice is forced by the fact that high-order tensorial quantities appear in the elasticity theory and, therefore, the use of explicit Cartesian indices (x, y, z) is unpractical. In any case, the new notation will not generate confusion since electrons, elsewhere in this volume labelled by Latin indices, will never appear in the discussion we are going to develop.

[2] In solid mechanics this result is known as the Cauchy polar decomposition theorem.

where $\mathbb{1}$ is the identity matrix. The product between the Jacobian matrix and its transpose matrix $\mathbb{J}^t$ is vanishingly small since the rotation is infinitesimal and, therefore, to a good approximation we can write

$$0 = \mathbb{J}\mathbb{J}^t = (\mathbb{R} - \mathbb{1})(\mathbb{R}^t - \mathbb{1}) = -\mathbb{J} - \mathbb{J}^t \quad \rightarrow \quad \mathbb{J} = -\mathbb{J}^t, \tag{5.4}$$

or, equivalently, we state that *the Jacobian matrix of a pure rotational deformation is antisymmetric*. Combining this result with equation (5.2) we conclude that in general *the deformation of any volume element consists in the combination of a local rotation and a local volume variation*, which can be either compressive or tensile, depending on the actual deformation considered.

This important result allows us to define the domain of *linear elasticity theory* we are developing in the present chapter: we will study the physics of *small volume variations of a solid body on the continuum scale under the action of sole deformation forces*, while pure rotations will not be considered. The reference to 'small' deformations is fully specified by the condition $\mathrm{Tr}(\mathbb{J}\mathbb{J}^t) \ll 1$.

5.1.2 The strain tensor

In the approximation of small deformations, the stretching, compression, bending or shearing of a solid body is entirely described by the symmetric part of the Jacobian matrix given in equation (5.2)

$$\epsilon_{ij} = \frac{1}{2}\left(\frac{\partial u_i}{\partial x_j} + \frac{\partial u_j}{\partial x_i} \right), \tag{5.5}$$

which is referred to as the *strain tensor* and, by its very definition, is symmetric $\epsilon_{ij} = \epsilon_{ji}$. Its physical meaning is revealed by the two examples shown in figure 5.1 where a simple tensile deformation (left) and a pure shear deformation (right) is applied to a homogeneous solid.

In the first case, a total length variation ΔL along the x_1 direction is applied without affecting the cross section. An infinitesimal slab originally placed in position x_1 is displaced by an amount[3] $\mathbf{u} = (x_1(\Delta L/L), 0, 0)$; we accordingly calculate the only non-zero component of the strain tensor $du_1/dx_1 = \Delta L/L$ and we get the

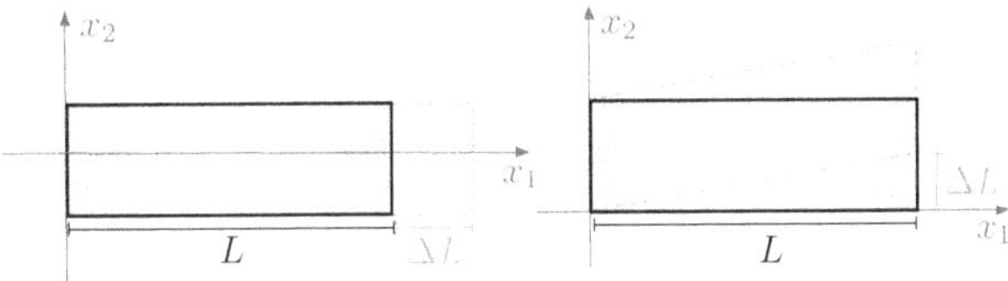

Figure 5.1. A simple tensile deformation (left) or pure shear deformation (right) is applied to a homogeneous solid. The deformation ΔL is applied along (left) or normal (right) to the x_1-axis. The third x_3-axis is perpendicular to the plane of the figure.

---

[3] This guarantees that the right terminal end of the sample is eventually found at position $x_1 = L + \Delta L$ as indicated in figure 5.1.

$$\text{strain tensor for a simple uniaxial tensile deformation} = \begin{pmatrix} \Delta L/L & 0 & 0 \\ 0 & 0 & 0 \\ 0 & 0 & 0 \end{pmatrix}, \quad (5.6)$$

This result holds also for the simple compressive deformation, but in this case obviously $\Delta L < 0$ so that $L + \Delta L < L$.

Let us now consider the case of the pure shear deformation: the face of the solid which lies at distance L for the origin is displaced along x_2 by an amount ΔL, while the total thickness along x_1 remains unchanged from its initial value L. In this case an infinitesimal slab originally placed in position x_1 is displaced by $\mathbf{u} = (0, (\Delta L/L)x_1, 0)$. Accordingly, only two off-diagonal elements of the strain tensor, namely ϵ_{12} and ϵ_{21}, are found to be non-zero and we get

$$\text{strain tensor for a pure shear deformation} = \begin{pmatrix} 0 & \Delta L/2L & 0 \\ \Delta L/2L & 0 & 0 \\ 0 & 0 & 0 \end{pmatrix}. \quad (5.7)$$

Equations (5.6) and (5.7) provide the most general form of a tensile/compressive and shear strain tensor, respectively. They can be used to describe any general deformation; for instance, if we non-isotropically disfigure a cubic solid with initial edge length L into an orthogonal prism, the strain tensor defining this deformation is

$$\text{cube} \longrightarrow \text{orthogonal prism} = \begin{pmatrix} \Delta L_1/L & 0 & 0 \\ 0 & \Delta L_2/L & 0 \\ 0 & 0 & \Delta L_3/L \end{pmatrix}, \quad (5.8)$$

where $\Delta L_1 \neq \Delta L_2 \neq \Delta L_3$ are the three unalike length variations along x_1, x_2, and x_3, respectively. This example illustrates an important property of the strain tensor

$$\sum_i \epsilon_{ii} = \frac{\Delta V}{V}, \quad (5.9)$$

or equivalently: *the trace of the strain tensor describes the fractional volume change of the system*. The strain tensor also describes the case of simple tensile deformation along x_1 with allowed lateral contractions/extensions (if the system is isotropic we have $\Delta L_2 = \Delta L_3$).

5.1.3 The stress tensor

So far we have discussed the mathematical entity in charge of describing any kind of small deformation, namely the strain tensor. We now need to address the forces acting on a material continuum and causing such deformation. It is customary to distinguish between two different categories: *body forces* and *surface forces*.

Body forces act throughout the volume of the solid and are typically due to an *external field* like a gravitational, electric, or magnetic one. They are formally described by means of their volume density $\mathbf{f}_v(\mathbf{r})$ so that the body force $d\mathbf{F}_v$ acting on the infinitesimal volume dV centred around the point $\mathbf{r}$ is $d\mathbf{F}_v = \mathbf{f}_v(\mathbf{r})dV$.

Surface forces, instead, are due to the *mechanical actions that each portion of the solid undergoes because of the remaining part of the continuum*[4]; they act across a surface within the solid body which ideally separates the selected portion from the surrounding distribution of continuum matter. If we identify the orientation of an infinitesimal portion dS of such a surface by means of its outward-pointing normal unit vector $\hat{n} = (n_1, n_2, n_3)$, then we understand that compressive and tensile surface forces are, respectively, negative and positive since they are antiparallel and parallel to $\hat{n}$, respectively. The infinitesimal surface force $d\mathbf{F}_s$ acting on dS is written as

$$d\mathbf{F}_s = \mathbf{f}_s \, dS, \tag{5.10}$$

where $\mathbf{f}_s$ has the physical meaning of an internal pressure and plays a key role in elasticity theory. It is often referred to as *tension*.

In order to calculate the internal force per unit area $\mathbf{f}_s$, let us consider the tetrahedrally shaped[5] infinitesimal volume dV shown in figure 5.2. In particular, we observe that the four faces of the infinitesimal tetrahedron have area dA_1, dA_2, dA_3 e dA_n and outward-pointing normal unit vectors $-\hat{i}, -\hat{j}, -\hat{k}$, and $\hat{n}$, respectively. The total force acting on the infinitesimal volume is

$$\begin{aligned}
\rho_m dV \, \mathbf{a} &= d\mathbf{F}_s + d\mathbf{F}_v \\
&= \mathbf{f}_{s,n} \, dA_n - \mathbf{f}_{s,1} dA_1 - \mathbf{f}_{s,2} dA_2 - \mathbf{f}_{s,3} dA_3 + \mathbf{f}_v dV,
\end{aligned} \tag{5.11}$$

where $\mathbf{a}$ is its acceleration, ρ_m the mass density of the solid body, and $\mathbf{f}_{s,1,2,3}$ are the three surface forces acting on the surfaces $dA_{1,2,3}$, respectively. Similarly, $\mathbf{f}_{s,n}$ acts on dA_n (here dh is the distance between the origin and the face dA_n). By dividing any term of this equation by dA_n and considering that $dV = dA_n dh/3$ we get[6]

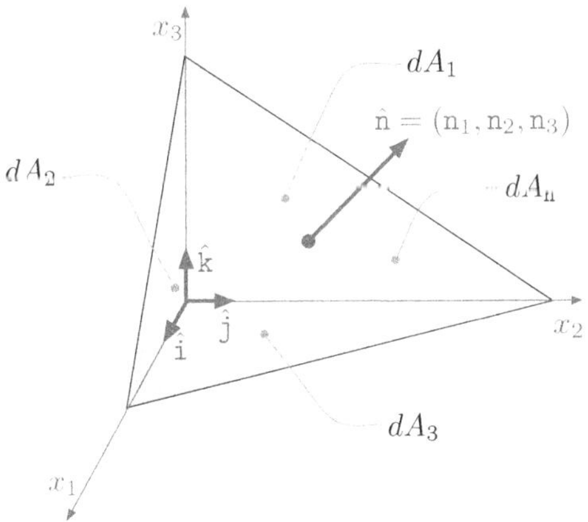

Figure 5.2. A tetrahedrally shaped infinitesimal volume. Its faces with area dA_1, dA_2, dA_3 and dA_n are, respectively, orthogonal to the unit vectors $\hat{i}$, $\hat{j}$, $\hat{k}$ and $\hat{n} = (n_1, n_2, n_3)$.

---

[4] This is clearly a macroscopic wording for the full set of ion–ion, electron–electron, and ion–electron interactions we have classified in the chapter 1.

[5] The final result will not depend on this arbitrary (but very convenient) assumption.

[6] By elementary geometry we have that $dA_i = n_i \, dA_n$ for any $i = 1, 2, 3$.

$$\mathbf{f}_{s,n} - \mathbf{f}_{s,1}\, n_1 - \mathbf{f}_{s,2}\, n_2 - \mathbf{f}_{s,3}\, n_3 + \frac{1}{3}\, \mathbf{f}_v\, dh = \frac{1}{3}\, \rho_m\, \mathbf{a}\, dh, \qquad (5.12)$$

which in limit $dh \to 0$ leads to

$$\mathbf{f}_s = \mathbf{f}_{s,1}\, n_1 + \mathbf{f}_{s,2}\, n_2 + \mathbf{f}_{s,3}\, n_3 \qquad (5.13)$$

This is a result of paramount importance: we proved that *the surface force $\mathbf{f}_s$ referred to any generic plane passing through a point inside the solid is fully determined by a set of three mutually orthogonal vector tensions*. This formal result, known as *Cauchy theorem*, states the existence anywhere within a solid body of a *stress tensor* $\mathbb{T}$ such that

$$\mathbf{f}_s = \mathbb{T}\, \hat{n}, \qquad (5.14)$$

where $\mathbb{T} = (\mathbf{f}_{s,1}, \mathbf{f}_{s,2}, \mathbf{f}_{s,3})$ or equivalently

$$f_{s,i} = \sum_j T_{ij}\, n_j, \qquad (5.15)$$

which corresponds to the mathematical entity in charge of describing the surface forces causing small deformations.

We illustrate this important achievement with two simple examples. Let us first suppose that the stress tensor is diagonal $T_{ij} = P\delta_{ij}$; then, it is easy to prove that $f_{s,i} = P\, n_i$ and we therefore understand that the matrix element P is nothing other than the hydrostatic pressure applied to the solid. On the other hand, if the stress tensor has the form

$$\mathbb{T} = \begin{pmatrix} 0 & \tau & 0 \\ \tau & 0 & 0 \\ 0 & 0 & 0 \end{pmatrix}, \qquad (5.16)$$

then we get

$$f_{s,1} = \tau\, n_2 \quad f_{s,2} = \tau\, n_1 \quad f_{s,3} = 0, \qquad (5.17)$$

which corresponds to a shear stress with two non-zero vector components acting along the x_1- and x_2-axis, respectively. We understand this physical situation by looking at figure 5.3(left): a force per unit area of magnitude τ is applied to the face 2 along the x_1 direction and a similar action is applied to face 1 along the x_2 direction. These tangential forces transform the square face 4 into a rhombus.

We conclude that *the physical meaning of the stress tensor is that of a vector pressure* whose geometrical representation is reported in figure 5.3(right) following the *Voigt notation*: T_{ij} represents the pressure applied to the jth face along the ith direction. It can be proved that *the stress tensor is symmetric* or, equivalently, $T_{ij} = T_{ji}$. While the formal details can be found elsewhere [3, 4], we can understand this symmetry property by developing a very simple argument with reference to figure 5.3(right): if we had $T_{12} \neq T_{21}$ then the volume element would undergo a

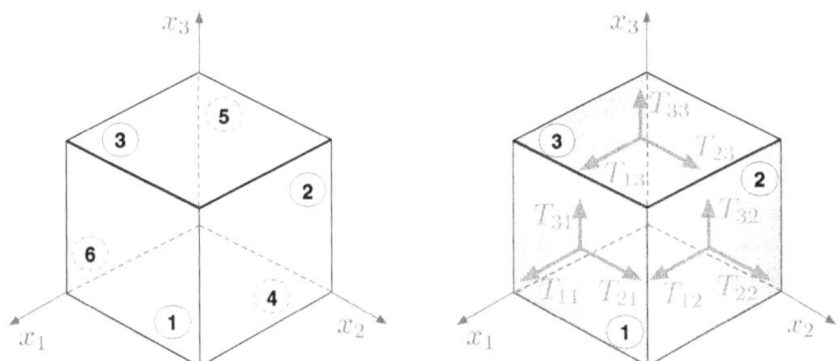

Figure 5.3. Left: the faces of an infinitesimal volume element are labelled according to the Voigt notation. Right: the corresponding nine components of the stress tensor T_{ij} with $i, j = 1, 2, 3$.

rotation around the x_3-axis; this would violate the local equilibrium condition and, therefore, it must be $T_{12} = T_{21}$.

We finally remark that, since the stress tensor is symmetric, it is always possible to define a frame of reference where it assumes a diagonal form. In order to distinguish its diagonal representation from the general one valid in any other frame of reference, we will indicate the matrix entries of a stress tensor in its diagonal form as T_{ij}^*.

5.2 Linear elasticity

5.2.1 The constitutive equation

So far we have prepared the formal environment to describe force actions and corresponding deformations. The next point is to look for the mathematical relationship linking the strain and stress tensors, which is usually referred to as the *constitutive equation* of elasticity.

In this regard, it must be preliminarily observed that *elasticity theory is unable to provide this relationship* which, instead, must be assumed 'a priori' of the problem we aim at investigating. Accordingly, any result of continuum elasticity will specifically depend on the adopted constitutive equation. This is the level at which the atomistic theory plays a major role since it provides the needed fundamental knowledge. In fact, once assigned the most appropriate model for the lattice many-body potential energy $U = U(\mathbf{R})$ governing the ion displacements (see sections 1.3.4 and 3.1), the constitutive stress–strain relationship is there contained, even if not always immediately apparent.

Consistently with the hypothesis of small deformations, we guess that *they are linearly dependent on their causing action*[7] and formally write

$$T_{ij} = \sum_{kh} C_{ijkh}\, \epsilon_{kh}, \tag{5.18}$$

---

[7] Elasticity theory can be formally developed under the more general assumption of arbitrary deformations. This is the realm of nonlinear elasticity [5] which requires much more sophisticated mathematical apparatus than here developed.

which can be simply looked at as the most general form of the Hooke law. By this *linear elastic constitutive equation* we introduce the fourth rank *elastic tensor* C_{ijkh}. In general, among its $3^4 = 81$ components *only 21 are independent* as determined by the symmetric character of the strain and stress tensor which imposes

$$C_{ijkh} = C_{jikh} \quad C_{ijkh} = C_{ijhk} \tag{5.19}$$

Another symmetry is imposed by the guessed constitutive equation (5.18) which in fact represents *the macroscopic counterpart of the harmonic crystal model* developed in chapter 3. Therefore, by analogy with equation (3.4) we can surely state that there exists a formal dependence of the elastic energy density $u = u(\epsilon_{ij})$ on the strain tensor which, within the adopted constitutive model, is cast in the harmonic form

$$u = \frac{1}{2} \sum_{ijkh} C_{ijkh}\, \epsilon_{ij}\, \epsilon_{kh}. \tag{5.20}$$

This allows us to identify the matrix entries of the elastic tensor as the macroscopic counterparts of the microscopic force constants appearing in equation (3.4) and accordingly write

$$C_{ijkh} = \frac{\partial^2 u}{\partial \epsilon_{ij}\partial \epsilon_{kh}}, \tag{5.21}$$

which leads to the third symmetry property

$$C_{ijkh} = C_{khij}, \tag{5.22}$$

and

$$\frac{\partial u}{\partial \epsilon_{ij}} = \frac{\partial}{\partial \epsilon_{ij}} \left[\frac{1}{2} \sum_{ijkh} C_{ijkh}\, \epsilon_{ij}\, \epsilon_{kh} \right] = \sum_{kh} C_{ijkh}\, \epsilon_{kh} = T_{ij}. \tag{5.23}$$

This result eventually establishes the full set of relations linking the most important quantities entering the linear elasticity theory.

5.2.2 The elastic tensor

The matrix entries of the elastic tensor are specific of any material. They are usually referred to as its *elastic (or stiffness) constants*: they determine, within the approximation of linear elasticity, the overall response of a solid to (small) deformation actions.

The symmetry properties of the strain and stress tensors, as well as equations (5.19) and (5.22), suggest we use the *Voigt compact notation* replacing second-rank and fourth-rank tensors with a vector and a square matrix, respectively. More specifically, by considering the six independent components of the strain and stress, the pairs of Cartesian indices $i, j = 1, 2, 3$ are grouped as[8]

$$
\begin{aligned}
&T_{11} \to T_1 \quad &&T_{22} \to T_2 \quad &&T_{33} \to T_3 \quad &&T_{12} \to T_4 \quad &&T_{23} \to T_5 \quad &&T_{13} \to T_6 \\
&\epsilon_{11} \to \epsilon_1 \quad &&\epsilon_{22} \to \epsilon_2 \quad &&\epsilon_{33} \to \epsilon_3 \quad &&2\epsilon_{12} \to \epsilon_4 \quad &&2\epsilon_{23} \to \epsilon_5 \quad &&2\epsilon_{13} \to \epsilon_6.
\end{aligned}
\tag{5.24}
$$

In this way, the stress–strain equation (5.18) is written in the compact matrix form

$$
\begin{pmatrix} T_1 \\ T_2 \\ T_3 \\ T_4 \\ T_5 \\ T_6 \end{pmatrix} =
\begin{pmatrix}
C_{11} & C_{12} & C_{13} & C_{14} & C_{15} & C_{16} \\
C_{12} & C_{22} & C_{23} & C_{24} & C_{25} & C_{26} \\
C_{13} & C_{23} & C_{33} & C_{34} & C_{35} & C_{36} \\
C_{14} & C_{24} & C_{34} & C_{44} & C_{45} & C_{46} \\
C_{15} & C_{25} & C_{35} & C_{45} & C_{55} & C_{56} \\
C_{16} & C_{26} & C_{36} & C_{46} & C_{56} & C_{66}
\end{pmatrix}
\begin{pmatrix} \epsilon_1 \\ \epsilon_2 \\ \epsilon_3 \\ \epsilon_4 \\ \epsilon_5 \\ \epsilon_6 \end{pmatrix},
\tag{5.25}
$$

which is a very effective way to highlight the fact that *there are only 21 independent components of the elastic tensor*, once its symmetry properties are properly taken into account.

Their number is further reduced if we take into consideration the symmetry of the underlying crystal lattice [3, 4]. In other words, *the number of independent elastic constants varies according to the crystal system actually considered*: the higher the symmetry of the lattice, the fewer of them are independent. For instance, it is found that the triclinic, monoclinic, orthorombic, and cubic lattice (order of increasing symmetry) have, respectively, 21, 13, 9, and 3 independent stiffness constants. In this latter case we write their matrix as

$$
\mathbb{C}^{\mathrm{Voigt}} =
\begin{pmatrix}
C_{11} & C_{12} & C_{12} & 0 & 0 & 0 \\
C_{12} & C_{11} & C_{12} & 0 & 0 & 0 \\
C_{12} & C_{12} & C_{11} & 0 & 0 & 0 \\
0 & 0 & 0 & C_{44} & 0 & 0 \\
0 & 0 & 0 & 0 & C_{44} & 0 \\
0 & 0 & 0 & 0 & 0 & C_{44}
\end{pmatrix},
\tag{5.26}
$$

where we added the label 'Voigt' to make it clear that this matrix representation is only valid in the compact notation.

Because of its linearity, the stress–strain equation (5.18) can be straightforwardly inverted

$$
\epsilon_{ij} = \sum_{kh} D_{ijkh} \, T_{kh},
\tag{5.27}
$$

where the *elastic compliance constants* D_{ijkh} are the matrix entries of the inverse elastic tensor. For a cubic crystal, it is easy to prove that

---

[8] Unfortunately, the compact notation is not always defined by the same grouping convention. This calls for some care when comparing the elastic equations reported in different textbooks.

Table 5.1. Room temperature stiffness constants (in 10^{11} N m^{-2} units) for some cubic crystalline solids with metallic, covalent, or ionic bonding.

	Metallic					Covalent			Ionic		
	Al	Cu	Au	Fe	Pb	C[a]	Si	Ge	NaCl	LiF	KBr
C_{11}	1.07	1.68	1.92	2.34	0.49	10.76	1.66	1.28	0.49	1.12	0.34
C_{12}	0.61	1.21	1.61	1.36	0.42	1.25	0.64	0.48	0.12	0.46	0.05
C_{44}	0.28	0.75	0.42	1.18	0.15	5.76	0.80	0.68	0.13	0.63	0.05

[a] Diamond crystalline structure.

$$D_{44} = \frac{1}{C_{44}}$$

$$D_{11} = \frac{C_{11} + C_{12}}{(C_{11} - C_{12})(C_{11} + 2C_{12})} \tag{5.28}$$

$$D_{12} = -\frac{C_{12}}{(C_{11} - C_{12})(C_{11} + 2C_{12})}.$$

In table 5.1 we report the elastic constants for some cubic crystalline systems.

5.2.3 Elasticity of homogeneous and isotropic media

We are now going to investigate the elasticity of a homogeneous and isotropic solid, that is a system where (i) the elastic constants are just the same everywhere and (ii) its elastic response is just the same along any direction. More formally, homogeneity and isotropicity are found whenever the elastic tensor is invariant upon translations and rotations. Although these features define a somewhat idealised situation, this case study is paradigmatically important and leads to very general results which, in fact, can be widely applied in practice.

By choosing a frame of reference where the stress tensor is diagonal (see section 5.1.3) and considering a uniaxial traction along the x_1 axis, it is empirically found that the system response is twofold: (i) it stretches along the x_1 direction and (ii) it shrinks in the (x_2, x_3) plane. Since the only non-zero stress component is T_{11}^*, we can formalise the observed phenomenology by defining the following strain tensor

$$\epsilon_{11}^* = +\frac{1}{E}T_{11}^* \quad \epsilon_{22}^* = -\frac{\nu}{E}T_{11}^* \quad \epsilon_{33}^* = -\frac{\nu}{E}T_{11}^* \quad \epsilon_{12}^* = \epsilon_{23}^* = \epsilon_{31}^* = 0, \tag{5.29}$$

where the two E and ν constants are introduced in the proposed combination for further convenience. It is important to remark that *just two constants are in fact needed to fully accomplish with this elastic problem* since we must only describe the observed material stretching along x_1 and its corresponding shrinking in a normal plane. We summarise the physical situation by saying that *a homogeneous and isotropic linear elastic medium has only two independent elastic moduli*, namely E and ν which are known as the *Young modulus* and the *Poisson ratio*, respectively. In table 5.2 we report their value for some elemental crystalline system.

Table 5.2. The Young modulus E, Poisson ratio ν, first λ and second μ Lamé coefficients for some elemental metallic solids.

	Al	Fe	Cu	Zn	Sn	W	Pb
E [GPa]	70	211	130	108	50	411	16
ν	0.35	0.29	0.34	0.25	0.36	0.28	0.44
λ [GPa]	34.3	82.4	28.3	57.1	98.9	192.9	11.7
μ [GPa]	26.9	82.4	46.2	38.1	16.1	151.6	6.0

If a uniaxial stress is similarly applied along the three axes we straightforwardly generalise equation (5.29) as follows

$$\epsilon_{kk}^* = \frac{1}{E}[(1 + \nu)T_{kk}^* - \nu(T_{11}^* + T_{22}^* + T_{33}^*)]$$

$$\epsilon_{ij}^* = 0 \quad \text{with} \ i \neq j, \tag{5.30}$$

which is convenient to write in matrix formalism

$$\mathbb{S}^* = \frac{1}{E}[(1 + \nu)\mathbb{T}^* - \nu \operatorname{Tr}(\mathbb{T}^*)\,\mathbb{I}], \tag{5.31}$$

where we have introduced the notation $\mathbb{S}^* = \{\epsilon_{ij}^*\}_{i,j=1,2,3}$. In the attempt to generalise this stress–strain equation, let us introduce the rotation matrix $\mathbb{R}$ transforming the frame of reference where the strain and stress tensors are diagonal into a generic one; accordingly, since

$$\mathbb{S} = \mathbb{R}'\,\mathbb{S}^*\,\mathbb{R} \quad \text{and} \quad \mathbb{T} = \mathbb{R}'\,\mathbb{T}^*\,\mathbb{R}, \tag{5.32}$$

we calculate that[9]

$$\mathbb{S} = \mathbb{R}'\left\{\frac{1}{E}[(1 + \nu)\mathbb{T}^* - \nu \operatorname{Tr}(\mathbb{T}^*)\,\mathbb{I}\,]\right\}\mathbb{R}$$

$$= \frac{1}{E}[(1 + \nu)\,\mathbb{R}'\mathbb{T}^*\mathbb{R} - \nu \operatorname{Tr}(\mathbb{T}^*)\,\mathbb{R}'\mathbb{I}\,\mathbb{R}\,] \tag{5.33}$$

$$= \frac{1}{E}[(1 + \nu)\,\mathbb{T} - \nu \operatorname{Tr}(\mathbb{T}^*)\,\mathbb{I}\,].$$

We must invert this equation in order to cast the stress–strain relation in the typical form given in linear elasticity by equation (5.18); this leads to

$$\mathbb{T} = \frac{E}{1 + \nu}\,\mathbb{S} + \frac{\nu}{1 + \nu}\,\operatorname{Tr}(\mathbb{T})\,\mathbb{I}. \tag{5.34}$$

---

[9] In this derivation we made of the identity $\operatorname{Tr}(\mathbb{T}^*) = \operatorname{Tr}(\mathbb{T})$ which is easily proved by considering the following results of matrix algebra: $\mathbb{R}'\mathbb{R} = \mathbb{I}$ and $\operatorname{Tr}(\mathbb{A}\mathbb{B}) = \operatorname{Tr}(\mathbb{B}\mathbb{A})$ for any pair of matrices $\mathbb{A}$ and $\mathbb{B}$.

The trace of the stress tensor is easily calculated from equation (5.33)

$$\mathrm{Tr}\,(\mathbb{T}) = \frac{E}{1 - 2\nu}\,\mathrm{Tr}\,(\mathbb{S}), \tag{5.35}$$

so that

$$\mathbb{T} = \frac{E}{1 + \nu}\,\mathbb{S} + \frac{\nu\,E}{(1 + \nu)(1 - 2\nu)}\,\mathrm{Tr}\,(\mathbb{S})\,\mathbb{I}. \tag{5.36}$$

By defining the two *Lamé coefficients* (whose values for some materials are reported in table 5.2)

$$\mu = \frac{E}{2(1 + \nu)} \tag{5.37}$$

$$\lambda = \frac{\nu\,E}{(1 + \nu)(1 - 2\nu)}, \tag{5.38}$$

we eventually write

$$\mathbb{T} = 2\mu\mathbb{S} + \lambda\mathrm{Tr}(\mathbb{S})\,\mathbb{I} \qquad \text{and} \qquad T_{ij} = 2\mu\epsilon_{ij} + \lambda\left(\sum_{k}\epsilon_{kk}\right)\delta_{ij}, \tag{5.39}$$

which represent *the linear elastic stress–strain equation for a homogeneous and isotropic medium* in matrix (left) or scalar (right) form. We derived it by a (robust) phenomenological approach, but it can be derived even more formally [3, 4].

5.3 Elastic moduli

We have so far introduced four different elastic parameters, namely the Young modulus, the Poisson ratio and the two Lamé coefficients. We previously introduced also the bulk modulus B when we investigated thermal expansion in section 4.2.1. Since B was defined as the inverse of the isothermal compressibility (see appendix C), that is it deals with volume variations, it is expected to be related to some elastic parameter. In order to elucidate this issue, let us consider a *hydrostatic stress* $T_{ij} = P\delta_{ij}$ (where P is the macroscopic hydrostatic pressure) and insert it into the constitutive equation (5.39) so as to get

$$\mathbb{S} = \frac{1}{3}\frac{1}{\lambda + \frac{2}{3}\mu}\,P\,\mathbb{I}. \tag{5.40}$$

The connection with equation (C.11) is established by defining

$$B = \lambda + \frac{2}{3}\mu, \tag{5.41}$$

so that

$$\mathbb{S} = \frac{1}{3B} \, P \, \mathbb{I} \qquad \rightarrow \qquad \text{Tr}\,(\mathbb{S}) = \sum_i \epsilon_{ii} = \frac{\Delta V}{V} = \frac{P}{B}, \tag{5.42}$$

which leads to the following definition

$$\frac{1}{B} = \frac{1}{V}\frac{\Delta V}{P}, \tag{5.43}$$

representing the finite difference counterpart of equation (C.11). This result reconciles the thermodynamical and elastic treatment of deformations affecting the system volume and it allows us to recast the stress–strain relation of a homogeneous and isotropic linear elastic medium in the form

$$\mathbb{T} = 2\mu\mathbb{S} + \left(B - \frac{2}{3}\mu\right)\text{Tr}(\mathbb{S})\,\mathbb{I}$$
$$= 3B\left[\frac{1}{3}\,\text{Tr}(\mathbb{S})\,\mathbb{I}\right] + 2\mu\left[\mathbb{S} - \frac{1}{3}\,\text{Tr}(\mathbb{S})\,\mathbb{I}\right], \tag{5.44}$$

where the first and second term on the right-hand side are, respectively, named *spherical part* and *deviatoric part* of the stress tensor: they describe the hydrostatic volume variation and the change in shape of the solid body subject to $\mathbb{T}$.

If we represent equation (5.44) in its matrix form and use the compact Voigt notation we immediately get

$$\mathbb{T} = \mathbb{C}^{\text{Voigt}}\,\mathbb{S} = \begin{pmatrix} B + 4\mu/3 & B - 2\mu/3 & B - 2\mu/3 & 0 & 0 & 0 \\ B - 2\mu/3 & B + 4\mu/3 & B - 2\mu/3 & 0 & 0 & 0 \\ B - 2\mu/3 & B - 2\mu/3 & B + 4\mu/3 & 0 & 0 & 0 \\ 0 & 0 & 0 & \mu & 0 & 0 \\ 0 & 0 & 0 & 0 & \mu & 0 \\ 0 & 0 & 0 & 0 & 0 & \mu \end{pmatrix}\mathbb{S}, \tag{5.45}$$

which, by comparison to equation (5.26), leads to the equation

$$2C_{44} = C_{11} - C_{12} \tag{5.46}$$

known as the *isotropy condition*, indeed a very useful result to assess to what extent a cubic crystal can be modelled as an isotropic elastic continuum. By inspection of table 5.1 we understand that a cubic crystal can be considered as elastically isotropic only in a first approximation. Instead, amorphous solids and polycrystals (see appendix C) fulfil equation (5.46) much more accurately and, therefore, they can be considered as the real counterpart of the ideal case of isotropic medium.

The stress–strain equation (5.44) has been written in terms of the bulk modulus and second Lamé coefficient. We can, however, write it in several equivalent ways by using any pair of elastic coefficients. In table 5.3 we report how to express any elastic coefficient in terms of any pair of the remaining parameters.

Table 5.3. Correspondence among the elastic moduli of linear elasticity.

	(λ, μ)	(B, μ)	(μ, ν)	(E, ν)	(E, μ)
λ		$B - \dfrac{2}{3}\mu$	$\dfrac{2\mu\nu}{1-2\nu}$	$\dfrac{\nu E}{(1+\nu)(1-2\nu)}$	$\dfrac{\mu(E-2\mu)}{3\mu-E}$
μ				$\dfrac{E}{2(1+\nu)}$	
B	$\dfrac{3\lambda+2\mu}{3}$		$\dfrac{2\mu(1+\nu)}{3(1-2\nu)}$	$\dfrac{E}{3(1-2\nu)}$	$\dfrac{E\mu}{3(3\mu-E)}$
E	$\dfrac{\mu(3\lambda+2\mu)}{\lambda+\mu}$	$\dfrac{9B\mu}{3B+\mu}$	$2(1+\nu)\mu$		
ν	$\dfrac{\lambda}{2(\lambda+\mu)}$	$\dfrac{3B-2\mu}{2(3B+\mu)}$			$\dfrac{E-2\mu}{2\mu}$

5.4 Thermoelasticity

We have so far implicitly assumed that the deformations are imposed to the system at zero temperature. While this assumption was useful to define a clean purely-elastic problem, we must duly generalise our theory to include stress actions applied at $T > 0$ K as well [6].

The starting point is of course the *energy balance* stated by the first law of thermodynamics (see equation (C.5)) which for a system with volume V in equilibrium at temperature T under some elastic action is written as

$$d\mathcal{U} = V \sum_{ij} T_{ij}d\epsilon_{ij} + T\,dS, \tag{5.47}$$

where the mechanical work contributing to the internal energy $\mathcal{U}$ has been written in terms of the stress tensor since we know that this latter describes any possible kind of volume and shape variation of the system. It is easy to reconcile equation (5.47) with the standard thermodynamical formulation by simply considering the case of a hydrostatic stress $T_{ij} = -P\,\delta_{ij}$, where P is the applied pressure whose negative sign indicates that the mechanical action is compressive. We assume it to operate quasi-statically, like anywhere else in the remaining of this chapter. By inserting the hydrostatic stress into equation (5.47) we get

$$\begin{aligned}
d\mathcal{U} &= V \sum_{ij} (-P\delta_{ij})d\epsilon_{ij} + T\,dS \\
&= -PV \sum_{i} d\epsilon_{ii} + T\,dS \\
&= -PV \frac{dV}{V} + T\,dS \\
&= -PdV + TdS,
\end{aligned} \tag{5.48}$$

consistently with the first law of thermodynamics. Equation (5.47) is valid for any arbitrary deformation and, therefore, it allows for a *thermodynamical definition of the stress tensor*

$$T_{ij} = \frac{1}{V} \left. \frac{\partial \mathcal{U}}{\partial \epsilon_{ij}} \right|_S = \frac{1}{V} \left. \frac{\partial \mathcal{F}}{\partial \epsilon_{ij}} \right|_T, \tag{5.49}$$

where $\mathcal{F} = \mathcal{U} - TS$ is the Helmholtz free energy, corresponding to the work exchanged quasi-statically during the constant-temperature deformation (see appendix C). This result defines the next task to accomplish, namely: deriving the explicit dependence $\mathcal{F} = \mathcal{F}(\epsilon_{ij})$.

Under the adopted assumption of small deformations we can expand the Helmholtz free energy in powers of the strain

$$\mathcal{F} = \mathcal{F}_0 + \sum_{ij} \left. \frac{\partial \mathcal{F}}{\partial \epsilon_{ij}} \right|_T \epsilon_{ij} + \frac{1}{2} \sum_{ijkh} \left. \frac{\partial^2 \mathcal{F}}{\partial \epsilon_{ij} \partial \epsilon_{kh}} \right|_T \epsilon_{ij}\epsilon_{kh} + \cdots$$
$$= \frac{1}{2} \sum_{ijkh} \left. \frac{\partial^2 \mathcal{F}}{\partial \epsilon_{ij} \partial \epsilon_{kh}} \right|_T \epsilon_{ij}\epsilon_{kh}, \tag{5.50}$$

where $\mathcal{F}_0$ is the free energy of the unstrained system at temperature T. As already done in many frameworks, we stopped the expansion to the second order[10] and gauged the free energy such that $\mathcal{F}_0 = 0$ with no loss of generality. Furthermore, we made use of the fact that at equilibrium there are no deformations or, equivalently, we set to zero the linear term in the power expansion thanks to equation (5.18).

We have come to an important result: *in the small deformation approximation, the Helmholtz free energy is quadratic in the strain tensor*. Since such an energy is of course a scalar quantity, we can easily guess its strain dependence: the mathematical expression for $\mathcal{F}$ can only contain terms depending on $[\mathrm{Tr}(\mathbb{S})]^2$ or $[\mathrm{Tr}(\mathbb{S}^2)]$. We accordingly write

$$\mathcal{F}(\epsilon_{ij}) = \frac{\lambda}{2} \sum_{ij} \epsilon_{ii}\epsilon_{jj} + \mu \sum_{ij} \epsilon_{ij}\epsilon_{ij}, \tag{5.51}$$

where the two Lamé coefficients are conveniently used to define the arbitrary weights of the two terms What it is really remarkable, however, is that *we have formally proved that a linear elastic homogeneous and isotropic medium has only two independent elastic moduli*, as anticipated by a phenomenological argument in section 5.2.3.

Equation (5.51) contains some far-reaching results. In the first instance, it dictates that *the elastic moduli must obey some fundamental conditions*. We known that the Helmholtz free energy is minimum at equilibrium (see appendix C); on the other hand, if we consider the zero-strain configuration (that is we set $\epsilon_{ij} = 0$ for any i, j in equation (5.51)) we find $\mathcal{F} = 0$. By combining these elementary observations, we

---

[10] Once again, this is basically equivalent to the harmonic approximation of lattice dynamics.

conclude that $\mathcal{F}$ *is a positive-definite function*. Accordingly, if we consider two specific deformations such that, respectively, $\epsilon_{ii} = 0$ and $\epsilon_{ij} = \epsilon\delta_{ij}$ (ϵ any real number), we easily get two conditions on the Lamé coefficients, namely: $\mu > 0$ and $3\lambda + 2\mu > 0$. By using the correspondences reported in table 5.3 we immediately obtain the noteworthy constraints on the bulk modulus $B > 0$ and Young modulus $E > 0$ which have an obvious physical interpretation: under a hydrostatic compression or an uniaxial tensile stress the volume and the length of a solid decreases or increases, respectively. The case of the Poisson ratio is more intriguing: since

$$\nu = \frac{3B - 2\mu}{2(3B + \mu)},\tag{5.52}$$

we easily obtain that $-1 < \nu < 1/2$. A negative Poisson ratio value implies that upon stretching a solid body along the x_1 direction, its section on the (x_2, x_3) plane increases. While this is not observed in most natural materials, this behaviour is observed in *auxetic systems* because of their internal structure and the way it replies to an uniaxial load: if stretched, they become thicker perpendicular to the tensile force. The auxetic behaviour is observed in macroscopically-engineered structures such as that shown in figure 5.4, as well as in some molecules (like paraffin, an acyclic chain hydrocarbon) or solid state systems (like cristobalite, a SiO_2 polymorph generated at very high temperature).

Next we observe that equation (5.51) provides the *thermodynamical foundation of the linear elastic constitutive relation*. In order to prove this, let us more conveniently rewrite the Helmholtz free energy as (the correspondences reported in table 5.3 are used)

$$\mathcal{F}(\epsilon_{ij}) = \frac{1}{2}B \sum_{ij} \epsilon_{ii}\epsilon_{jj} + \mu \sum_{ij}\left(\epsilon_{ij}\epsilon_{ij} - \frac{1}{3}\epsilon_{ii}\epsilon_{jj}\right),\tag{5.53}$$

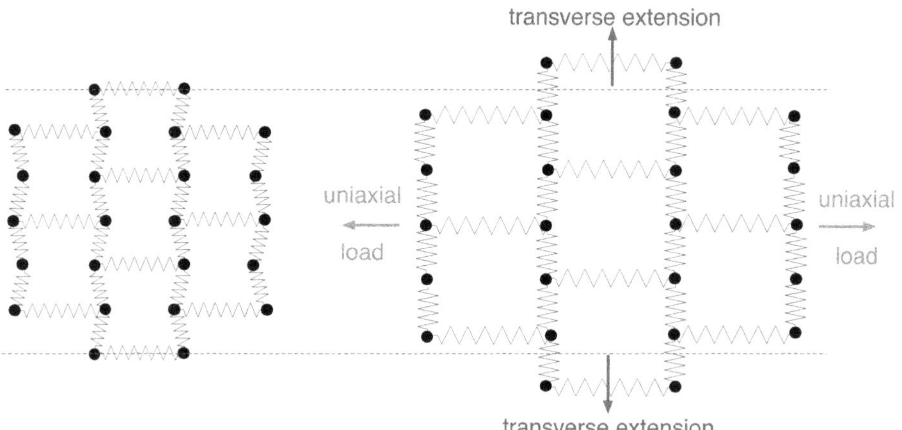

Figure 5.4. A mass-spring auxetic structure made with negative Poisson ratio: upon uniaxial loading (red), a transverse expansion (blue) is observed.

and use the stress definition given in equation (5.49)

$$
\begin{aligned}
T_{kh} &= \frac{1}{V} \left. \frac{\partial \mathcal{F}}{\partial \epsilon_{kh}} \right|_T \\
&= \frac{1}{2} B \sum_{ij} \left(\frac{\partial \epsilon_{ii}}{\partial \epsilon_{kh}} \epsilon_{jj} + \epsilon_{ii} \frac{\partial \epsilon_{jj}}{\partial \epsilon_{kh}} \right) \\
&\quad + \mu \sum_{ij} \left(2\frac{\partial \epsilon_{ij}}{\partial \epsilon_{kh}} \epsilon_{ij} - \frac{1}{3} \frac{\partial \epsilon_{ii}}{\partial \epsilon_{kh}} \epsilon_{jj} - \frac{1}{3} \epsilon_{ii}\frac{\partial \epsilon_{jj}}{\partial \epsilon_{kh}} \right) \\
&= 2\mu\epsilon_{kh} + \left(B - \frac{2}{3}\mu \right)\left(\sum_i \epsilon_{ii} \right) \delta_{kh},
\end{aligned}
\tag{5.54}
$$

which, as anticipated, represents the counterpart of the stress–strain relation given in equation (5.44).

In concluding this section, we aim at defining the general framework to deal with *deformations occurring at variable temperature*. This class of phenomena includes two different situations: either a deformation induced by a temperature variation[11] or a change in temperature generated by a deformation caused by some mechanical action. Their formal theory is not trivial at all and it is developed in more advanced textbooks. Here we rather follow a phenomenological argument. The key point is to work out a model describing the *coupling between temperature and strain*: to this aim, we will rely on the general assumption of small deformations and, furthermore, we will consider just small temperature variations ΔT. These approximations make it physically sound to consider a *coupling term linear in both temperature and strain*. By following the same reasoning that led to equation (5.51), we understand that such a coupling term can only depend on strain through its trace[12]. All together these arguments suggest writing the Helmholtz free energy as[13]

$$
\mathcal{F} = \frac{1}{2}B \sum_{ij} \epsilon_{ii}\epsilon_{jj} + \mu \sum_{ij} \left(\epsilon_{ij}\epsilon_{ij} - \frac{1}{3}\epsilon_{ii}\epsilon_{jj} \right) - B\beta \left(\sum_i \epsilon_{ii} \right) \Delta T
\tag{5.55}
$$

where for further convenience we replaced in equation (5.51) the two Lamé coefficients with the (B, μ) pair (see table 5.3). The last term on the right-hand side represents the combined effect of the strain field and temperature variation. The corresponding coupling coefficient has been conveniently written as the product between the bulk modulus B and the thermal expansion coefficient β, as proved by the following simple argument. Let us consider a *free thermal expansion*, occurring in absence of any applied stress: when the system has equilibrated after expansion, it

---

[11] This is the thermal expansion phenomenon, already discussed in section 4.2.1.

[12] More formally: since the strain tensor is a second-rank symmetric tensor, we can build out of its elements only one scalar term, namely $\mathrm{Tr}(\mathbb{S})$.

[13] As previously, we gauged to zero the free energy of the unstrained system at temperature T.

reaches a new stress-free $T_{ij} = 0$ configuration for which, according to equation (5.49), we can write

$$T_{ij} = 0 = \frac{\partial \mathcal{F}}{\partial \epsilon_{ij}} = B \left(\sum_i \epsilon_{ii} \right) \delta_{ij} + 2\mu \left[\epsilon_{ij} - \frac{1}{3} \left(\sum_i \epsilon_{ii} \right) \delta_{ij} \right] - B\beta \, \Delta T \, \delta_{ij}, \qquad (5.56)$$

and by considering the diagonal element $i = j$ we easily obtain[14]

$$0 = B \left(\sum_i \epsilon_{ii} \right) - B \beta \, \Delta T, \qquad (5.57)$$

or equivalently

$$\beta = \frac{1}{\Delta T} \left(\sum_i \epsilon_{ii} \right) = \frac{1}{V} \frac{\Delta V}{\Delta T}, \qquad (5.58)$$

which in fact represents the finite difference counterpart[15] of equation (C.9).

References

[1] Gurtin N E 1981 *An Introduction to Continuum Mechanics* (New York: Academic)
[2] Asaro R and Lubarda V 2006 *Mechanics of Solids and Materials* (Cambridge: Cambridge University Press)
[3] Love A E H 2002 *A Treatise on the Mathematical Theory of Elasticity* (New York: Dover)
[4] Atkin R J and Fox N 1980 *An Introduction to the Theory of Elasticity* (New York: Dover)
[5] Novozhilov V V 1999 *Foundations of the Nonlinear Theory of Elasticity* (New York: Dover)
[6] Weiner J H 2002 *Statistical Mechanics of Elasticity* (New York: Dover)

---

[14] Consistently with the assumed isotropic elastic behaviour, we have $\epsilon_{ii} - (\sum_i \epsilon_{ii})/3 = 0$.
[15] In this case the constant-pressure condition required by the thermodynamic definition of the thermal expansion coefficient is rigorously fulfilled, since the system is supposed to be stress-free before and after the expansion.

Part III

Electronic structure

Solid State Physics
A primer
Luciano Colombo

Chapter 6

Electrons in crystals: general features

Syllabus—*We establish the conceptual framework and some general results useful in addressing the electronic structure of crystalline solids. In particular, based on the general discussion developed in chapter 1, we outline three different approaches based on the single-particle approximation, namely the free electron theory, the tight-binding theory and the density functional theory, which will be thoroughly discussed in the following chapters. Next, we elaborate the Fermi–Dirac distribution law (describing the probability that a given electron energy level is occupied), the Bloch theorem (defining the most general form of the one-electron crystalline wavefunction), and the general features of the energy spectrum of an electron which is subject to a periodic potential. This will naturally lead to the concept of allowed energy bands and forbidden energy gaps which is the essence of the band theory.*

6.1 The conceptual framework

In developing an atomistic and quantum picture of the solid state we had to admit that it is necessary to adopt approximations, even very drastic ones as thoroughly reported in section 1.3. As far as electrons are concerned, this has led to equation (1.15) which defines the constitutive eigenvalue problem for the total crystalline electron wavefunction[1] $\Psi_e^{(R)}(\mathbf{r})$: by solving it, we ultimately define the *electronic structure* of the crystal.

Despite the remarkable number of adopted simplifications already introduced[2], equation (1.15) still represents a formidably complicated many-body quantum problem. The search for general methods able to solve it, either analytically or numerically, represents an advanced topic in theoretical condensed matter physics, ranging from single-particle theories to many-body formalisms to quantum field

---

[1] Nomenclature, notation, and symbols hereafter used have been previously defined in chapter 1.

[2] Namely: the semi-classical, frozen-core, non-magnetic, non-relativistic, and adiabatic approximations which have been introduced, motivated, and discussed in section 1.3.

theories [1–6]. These methods are currently the topic of active research and mostly fall well beyond the level (and the scope) of this introductory textbook. In our attempt to elaborate a more elementary theory we will follow a simplified approach, which basically consists in two logical steps: first, we will root our theory in the *single-particle approximation*; second, we will take profit from some materials-specific characteristics so as to further reduce the mathematical complexity of the physical problem.

The first step was already introduced in section 1.4.1: electron–ion and electron–electron interactions are replaced by the crystal field potential $V_{\mathrm{cfp}}(\mathbf{r})$, namely *a local one-electron potential* with the same periodicity of the underlying crystal lattice and effectively describing the most relevant many-body features. Hereafter, we will assume that such a potential is known, for instance through a self-consistent calculation, as outlined in section 1.4.1. Its practical implementation is an advanced topic of computational solid state theory [1–3].

The second step is by its nature less general and takes a different form depending on which class of solids, either metals or insulators, is considered. Let us start by considering metals since, while most of ordinary solids are insulators, they deserve special attention because of their unique physical properties. *In addressing a metal we will draw the very drastic approximation to completely overlook ions.* In other words, we will neglect any effect due to the lattice on the electron system. This is indeed a much stronger approximation than the adiabatic one, since it basically states that electrons do not undergo the action of any external potential[3]: they are considered as *free particles.* In addition, the problem is further simplified by neglecting as well the internal interactions among electrons: they are further treated as *independent particles.* There is no need, therefore, to introduce a suitable crystal field: electron–electron Coulomb interactions as well as exchange and correlation effects will not be explicitly taken into account and, accordingly, we will set $V_{\mathrm{cfp}}(\mathbf{r}) = 0$ anywhere within the metal. The resulting picture is referred to as the *free electron theory*[4]. We will make use of the free electron theory in chapter 7 to describe the main features of metals: basically, their system of conduction electrons will be treated as a gas of free and independent quantum particles, confined within the volume of the solid. The validity of the model will be heuristically proved by the very good agreement between its predictions on thermal and electric transport in metals and the experimental data. Also, through this model we will introduce in a very simple way some fundamental concepts largely used in electronic structure theory. In this respect, the free electron theory is still a mostly useful pedagogical tool.

In the opposite case of insulating solids, *valence electrons will be treated as a system of single-particles, individually subject to the action of the external crystal field potential* $V_{\mathrm{cfp}}(\mathbf{r})$ and the corresponding eigenvalue problem will be formulated as in equation (1.22). The true existence of such a periodic potential allows the internal

---

[3] In this case the external potential would be due to the electron–ion Coulomb interactions.

[4] Historically, this was the first model developed by scientists in addressing the electronic structure of crystalline solids [7, 8]: it was originally elaborated by Drude in 1900 in order to explain the electrical conduction in metals and it was entirely based on the classical kinetic theory of gases; next, it was generalised by Sommerfeld so as to include quantum features.

volume of the crystal to be periodically separated into *core regions* and *interstitial regions*. With reference to figure 6.1, the former ones correspond to the crystal portions nearest to the ion positions[5]: here $V_{\rm cfp}(\mathbf{r})$ is expected to be very similar to the potential generated by a single isolated atom. Correspondingly, within each core region the crystalline wavefunction is expected to be very similar to an atomic wavefunction centred on the selected lattice site. On the other hand, in the interstitial regions $V_{\rm cfp}(\mathbf{r})$ is substantially different from the atomic case[6] and, therefore, it is quite reasonable to state that here the crystalline wavefunction is no longer atomic-like: as a matter of fact, in these crystal portions there is naturally expected to be some overlap among atomic wavefunctions centred on neighbouring lattice sites. In brief, we can summarise the physical situation by saying that (i) some overlap between atomic wavefunctions does occur, but (ii) it is not strong enough to fully invalidate the atomic description of electron quantum states. This argument, as qualitative as nevertheless firmly rooted in some fundamental quantum features of the single-particle picture, suggests that in a non-metallic solid[7] the electron wavefunction can be cast in the form of *a linear combination of localised atomic orbitals*. This is the essence of the *tight-binding theory*, which for obvious reasons is alternatively referred to as the *linear combination of atomic orbitals* (LCAO) method. We will make use of the tight-binding theory in chapter 8 to describe the main features of insulators; in particular, we will rely on it in investigating the so-called 'band structure' of semiconductor materials (chapter 9): a class of non-metallic systems with several intriguing physical properties, making them of paramount importance for technological applications.

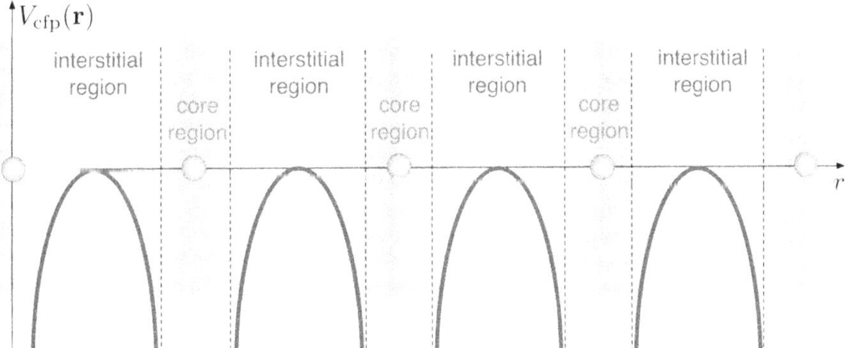

Figure 6.1. Schematic representation of the crystal field potential $V_{\rm cfp}(\mathbf{r})$ acting on a valence electron in a one-dimensional solid. Core and interstitial regions are shown: the darker the grey shadowing, the more atomic-like is the character of the electron wavefunction.

---

[5] For further convenience, we have gauged to zero the topmost value of $V_{\rm cfp}(\mathbf{r})$ in the interstitial regions. This convention will be fully exploited in section 6.4.

[6] In these regions the net potential is given by the overlap of the tails of the single-atom potentials centred on the various lattice sites.

[7] In fact, it is also valid to describe the electronic structure of transition metals, like Ni or Cu, with partially filled d-subshells [9].

A third completely different approach is indeed possible, no longer based on the knowledge of the single-electron wavefunctions (either free-particle-like or atomic-like) but rather on the electron density [10]. This approach, named *density functional theory*, entirely rests on two fundamental quantum mechanical theorems, respectively, stating that (i) the ground-state energy of an electron system is a unique functional of the electron density and that (ii) the density minimising the total energy of the system is actually the true electron density than one would calculate by solving the full Schrödinger problem. This theory fully incorporates any kind of interactions ruling over the physics of crystalline electrons, also including exchange and correlation effects. Once the electron density is known, most of the ground-state properties of a crystal can be calculated, sometimes straightforwardly and sometimes by more sophisticated methodologies. The density functional theory is considered as the 'standard model' of modern solid state physics since it is not only more fundamental and superior to the free electron or tight-binding theories, but also because it can be implemented in a totally parameter-free way and it provides the most accurate predictions on the electronic structure properties. This theory, which is in fact an advanced topic of solid state physics, will be outlined in chapter 10.

In the remaining part of this chapter we discuss some general results that will represent as many cornerstones for the electronic structure theory of crystalline solids. We will not refer to any specific kind of solids and we will instead develop our arguments at a very general level.

6.2 The Fermi–Dirac distribution function

Let us consider a gas of N electrons, confined within a volume V in equilibrium at temperature T. Quantum mechanics provides the energy spectrum of this system, as we will extensively discuss in chapters 7 and 8; we label by E_i with $i = 1, 2, 3, \dots$ the energies of the single-electron levels[8]. Hereafter in this section we assume that such energies are known and ask ourselves *how likely each level is to be occupied*.

Let us assume that two energy levels E_1 e E_2 are occupied with probability $n_{FD}(E_1, T)$ and $n_{FD}(E_2, T)$, respectively. Since the system is in equilibrium, on average the number of electrons undergoing the transition $E_1 \rightarrow E_2$ in the unit time equals the number of electrons undergoing the inverse transition $E_2 \rightarrow E_1$. This statement represents a specific formulation of the more general principle of micro-reversibility[9]. Such transitions could be generated by any kind of mechanism, like electron–electron or electron-defect scattering.

---

[8] We remark that this simplified labelling convention is only valid to count the different states; more rigorously, they should be indicated as E_{m_i} as discussed in section 1.4.1. However, for the present discussion the simplified notation is sufficient and will not generate any ambiguity.

[9] This is tantamount to stating that in the guessed equilibrium condition, each elementary electron process is as equally probable as its reverse process [11]: since microscopic physical laws are ultimately invariant under time reversal, a transition between two microscopic quantum states and its inverse transition are equally probable. Microreversibility is, in turn, the foundation of the detailed balance principle [11, 12] which is widely used in statistical mechanics.

The number of electrons undergoing the transition $E_1 \rightarrow E_2$ per unit time is calculated as the product between the number of electrons initially occupying the level E_1 and the rate $R_{1\rightarrow2}$ of occurrence of such transition. The initial number of electrons is in turn given by the product between the total number of particles N and the probability that the initial energy level is occupied. Similar definitions hold for the inverse transition. If electrons were classical particles, the occupation probability of any given level with energy E would be proportional to the Boltzmann factor $\exp(-E/k_BT)$ [11] and, therefore, we could write the *classical version of micro-reversibility* as

$$\exp(-E_1/k_BT)\ R_{1\rightarrow2}^{\text{classical}} = \exp(-E_2/k_BT)\ R_{2\rightarrow1}^{\text{classical}}, \tag{6.1}$$

where $R_{1\rightarrow2}^{\text{classical}}$ and $R_{1\rightarrow2}^{\text{classical}}$ are the transition rates calculated according to classical physics. Since, however, electrons are identical and indistinguishable quantum particles they must obey the Pauli principle (see appendix E): two electrons cannot have the same set of quantum numbers. This suggests the most general way to switch from classical to quantum transition rates

$$R_{1\rightarrow2}^{\text{quantum}} = R_{1\rightarrow2}^{\text{classical}}\ [1 - n_{\text{FD}}(E_2, T)] \quad \text{and} \quad R_{2\rightarrow1}^{\text{quantum}} = R_{2\rightarrow1}^{\text{classical}}\ [1 - n_{\text{FD}}(E_1, T)], \tag{6.2}$$

where the terms in square parenthesis $[\cdots]$ define the probability that the final state is not previously occupied. *Quantum microreversibility* is straightforwardly cast in the form

$$n_{\text{FD}}(E_1, T)\ R_{1\rightarrow2}^{\text{quantum}} = n_{\text{FD}}(E_2, T)\ R_{2\rightarrow1}^{\text{quantum}}, \tag{6.3}$$

which, by inserting equation (6.1), leads to

$$\frac{R_{1\rightarrow2}^{\text{classical}}}{R_{2\rightarrow1}^{\text{classical}}} = \frac{\exp(-E_2/k_BT)}{\exp(-E_1/k_BT)} = \frac{n_{\text{FD}}(E_2, T)}{n_{\text{FD}}(E_1, T)}\ \frac{1 - n_{\text{FD}}(E_1, T)}{1 - n_{\text{FD}}(E_2, T)}, \tag{6.4}$$

or equivalently

$$\frac{1 - n_{\text{FD}}(E_1, T)}{n_{\text{FD}}(E_1, T)}\ \exp(-E_1/k_BT) = \frac{1 - n_{\text{FD}}(E_2, T)}{n_{\text{FD}}(E_2, T)}\ \exp(-E_2/k_BT). \tag{6.5}$$

Since this result applies to each arbitrary pair of energies E_1 and E_2, we logically draw the conclusion that the two expressions appearing on the left and right side of this equation must be equal to an identical function, which only depends on temperature. Following the standard convention [11, 13, 14], we name this function $\exp(-\mu_c/k_BT)$ where μ_c is the *chemical potential*. For any arbitrary value of the electron energy E we obtain

$$\frac{1 - n_{\text{FD}}(E, T)}{n_{\text{FD}}(E, T)}\ \exp(-E/k_BT) = \exp(-\mu_c/k_BT), \tag{6.6}$$

which leads to

$$n_{\text{FD}}(E, T) = \frac{1}{1 + \exp[(E - \mu_c)/k_BT]}, \tag{6.7}$$

describing *the probability that the quantum level E is occupied when the electron system is in equilibrium at temperature T*. As indicated by the subscript FD this probability corresponds to the *Fermi–Dirac distribution law* obtained in appendix E by means of a very general statistical argument: this provides evidence that electrons are fermions, consistently with the fact that they have half odd spin $\hbar/2$.

By comparing equation (6.7) with equation (E.11), we realise that apparently only the latter contains information about the degeneracy of the single-particle quantum level. This could be confusing at first glance and, therefore, it deserves a further comment. In non-magnetic approximation, spin-related interactions are strictly speaking neglected: electron energies are independent of the spin (in the absence of an external magnetic field) and the Pauli principle allows up to two electrons to occupy the same single-particle quantum state, provided that they have opposite spin. This corresponds to a twofold degeneracy. On the other hand, if spin-related interactions are included, each single-particle level can only be occupied by a sole electron. In conclusion, equation (6.7) must be understood as the occupation probability of spin-resolved single-particle energy levels. Instead, a factor 2 should be added in case the spin is fully neglected, in order to properly take into account that each level with energy E actually represents two possible states, one for each electron spin polarisation.

It is easy to recognise that the chemical potential μ_c appearing in equation (6.7) corresponds to the Lagrange multiplier introduced in appendix E to enforce the conservation of the total number of electrons; by the arguments developed in appendix C, μ_c is therefore understood as the thermodynamical parameter ruling over the chemical equilibrium condition between systems of electrons: if two electron reservoirs are put into contact, then some electrons will flow from one to the other until the chemical potentials of the two systems are the same. Its physical meaning is so clarified by thermodynamics: μ_c *provides the work needed to add one electron to the system at constant temperature*, or equivalently:

$$\mu_c(T) = \mathcal{F}(T,\, V,\, N + 1) - \mathcal{F}(T,\, V,\, N), \tag{6.8}$$

where $\mathcal{F}(T,\, V,\, N)$ is the free energy of the system of N electrons.

The discussion so far developed clearly indicates that both the Fermi–Dirac distribution and the chemical potential depend on temperature. More specifically, $\mu_c(T)$ must be adjusted[10] so that at any temperature T the total number of electrons remains constant (for an isolated system). Also, it is easy to see from equation (6.7) that

$$\lim_{E \to -\infty} n_{\mathrm{FD}}(E,\, T) = 1 \quad \text{and} \quad \lim_{E \to +\infty} n_{\mathrm{FD}}(E,\, T) = 0, \tag{6.9}$$

at any non-zero temperature; the width of the energy range over which the Fermi–Dirac distribution is non-constant is of the order of $\sim k_{\mathrm{B}}T$. Furthermore, by calculating the derivative

---

[10] This procedure will be fully exploited in section 7.3.2.

$$\frac{\partial n_{\mathrm{FD}}(E,\,T)}{\partial E} = -\frac{1}{k_{\mathrm{B}}T}\frac{\exp\left[(E-\mu_{\mathrm{c}})/k_{\mathrm{B}}T\right]}{\left\{1+\exp[(E-\mu_{\mathrm{c}})/k_{\mathrm{B}}T]\right\}^{2}}, \tag{6.10}$$

we get the maximum value of the slope of the Fermi–Dirac distribution law

$$\left.\frac{\partial n_{\mathrm{FD}}(E,\,T)}{\partial E}\right|_{E=\mu_{\mathrm{c}}} = -\frac{1}{4k_{\mathrm{B}}T}, \tag{6.11}$$

proving that as the temperature decreases, $n_{\mathrm{FD}}(E,\,T)$ gets closer and closer to a step-like function centred at energy $E = \mu_{\mathrm{c}}$. In particular, at zero temperature we have

$$n_{\mathrm{FD}}(E,\,0) = \begin{cases} 1 & \text{for } E < \mu_{\mathrm{c}} \\ 1/2 & \text{for } E = \mu_{\mathrm{c}} \\ 0 & \text{for } E > \mu_{\mathrm{c}}, \end{cases} \tag{6.12}$$

We can generalise this result by observing that at any finite temperature it is always found that $n_{\mathrm{FD}}(\mu_{\mathrm{c}},\,T) = 1/2$: in other words, *the chemical potential corresponds to that energy value which selects the electron state occupied with 50% probability*. Since $n_{\mathrm{FD}}(E,\,T)$ is smoothed by increasing the temperature, the chemical potential is correspondingly shifted, as shown in figure 6.2. The specific temperature-dependence of $\mu_{\mathrm{c}}(T)$ will be addressed in chapter 7 in the framework of the free electron theory and in chapter 9 when we will discuss the case of extrinsic semiconductors.

In concluding this section we anticipate a nomenclature widely used in the following: the zero-temperature value of the chemical potential is usually referred to as the *Fermi energy* E_{F} which is therefore formally defined as $E_{\mathrm{F}} = \lim_{T \to 0} \mu_{\mathrm{c}}(T)$.

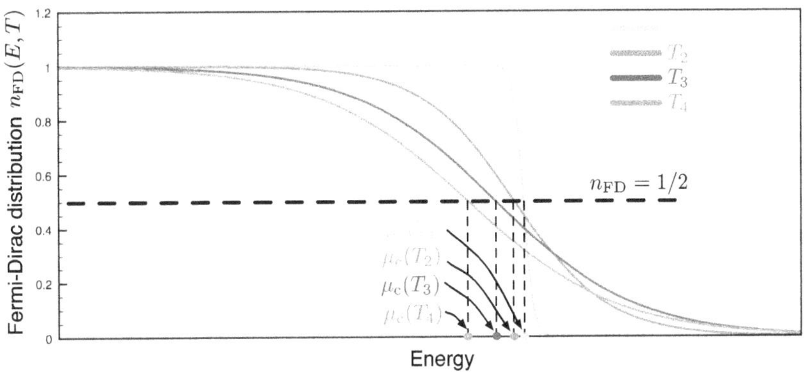

Figure 6.2. The Fermi–Dirac distribution law $n_{\mathrm{FD}}(E,\,T)$ calculated at four different temperatures $T_1 < T_2 < T_3 < T_4$. The 50% occupation probability (corresponding to $n_{\mathrm{FD}} = 1/2$) is shown by a black dashed line: its intercepts with the curves $n_{\mathrm{FD}}(E,\,T)$ define the positions of the chemical potential $\mu_{\mathrm{c}}(T)$ at the different temperatures.

6.3 The Bloch theorem

We now derive a formal result due to the translational invariance of any crystal lattice discussed in chapter 2.

Let us start from the single electron approximation developed in section 1.4.1, where we proved that the Schrödinger problem given by equation (1.22) must be solved for each crystalline electron. The electron Hamiltonian operator $\hat{H}(\mathbf{r})$

$$\hat{H}(\mathbf{r}) = -\frac{\hbar^2}{2m_e}\nabla^2 + \hat{V}_{\text{cfp}}(\mathbf{r}), \tag{6.13}$$

is obviously depending upon the position $\mathbf{r}$ of the particle within the crystal and, because of the property of translational invariance of the lattice, we have

$$\hat{H}(\mathbf{r}) = \hat{H}(\mathbf{r} + \mathbf{R}_l), \tag{6.14}$$

as shown in figure 2.3. In order to formally treat such an invariance, it is useful to introduce the *translation operator* $\hat{T}_{\mathbf{R}_l}$ whose action on a generic space function $f(\mathbf{r})$ is defined as

$$\hat{T}_{\mathbf{R}_l} f(\mathbf{r}) = f(\mathbf{r} + \mathbf{R}_l). \tag{6.15}$$

The translational invariance is revealed by stating that *the one-electron Hamiltonian operator commutes with the translation operator*, that is: $[\hat{H}(\mathbf{r}), \hat{T}_{\mathbf{R}_l}] = 0$. Therefore, the solutions $\psi(\mathbf{r})$ of equation (1.22) are also eigenfunctions of the translation operator

$$\hat{T}_{\mathbf{R}_l}\psi(\mathbf{r}) = t(\mathbf{R}_l)\,\psi(\mathbf{r}), \tag{6.16}$$

where the number $t(\mathbf{R}_l)$ is the eigenvalue of $\hat{T}_{\mathbf{R}_l}$; it is intuitive to figure out that, according to equation (2.1), $t(\mathbf{R}_l)$ depends on the set $\{n_1, n_2, n_3\}$. Furthermore, by composing two translations

$$\hat{T}_{\mathbf{R}_{l'}}\hat{T}_{\mathbf{R}_l}\psi(\mathbf{r}) = \hat{T}_{\mathbf{R}_l+\mathbf{R}_{l'}}\psi(\mathbf{r}) = t(\mathbf{R}_l + \mathbf{R}_{l'})\,\psi(\mathbf{r}), \tag{6.17}$$

we understand that $t(\mathbf{R}_l + \mathbf{R}_{l'}) = t(\mathbf{R}_l)\,t(\mathbf{R}_{l'})$.

We assume that the electron wavefunctions have been properly normalised

$$\int_V |\psi(\mathbf{r})|^2 d\mathbf{r} = 1, \tag{6.18}$$

over the finite volume V of the crystalline sample we are studying. Since such a normalisation condition is not affected by translations, we conclude that $|t(\mathbf{R}_l)|^2 = 1$ or, equivalently, that $t(\mathbf{R}_l) = \exp(i\alpha)$ where α is a still undetermined real number. By combining this result with our knowledge about the composition of $t(\mathbf{R}_l)$ numbers, we understand that α must proportional to $\mathbf{R}_l$ and, therefore, we can write

$$\alpha = \mathbf{k} \cdot \mathbf{R}_l, \tag{6.19}$$

where it is straightforward to identify $\mathbf{k}$ with the wavevector of the matter wave that quantum mechanically describes the crystalline electron. In solid state physics $\mathbf{k}$ is looked at as the counterpart of the quantum numbers used in atomic physics to label electronic states.

By combining the results we obtained, we can eventually state that the *wave-function of a crystalline electron must always have the following property*

$$\psi(r + \mathbf{R}_l) = e^{i\mathbf{k}\cdot\mathbf{R}_l}\,\psi(\mathbf{r}), \tag{6.20}$$

a fundamental result known as the *Bloch theorem*. It is really hard to underestimate the importance of such a result: under the sole single-particle approximation and by only exploiting the inherent symmetry property of any crystal systems (namely, its translational invariance), we have derived *a general common feature of any electron wavefunction*: a translation only affects the eigenfunction of the single-particle Hamiltonian by a phase change.

The Bloch theorem can be cast in a different and similarly useful form. Let us introduce a new function $u_\mathbf{k}(\mathbf{r})$ such that

$$u_\mathbf{k}(\mathbf{r}) = e^{-i\mathbf{k}\cdot\mathbf{r}}\,\psi(\mathbf{r}). \tag{6.21}$$

The translation operator transforms such a function as

$$\hat{T}_{\mathbf{R}_l}u_\mathbf{k}(\mathbf{r}) = u_\mathbf{k}(\mathbf{r} + \mathbf{R}_l) = e^{-i\mathbf{k}\cdot(\mathbf{r}+\mathbf{R}_l)}\,\psi(\mathbf{r} + \mathbf{R}_l). \tag{6.22}$$

If we now make use of equation (6.20) we can write

$$\begin{aligned}
\hat{T}_{\mathbf{R}_l}u_\mathbf{k}(\mathbf{r}) &= e^{-i\mathbf{k}\cdot(\mathbf{r}+\mathbf{R}_l)}\,\psi(\mathbf{r} + \mathbf{R}_l) \\
&= e^{-i\mathbf{k}\cdot(\mathbf{r}+\mathbf{R}_l)}\,e^{i\mathbf{k}\cdot\mathbf{R}_l}\psi(\mathbf{r}) \\
&= e^{-i\mathbf{k}\cdot\mathbf{r}}\psi(\mathbf{r}) \\
&= u_\mathbf{k}(\mathbf{r}),
\end{aligned} \tag{6.23}$$

which leads to the conclusion that *the function $u_\mathbf{k}(\mathbf{r})$ is a periodic function*. By inverting equation (6.21), we immediately obtain the noteworthy result

$$\psi(\mathbf{r}) = e^{i\mathbf{k}\cdot\mathbf{r}}\,u_\mathbf{k}(\mathbf{r}), \tag{6.24}$$

stating that *any crystalline electron wavefunction is always the product between a plane wave and a second function with the same periodicity of the lattice*. Wavefunctions of this form are referred to as *Bloch functions*. This result is again known as the *Bloch theorem* which, therefore, can be formulated in the two alternative forms provided in equations (6.20) and (6.24). Even in this case the result is of general validity since it is only based on the translational invariance of the lattice.

The importance of the Bloch theorem will be extensively exploited in the next chapters. However, we conclude this section by a simplified argument which should help in catching the paradigmatic relevance of the theorem. Let us consider the form of the theorem given in equation (6.24): a special case of periodic function is when $u_{\mathbf{k}}(\mathbf{r}) = $ constant. Under this condition, the Bloch functions reduce to plane waves: crystalline electrons described by such wavefunctions grossly behave as free particles[11]. Crystals where $u_{\mathbf{k}}(\mathbf{r}) = $ constant are accordingly expected to show a metallic behaviour. On the other hand, whenever $u_{\mathbf{k}}(\mathbf{r})$ is not a constant, the overall behaviour of the electron system will be qualitatively different and, therefore, addressed as non-metallic. Although very qualitative, this argument makes it clear that the charge transport properties of condensed matter systems are solely explained by quantum mechanical arguments, where the actual mathematical form of the Bloch functions plays a major role.

6.4 Electrons in a periodic potential

Just as it has been possible to obtain the general form of the wavefunction of crystalline electrons without taking into consideration any materials-specific property, so we are about to derive *the general structure of the energy spectrum for valence electrons* by only considering the periodicity of the single-particle potential $\hat{V}_{\mathrm{cfp}}(\mathbf{r}) = \hat{V}_{\mathrm{cfp}}(\mathbf{r} + \mathbf{R}_{\mathrm{l}})$.

To this aim we discuss the simple case of a one-dimensional monoatomic crystal under the construction usually referred as *Kronig–Penney model*. The situation sketched in figure 6.1 is further simplified by approximating the crystal field potential with a function $V(x)$ which consists in a periodic sequence of *potential wells* spanning the core regions, separated by *finite barriers* occupying the interstitial ones. This idealised situation is represented in figure 6.3. We remark that we have for convenience set the zero of the potential at the bottom of the wells, while a and b, respectively, indicate the width of the wells and barriers. Therefore, the lattice periodicity is $a + b$ or, equivalently, in this case the lattice vectors are written as $R_{\mathrm{l}} = n(a + b)$ with $n = 0, \pm 1, \pm 2, \pm 3, \ldots$ (see equation (2.1)). We understand that $a < b$ by guessing that interstitial regions are larger than core ones[12].

Thanks to the crystal periodicity, it is sufficient to solve the quantum problem of a valence electron under the action of the Kronig–Penney potential $V(x)$ only for a single pair of adjacent core and interstitial regions. With reference to figure 6.3 we write

$$V(x) = \begin{cases} 0 & 0 < x < a \quad \text{core region} \\ V_0 & -b \leqslant x \leqslant 0 \quad \text{interstitial region,} \end{cases} \tag{6.25}$$

so that

$$\begin{cases} \dfrac{d^2\psi(x)}{dx^2} + \dfrac{2m_{\mathrm{e}}}{\hbar^2}E\psi(x) = 0 & \text{core region} \\[4mm] \dfrac{d^2\psi(x)}{dx^2} + \dfrac{2m_{\mathrm{e}}}{\hbar^2}(E - V_0)\psi(x) = 0 & \text{interstitial region,} \end{cases} \tag{6.26}$$

---

[11] In fact, the real situation is more subtle, as discussed when treating charge transport in solids. The argument we are here developing is just qualitative.

[12] This justifies the graphical representation reported in figure 6.3, but neither the mathematics developed below nor the general results we are going to obtain actually depend on this choice.

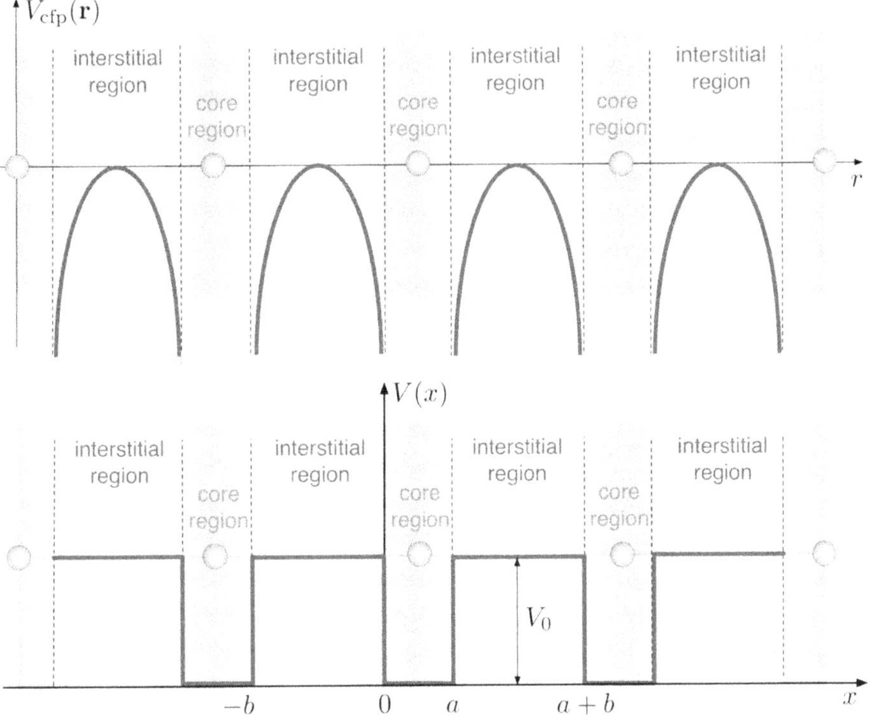

Figure 6.3. Top: schematic representation of the crystal field potential $V_{\text{cfp}}(\mathbf{r})$ acting on a valence electron in a one-dimensional monoatomic crystal. Bottom: the corresponding Kronig–Penney model. The function $V(x)$ consists in a periodic sequence of potential wells of width a centred on the lattice ions, separated by finite barriers of thickness b and height V_0 spanning the interstitial regions.

where we have indicated by $\psi(x)$ and E the electron wavefunction and energy, respectively. This equation is the counterpart of equation (1.22) in a simplified notation. By introducing the shortcuts

$$\alpha^2 = \frac{2m_e}{\hbar^2}E \qquad \text{and} \qquad -\beta^2 = \frac{2m_e}{\hbar^2}(E - V_0), \tag{6.27}$$

we can rewrite equations (6.26) in a more convenient way

$$\begin{cases} \dfrac{d^2\psi(x)}{dx^2} + \alpha^2\psi(x) = 0 & \text{core region} \\[2mm] \dfrac{d^2\psi(x)}{dx^2} - \beta^2\psi(x) = 0 & \text{interstitial region.} \end{cases} \tag{6.28}$$

Since by construction the Kronig–Penney potential is periodic, the wavefunction $\psi(x)$ must be a Bloch function or, equivalently

$$\psi(x) = e^{ikx}\, u(x) \quad \text{with} \quad u(x) = u(x + R_1). \tag{6.29}$$

By inserting this form of the wavefunction into equations (6.28) we easily obtain

$$
\begin{cases}
\dfrac{d^2u(x)}{dx^2} + 2ik\dfrac{du(x)}{dx} + (\alpha^2 - k^2)u(x) = 0 & \text{core region} \\[2ex]
\dfrac{d^2u(x)}{dx^2} + 2ik\dfrac{du(x)}{dx} + (\beta^2 + k^2)u(x) = 0 & \text{interstitial region,}
\end{cases}
\tag{6.30}
$$

whose solutions are

$$
\begin{cases}
u_1(x) = A\exp[i(\alpha - k)x] + B\exp[-i(\alpha + k)x] & \text{core region} \\[1ex]
u_2(x) = C\exp[(\beta - ik)x] + D\exp[-(\beta + ik)x] & \text{interstitial region.}
\end{cases}
\tag{6.31}
$$

The matching between $u_1(x)$ and $u_2(x)$ at the boundaries between core and interstitial regions is obtained as customarily in quantum mechanics by imposing the continuity of the wavefunction and of its first derivative[13]; these conditions lead to the following set of four equations

$$
\begin{aligned}
&u_1(x = 0) = u_2(x = 0) &\quad& \left.\frac{du_1(x)}{dx}\right|_{x=0} = \left.\frac{du_2(x)}{dx}\right|_{x=0} \\[2ex]
&u_1(x = a) = u_2(x = -b) &\quad& \left.\frac{du_1(x)}{dx}\right|_{x=a} = \left.\frac{du_2(x)}{dx}\right|_{x=-b}.
\end{aligned}
\tag{6.32}
$$

By inserting equations (6.31) into the boundary conditions provided by equations (6.32), we easily obtain an algebraic system where the unknowns are the four coefficients A, B, C and D. The system has a non-trivial solution only if the determinant of such coefficients vanishes, that is only if

$$
\frac{\beta^2 - \alpha^2}{2\alpha\beta}\sinh(\beta b)\sin(\alpha a) + \cosh(\beta b)\cos(\alpha a) = \cos[k(a + b)],
\tag{6.33}
$$

which represents *the fundamental equation of the Kronig–Penney model*: it is a transcendental function with no closed-form solution; it can only be solved numerically.

In order to proceed in the most straightforward way, it is useful to replace the physical barriers with infinitesimally-thin ones, while keeping unaffected their strength[14]. This is tantamount to saying that we describe the interstitial regions by means of barriers such that

$$
\begin{cases}
V_0 \to +\infty \\[2ex]
b \to 0
\end{cases}
\quad \text{with } V_0 b = \text{constant},
\tag{6.34}
$$

---

[13] We are considering barriers with finite height: accordingly, the second derivative of the wavefunction is well defined and it is finite. Therefore, the first derivative is a continuous function.

[14] It is formally proved in quantum mechanics that the penetration through a potential barrier only depends on its strength $V_0 b$, namely on the product between the barrier height and width [7, 15, 16].

This is clearly a choice of mathematical convenience, which nevertheless leaves the physical nature of the problem unchanged. The convenience lies in the fact that, under the conditions given in equation (6.34), we have $\cosh(\beta b) \sim 1$, $\sinh(\beta b) \sim \beta b$ and $V_0 \gg E$ so that equation (6.33) can be replaced by

$$P\frac{\sin(\alpha a)}{\alpha a} + \cos(\alpha a) = \cos(ka) \quad \text{with } P = \frac{m_e V_0 \, ba}{\hbar^2}, \tag{6.35}$$

which represents a really interesting result: since the right-hand side of this equation is an ordinary trigonometric cosine function, then the only allowed values of the variable (αa) are those for which $-1 \leqslant P\sin(\alpha a)/(\alpha a) + \cos(\alpha a) \leqslant +1$. We graphically solve equation (6.35) as shown in figure 6.4, where we have set $F(\xi) = P\sin(\xi)/\xi + \cos(\xi)$ and used the shortcut $\xi = \alpha a$.

The main feature of this graphical procedure is that the solutions of equation (6.35) are grouped into intervals of allowed values, separated by intervals of forbidden ones. They will be hereafter referred to as *allowed bands* and *forbidden gaps*, respectively. Given the definition of α reported in equation (6.27), this leads into a remarkably important result: *the Kronig–Penney model proves that the energy spectrum of a crystalline electron consists in allowed bands separated by forbidden gaps*. Although this result has been obtained by approximating the crystal field potential with a periodic sequence of rectangular potential wells separated by finite barriers, it is generally valid for any type of periodic potential: therefore, *the existence of bands and gaps will be the true dominant feature of the electronic structure of all periodically-invariant crystalline solids*.

A number of interesting features are found in figure 6.4. First of all, we remark that the higher the energy E of the electron (or, equivalently, the larger the value of the variable ξ), the larger are the band widths and the smaller are the extensions of the gaps. Next, we observe that as P increases, the allowed bands become less and

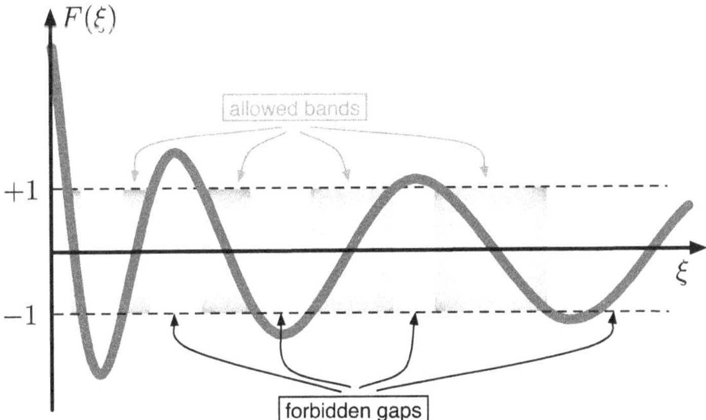

Figure 6.4. Graphical solution of equation (6.35) (Kronig–Penney model). The only allowed values of the variable $\xi = (\alpha a)$ are those for which $-1 \leqslant F(\xi) \leqslant +1$ with $F(\xi) = P\sin(\xi)/\xi + \cos(\xi)$. The parameters α and a are defined in equation (6.27) and figure 6.3, respectively. The bands of allowed ξ values are shaded in light red colour.

less extended; in particular, in the limit $P \to +\infty$ the bands transform into a sequence of single energy levels: the corresponding spectrum is nothing other than that of an electron confined within an infinitely-deep potential well. Finally, as P decreases the allowed bands become larger and larger; in the limit $P \to 0$ all bands merge into an interrupted continuum of allowed states, corresponding to the free electron case.

References

[1] Martin R M 2012 *Electronic Structure–Basic Theory and Practical Methods* (Cambridge: Cambridge University Press)

[2] Grosso G and Pastori Parravicini G 2014 *Solid State Physics* 2nd edn (Oxford: Academic)

[3] Cohen M L and Louie S G 2016 *Fundamentals of Condensed Matter Physics* (Cambridge: Cambridge University Press)

[4] Bruus H and Flensberg K 2004 *Many-body Quantum Theory in Condensed Matter Physics: An Introduction* (Oxford: Oxford University Press)

[5] Inkson C J 1984 *Many-body Theory of Solids–An Introduction* (New York: Springer)

[6] Tsvelik A M 2002 *Quantum Field Theory in Condensed Matter Physics* (Cambridge: Cambridge University Press)

[7] Eisberg R and Resnick R 1985 *Quantum Physics of Atoms, Molecules, Solids, Nuclei, and Particles* 2nd edn (Hoboken, NJ: Wiley)

[8] Ashcroft N W and Mermin N D 1976 *Solid State Physics* (London: Holt-Saunders)

[9] Colombo L 2019 *Atomic and molecular physics: a primer* (Bristol: IOP Publishing Ltd)

[10] Sholl D S and Steckel J A 2009 *Density Functional Theory–A Practical Introduction* (New York: Wiley)

[11] Reif F 1987 *Fundamentals of Statistical and Thermal Physics* (New York: McGraw-Hill)

[12] Lebon G, Jou D and Casas-Vázquez J 2008 *Understanding Non-equilibrium Thermodynamics* (Berlin: Springer)

[13] Glazer M and Wark J 2001 *Statistical Mechanics–A Survival Guide* (Oxford: Oxford University Press)

[14] Swendsen R H 2012 *An Introduction to Statistical Mechanics and Thermodynamics* (Oxford: Oxford University Press)

[15] Miller D A B 2008 *Quantum Mechanics for Scientists and Engineers* (New York: Cambridge University Press)

[16] Griffiths D J and Schroeter D F 2018 *Introduction to Quantum Mechanics* 3rd edn (Cambridge: Cambridge University Press)

IOP Publishing

Solid State Physics
A primer
Luciano Colombo

Chapter 7

Free electron theory

Syllabus—The classical and quantum theory of the free electron gas are developed to predict its transport (both charge and heat), thermal, and optical properties. They provide the Drude and Sommerfeld theory of metals, respectively. In the quantum case, we also elaborate a complete description of the electron states, entirely based on the Fermi–Dirac statistics for fermions. Merits and failures of both models are extensively discussed by comparing their predictions to experimental evidence. Eventually we motivate the need of a more fundamental description of crystalline quantum electron states, which will be developed in the next chapter.

7.1 General features of the metallic state

Metals are characterised at the macroscopic level by the ability to conduct electricity. Phenomenologically, the charge transport properties are defined by their resistivity which typically ranges in between 10^{-8} and 10^{-6} Ωm at $T - 300$ K. The presence of impurities detrimentally affects the charge transport in these materials and, therefore, their conductivity is typically lowered by increasing the concentration of defects. Finally, the resistivity is found to decrease monotonically with decreasing temperature[1].

The metallic state is very common in Nature, since more than two thirds of the elements are in fact good conductors. They are preferentially found on the left-hand side of the periodic table; accordingly, their atomic ground-state configuration typically consists in a large majority of electrons hosted by core states and just a few others found in valence states, as shown in appendix A. The number n_e of valence

---

[1] Metals are not the only solids able to conduct electricity: semiconductors do it as well. However, their charge transport properties are remarkably different: while their native resistivity is much larger than in metals, it can be dramatically reduced by adding impurities, a process named 'doping' which enables the microelectronics technology. Furthermore, the resistivity of a semiconductor varies with temperature in a complicated way, but it becomes very large at low temperature: strictly speaking, at $T = 0$ K an undoped semiconductor is in fact an insulator. For these reasons, the free electron theory cannot be applied to semiconductors.

electrons per cm^3 is given by the product (number of atoms per mole) × (number of moles per cm^3) × (number of valence electrons per atom) or equivalently

$$n_e = \mathcal{N}_A \, \frac{d_m}{A} \, Z_v, \tag{7.1}$$

where d_m is the mass density of the metal, while the symbols $\mathcal{N}_A$, Z_v, and A are the Avogadro number, the number of valence electrons per atom (chemical valence), and the atomic mass number, respectively, previously defined in sections 1.2.1 and 1.3.2. As reported in table 7.1 *this corresponds to a typical number density of the order of 10^{22} electrons cm^{-3}*, which is much larger than found in any ordinary atomic or molecular gas in normal conditions of temperature and pressure[2]. We can also assign a *volume per electron*, which corresponds to a sphere of radius r_e defined so that

$$\frac{4}{3}\pi \, r_e^3 = \frac{1}{n_e}. \tag{7.2}$$

If we compare the calculated values of r_e with the typical interatomic distances in crystals (which are of the order of few Å), we come to the conclusions that *in metals there is plenty of room available to valence electrons*. Finally, we take into consideration that they are only weakly bound to their ion core: therefore, it is quite reasonable to assume that, upon collecting many atoms to form the crystal, they homogeneously delocalise throughout the interstitial regions, thus giving rise to unidirectional metal bonds, as anticipated in figure 2.22 and related discussion.

This body of phenomenological evidence supports the idea of modelling the conduction gas of a metal as a homogeneous gas of delocalised, free, independent, and charged particles. Although based on very drastic approximations, this picture is nevertheless promising to describe at least the main features of metals. The key question is: which is the most accurate level to implement this *free electron gas model*?

Table 7.1. Number Z_v of valence electrons, electron density n_e (units 10^{22}/cm^3), radius r_e (units Å) of the sphere defining the volume per electron, and room-temperature resistivity ρ_e (units 10^{-8} Ω m) for some metallic elements. The calculated Drude relaxation time τ_e (units 10^{-14} s) is also reported, as calculated by equation (7.7).

	Li	Na	Cu	Ag	Au	Mg	Ca	Fe	Zn	Al	Sn	Pb
Z_v	1	1	1	1	1	2	2	2	2	3	4	4
n_e	4.7	2.6	8.5	5.9	5.9	8.6	4.6	17.0	13.2	18.1	14.8	13.2
r_e	1.7	2.1	1.4	1.6	1.6	1.4	1.7	1.1	1.2	1.1	1.2	1.2
ρ_e	9.5	4.9	1.7	1.6	2.2	4.4	3.4	10.0	6.0	2.7	11.5	21
τ_e	0.79	2.79	2.46	3.76	2.73	0.94	2.27	0.21	0.45	0.73	0.21	0.13

---

[2] This fact should be read as a first indication that the system of valence electrons is not at all an ordinary gas and that it requires the development of a specific theory.

7.2 The classical (Drude) theory of the conduction gas

A first simple approach to the physics of the free electron gas is purely classical, mostly based on the kinetic theory of gases [1]. In the *Drude theory of the metallic state* [2–4] electrons are described as point-like charged particles, confined within the volume of a solid specimen. The very drastic approximations of free and independent particles outlined in the previous section are slightly corrected by assuming that *electrons occasionally undergo collisions with ion vibrations, with other electrons and with lattice defects* possibly hosted by the sample; the key simplifying assumption is that we define a unique *relaxation time* τ_e (thus averaging among all possible scattering mechanisms) defined such that $1/\tau_e$ is the probability per unit time for an electron to experience a collision of whatever kind[3]. This approach is usually referred to as the *relaxation time approximation*. The free-like and independent-like characteristics of the particles of the Drude gas are instead exploited by assuming that between two collisions electrons move according to the Newtons equations of motion, that is uniformly and in straight lines. Collisions are further considered as instantaneous events which abruptly change the electron velocities; also, they are assumed to be the only mechanism by which the Drude gas is able to reach the thermal equilibrium. In other words, the velocity of any electron emerging from a scattering event is randomly distributed in space, while its magnitude is related to the local value of the temperature in the microscopic region of the sample close to the scattering place (local equilibrium).

7.2.1 Electrical conductivity

The first application of the Drude theory is to predict the direct-current electrical conductivity of a metal. Let $\mathbf{v}_d$ be the *electron drift velocity* under the action of an externally-applied uniform and constant electric field $\mathbf{E}$. The overall dynamical effect of the collisions experienced by the accelerated electrons is described as a frictional term in their Newton equation of motion

$$-e\mathbf{E} = m_e\dot{\mathbf{v}}_d + \beta\mathbf{v}_d,\tag{7.3}$$

where β is a coefficient to be determined. Basically, the added frictional term forces the electron distribution to relax towards the equilibrium Fermi–Dirac one when the external electric field is removed. In a steady-state condition we have $d\mathbf{v}_d/dt = 0$ and therefore

$$-\frac{e}{m_e}\mathbf{E} = \frac{\beta}{m_e}\mathbf{v}_d,\tag{7.4}$$

which naturally[4] leads to defining $\beta = m_e/\tau_e$. This allows us to calculate the electron drift velocity as

---

[3] This notion will be reconsidered in greater detail in section 7.3.3.
[4] This result directly comes from the dimensional analysis of equation (7.4): β/m_e must be the inverse of a time which is straightforward to identify in the relaxation time.

$$\mathbf{v}_{\mathrm{d}} = -\frac{e\tau_{\mathrm{e}}}{m_{\mathrm{e}}}\,\mathbf{E}, \tag{7.5}$$

from which we obtain the steady-state charge current density $\mathbf{J}_{\mathrm{q}}$

$$\mathbf{J}_{\mathrm{q}} = -n_{\mathrm{e}}e\mathbf{v}_{\mathrm{d}} = \frac{n_{\mathrm{e}}e^2\tau_{\mathrm{e}}}{m_{\mathrm{e}}}\,\mathbf{E}, \tag{7.6}$$

and the *Drude expression for the direct-current conductivity* σ_{e}

$$\sigma_{\mathrm{e}} = \frac{n_{\mathrm{e}}e^2\tau_{\mathrm{e}}}{m_{\mathrm{e}}}, \tag{7.7}$$

which links this quantity to few microscopic physical parameters associated either with the charge carriers (e and m_{e}) or to the specific material (n_{e} and τ_{e}). The conductivity is the inverse of the *electrical resistivity* $\rho_{\mathrm{e}} = 1/\sigma_{\mathrm{e}}$, a physical property which is easily measured: therefore, the Drude theory allows for a direct estimation of the order of magnitude of the relaxation time related to the charge current[5] which turns out to be as small as $\tau_{\mathrm{e}} \sim 10^{-14}$ s; its predicted value is reported in table 7.1 for some selected metallic elements. By applying the kinetic theory to the (classical) electron gas, we can estimate the electron thermal velocity $v_{\mathrm{e}}^{\mathrm{th}}$ by means of the equipartition theorem[6] and accordingly define the *electron mean free path* $\lambda_{\mathrm{e}} \sim 1-10\text{\AA}$ which represents the average distance covered by an electron between two successive collisions. It is reassuring to get a number which is comparable with the typical interatomic distance in a crystalline solid: this supports the robustness of the Drude model.

7.2.2 Optical properties

Another success of the classical free electron gas theory is that it correctly predicts the optical properties of metals, which are found to strongly reflect any electromagnetic radiation in the visible spectrum, while at higher frequency they are able to absorb [5], as shown in figure 7.1 in the paradigmatic case of aluminium.

In order to estimate the optical reflectivity of a free electron gas, we need to evaluate its frequency-dependent refractive index $\sqrt{\epsilon_{\mathrm{r}}}$, where ϵ_{r} is the relative permittivity of the metal [5]. Let $\mathbf{E}(t) = \mathbf{E}_0 \exp(-i\omega t)$ be a time-varying and uniform electric field applied to a metallic sample, where $\mathbf{E}_0$ and ω are its amplitude and frequency, respectively. Following the same path which led to equation (7.3), we write the electron equation of motion as

$$-e\mathbf{E}_0 \exp(-i\omega t) = m_{\mathrm{e}}\dot{\mathbf{v}}_{\mathrm{d}}(t) + \frac{m_{\mathrm{e}}}{\tau_{\mathrm{e}}}\mathbf{v}_{\mathrm{d}}(t), \tag{7.8}$$

---

[5] The reason why we are underlying that the present τ_{e} is related to the charge current will be clear in section 7.2.3 where a second relaxation time for electrons will be introduced to discuss the heat current.

[6] Assuming that the electron gas is a classical gas at equilibrium, we can write $m_{\mathrm{e}}(v_{\mathrm{e}}^{\mathrm{th}})^2/2 = 3k_{\mathrm{B}}T/2$ since its free and independent particles have only three translational degrees of freedom [1].

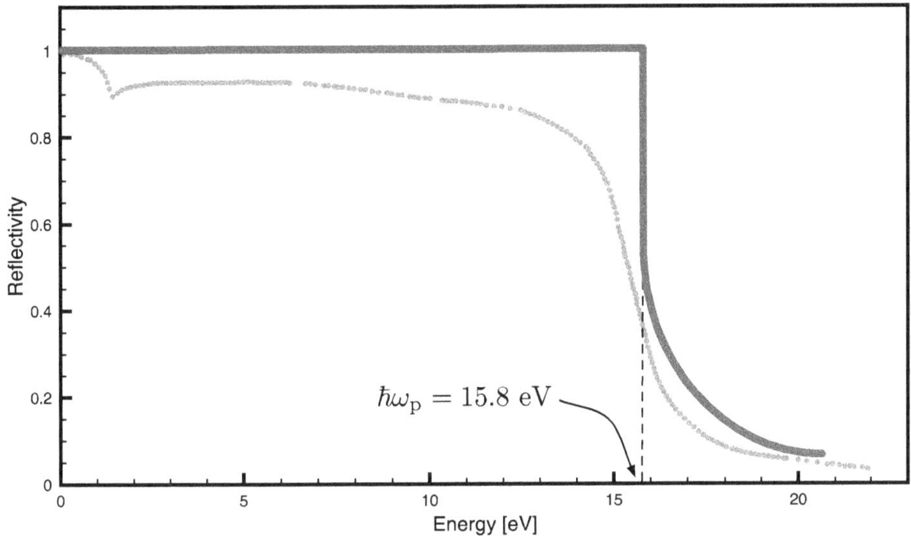

Figure 7.1. The reflectivity of aluminium. Red dots: experimental data taken from [6]. Full blue line: reflectivity predicted by the Drude theory for $\hbar\omega_p = 15.8$ eV, where ω_p is aluminium plasma frequency.

where we have introduced the time dependence in $\mathbf{v}_d(t)$ since we understand that, under the action of an oscillating electric field, the drift velocity of a free electron also follows a periodic variation with the same frequency. More specifically, we write $\mathbf{v}_d(t) = \mathbf{v}_{d,0}\exp(-i\omega t)$. From equation (7.8) we easily get the drift velocity[7]

$$\mathbf{v}_{d,0} = -\frac{e\tau_e}{m_e}\frac{1}{1 - i\omega\tau_e}\mathbf{E}_0, \tag{7.9}$$

and by integration we obtain the time-dependent displacement $\mathbf{s}(t)$ of the electron

$$\mathbf{s}(t) = \int_0^t \mathbf{v}_d(t')dt' = \frac{e\tau_e}{i\omega m_e}\frac{1}{1 - i\omega\tau_e}\mathbf{E}_0\exp(-i\omega t), \tag{7.10}$$

where with no loss of generality we have set $\mathbf{s}(0) = 0$ for convenience. We can now calculate the polarisation (that is the induced electric dipole moment per unit volume) $\mathbf{P} = -n_e e\mathbf{s}(t)$ and, through the standard relation $\epsilon_0\epsilon_r\mathbf{E} = \epsilon_0\mathbf{E} + \mathbf{P}$, eventually obtain the relative permittivity of the metal as

$$\epsilon_r = 1 - \frac{n_e e^2}{\epsilon_0 m_e}\frac{\tau_e}{i\omega}\frac{1}{1 - i\omega\tau_e} = 1 - \frac{\tau_e}{i\omega}\frac{1}{1 - i\omega\tau_e}\omega_p^2, \tag{7.11}$$

---

[7] Incidentally, through the current density $\mathbf{j} = -n_e e\mathbf{v}_d$, we can also obtain the Drude expression for the alternate-current conductivity $\sigma_e(\omega) = (n_e e^2\tau_e)/m_e(1 - i\omega\tau_e)$ which, as expected, reduces to equation (7.7) in the $\omega \to 0$ limit.

where we have introduced the *plasma frequency* $\omega_{\mathrm{p}}^2 = n_e e^2/\epsilon_0 m_e$ of the electron gas. In the limit $\omega\tau_e \gg 1$ (that is, for any electromagnetic wave with frequency well within the visible spectrum or higher) this expression reduces to

$$\epsilon_{\mathrm{r}}(\omega) = 1 - \frac{\omega_{\mathrm{p}}^2}{\omega^2}, \qquad (7.12)$$

which indicates that whenever $\omega < \omega_{\mathrm{p}}$ the refractive index $\sqrt{\epsilon_{\mathrm{r}}}$ becomes imaginary: therefore, the electric field becomes evanescent within the material. In other words, we have proved that *no propagation of an electromagnetic wave in a metal is possible at frequencies below the plasma frequency*. The situation is summarised by stating that for $\omega < \omega_{\mathrm{p}}$ a metal is totally reflecting, while for $\omega > \omega_{\mathrm{p}}$ it is able to transmit an electromagnetic wave. The frequency-dependent reflectivity $R(\omega)$ is defined as [5]

$$R(\omega) = \left| \frac{\sqrt{\epsilon_{\mathrm{r}}(\omega)} - 1}{\sqrt{\epsilon_{\mathrm{r}}(\omega)} + 1} \right|^2, \qquad (7.13)$$

and eventually calculated by using equation (7.12). The result for $R(\omega)$ is pretty well confirmed by the experimentally observed optical reflectivity of metals, as shown in figure 7.1 in the case of aluminium. The plasma frequency values are easily calculated within the Drude model by simply taking the density data from table 7.1: they are found to be of the order of $\sim 10^{-15}\mathrm{s}^{-1}$ for most metals.

The true existence of the plasma frequency has the important consequence that *a free electron gas can sustain a charge density oscillation*: it corresponds to a back-and-forth longitudinal displacement of the electron gas as a whole with respect to the background of the positive ions (which remain clamped at their lattice positions), as schematically shown in figure 7.2. If the electron gas is displaced by an amount d, then a total electric field $E = en_e d/\epsilon_0$ is established within the

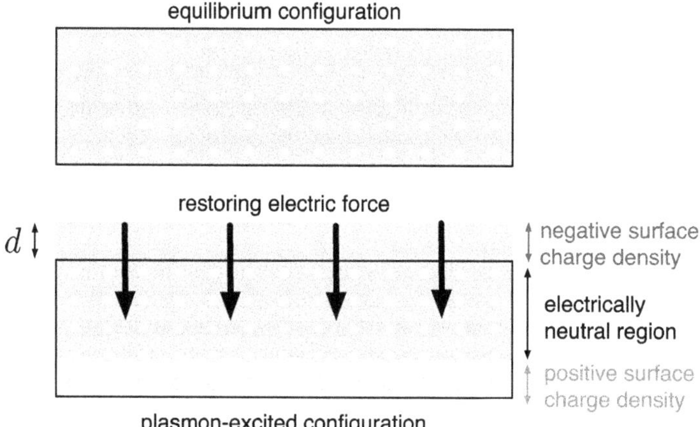

Figure 7.2. Top: the electron gas (light blue continuum) is displaced as a whole by d, while the ions (red dots) are kept at their lattice positions. Bottom: an electrostatic restoring electric force is established, causing plasma oscillations.

sample, acting as a restoring force. The resulting equation of motion of the gas as a whole is written as

$$Nm_e\,\ddot{d} = -Ne\,E = -\frac{Ne^2n_e}{\epsilon_0}d, \qquad (7.14)$$

where N is the total number of electrons. This is the equation of motion of a harmonic oscillator which vibrates at the plasma frequency $\omega_p^2 = n_e e^2/\epsilon_0 m_e$. The quantum counterpart of this collective oscillation is introduced as in the phonon case (see section 3.5): a *plasmon* is therefore defined as a pseudo-particle (a collective excitation, in fact) of energy $\hbar\omega_p$ which can be created or annihilated under suitable conditions. Because of the typical values of the plasma frequency, plasmons have energy of the order of several eV and, therefore, they cannot be thermally excited; this implies that, as far as its collective longitudinal oscillations are concerned, the conduction gas of a metal lies in its ground-state at any ordinary temperature. On the other hand, we could make a beam of fast electrons[8] to pass through a thin metallic film, collecting the energies of the emerging ones. Experiments provide evidence that the transmitted beam shows energy losses at discrete values, corresponding to the generation of one or more plasmons, in excellent agreement with the prediction of the Drude theory [2, 5].

7.2.3 Thermal transport

We finally discuss the thermal conduction in metals. The basic assumption in this case is that *most of the heat current is carried by the gas of conduction electrons,* consistently with the empirical observation that metals are as good electrical conductors as they are good thermal conductors: it is therefore natural to look at *electrons as the primary microscopic carriers for both transport phenomena.* More specifically, we understand that electrons are able to transport heat since those coming from the hotter region of the sample carry a larger amount of thermal energy than the electrons moving from the colder region.

The approximation to consider only electrons as heat carriers implies that in the Drude theory we formally assume $\kappa_{tot} = \kappa_e + \kappa_l \sim \kappa_e$, where κ_e and κ_l are the separate contributions due to the electronic and ionic degrees of freedom, respectively. In order to calculate the actual form of κ_e, we can proceed by analogy[9] with the lattice case discussed in section 4.3 by setting

$$\kappa_e^{\text{Drude}} = \frac{1}{3}\,\tau_e\,\langle(v_e^{\text{th}})^2\rangle c_V^e(T), \qquad (7.15)$$

where $c_V^e(T)$ is the constant-volume specific heat of the electron gas. It is important to remark that in this equation we used the same symbol τ_e as before, but with a

---

[8] In this kind of experiment incident electrons typically have a kinetic energy in the range 1–10 keV.

[9] The analogy is supported by the fact that, in order to calculate the lattice thermal conductivity as in equation (4.40), we used the corpuscular language of phonons. So, we can simply replace any phonon-like quantity by its electron-like counterpart.

subtly different meaning: in equation (7.15) it is understood as the relaxation time occurring in the heat current phenomenon, while in equation (7.7) it was associated with the charge current one. This abuse of notation suggests that within the present free electron theory *we are assuming the charge current and heat current relaxation times to be just the same*: indeed a useful approximation, which however will be revised in section 7.3.4. By describing the conduction gas classically, we calculate[10] $\langle (v_e^{th})^2 \rangle = 3k_B T / m_e$ and $c_V^e(T) = 3n_e k_B/2$ and, by means of equation (7.7), we predict the ratio between its thermal and electrical conductivity to be

$$\frac{\kappa_e^{\text{Drude}}}{\sigma_e} = \frac{3}{2}\left(\frac{k_B}{e}\right)^2 T, \tag{7.16}$$

which is in remarkable good agreement with the *experimental Wiedemann-Franz law* stating that in most metals the ratio between the thermal and electric conductivities of the conduction gas is proportional to T at sufficiently high temperatures (room temperature belongs to this range). To a very good extent, the proportionality constant is found to be about the same for all metals. Equation (7.16) not only accurately predicts the high temperature trend of the κ_e/σ_e ratio, but it also provides the estimation of the constant

$$\frac{\kappa_e^{\text{Drude}}}{\sigma_e T} = \frac{3}{2}\left(\frac{k_B}{e}\right)^2 = 1.11 \cdot 10^{-8} \text{ W } \Omega \text{ K}^{-2}, \tag{7.17}$$

which is known as the *Lorenz number*. Finally, we remark that this result critically depends on the assumption that the relaxation times limiting charge and heat currents are the same.

7.2.4 Failures of the Drude theory

From the discussion developed so far, one could draw the conclusion that Drude theory is basically sufficient to explain the main physical features of electron gas in a metal, given its apparent success in all topics where we have applied it. Unfortunately, this is not true.

First of all, we observe that *the predicted value for the Lorenz number is (roughly) only half of the measured one* [2, 3]. Second, *the ability to explain the phenomenological Wiedemann–Franz law is just casual* since it is based on two large numerical errors pointing in opposite directions: (i) the actual specific heat of an electron gas at the typical metallic density is much smaller than predicted classically and (ii) the typical electron velocities are much larger than calculated by equipartition. By chance these two rather inaccurate estimations almost perfectly compensate for each other, thus giving an illusory impression of robustness to the Drude theory. In addition, *the Lorenz number is not a constant at low temperatures* [7], mainly because it is found that κ_e is a function of temperature. As a matter of fact, only a full

---

[10] As before, these results directly derive from the application of the equipartition theorem to the electron gas.

quantum theory is able to correct such discrepancies so as to match the experimental evidences.

In addition to the above failures, the Drude theory is also unsuccessful in describing the electron gas under the action of both an electric and magnetic field, as found when investigating the *Hall effect* [2, 3]. Let us consider the situation shown in figure 7.3 where a constant and uniform magnetic field $\mathbf{B} = (0, 0, B)$ is applied normal to the current density $\mathbf{j}$ generated by a constant and uniform electric field $\mathbf{E} = (E_x, E_y, 0)$. As before, the equation of motion for the electrons of the conducing gas are Newton-like

$$-e\mathbf{E} - e\mathbf{v}_\mathrm{d} \times \mathbf{B} = m_\mathrm{e}\dot{\mathbf{v}}_\mathrm{d} + \frac{m_\mathrm{e}}{\tau_\mathrm{e}}\mathbf{v}_\mathrm{d}, \tag{7.18}$$

where the left-hand side is now given by the Lorentz force $-e(\mathbf{E} + \mathbf{v}_\mathrm{d} \times \mathbf{B})$. If we consider a steady-state regime, we have $\dot{\mathbf{v}}_\mathrm{d} = 0$ and the equation of motion leads to

$$v_{\mathrm{d},x} = -\frac{e\tau_\mathrm{e}}{m_\mathrm{e}}E_x - \frac{e\tau_\mathrm{e}}{m_\mathrm{e}}B\,v_{\mathrm{d},y} \quad \text{and} \quad v_{\mathrm{d},y} = -\frac{e\tau_\mathrm{e}}{m_\mathrm{e}}E_y + \frac{e\tau_\mathrm{e}}{m_\mathrm{e}}B\,v_{\mathrm{d},x}, \tag{7.19}$$

By imposing the condition that no transverse current flows in the y direction (an open-circuit configuration corresponding to the condition $v_y = 0$) we get

$$\frac{E_y}{E_x} = -\frac{e\tau_\mathrm{e}}{m_\mathrm{e}}B, \tag{7.20}$$

which allows us to define the *Hall coefficient* R_H as

$$R_\mathrm{H} = \frac{E_y}{j_x B} = \frac{E_y}{\sigma_\mathrm{e}E_x B} = -\frac{1}{n_\mathrm{e}e}, \tag{7.21}$$

where j_x is the current density along the x direction and equation (7.7) has been used for the conductivity σ_e. Therefore, Drude theory predicts that *the Hall coefficient is independent of the applied magnetic field and, in any case, negative.* Unfortunately, both conclusions are wrong!

In order to enlighten these failures, let us plot as shown figure 7.4 the quantity $-1/n_\mathrm{e}eR_\mathrm{H}^*$ versus the applied magnetic field, where R_H^* is the result of a direct experimental measurement of the Hall coefficient. Plotting this quantity, we directly measure the deviation of the Drude theory from the real physical situation. This is

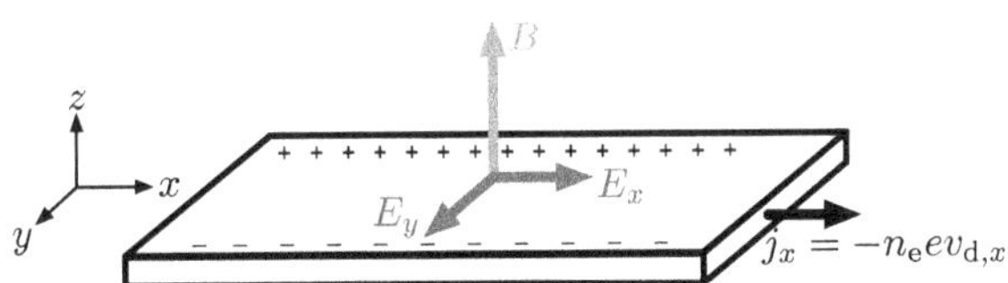

Figure 7.3. Left: schematic representation of the Hall effect. The + and − signs respectively indicate the depletion/accumulation of electrons on the opposite sides of the metallic sample (shown in grey) caused by the magnetic field $\mathbf{B} = (0, 0, B)$ (red arrow). The applied electric field $\mathbf{E} = (E_x, E_y, 0)$ is shown by blue arrows. When the circuit is open in the y direction, only the current density $j_x = -en_\mathrm{e}v_{\mathrm{d},x}$ is observed (black arrow).

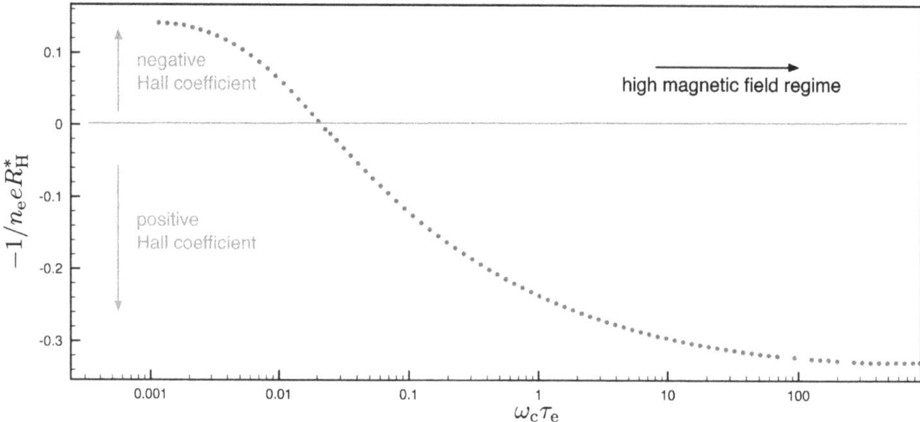

Figure 7.4. The experimentally observed dependence of the Hall coefficient R_H^* (blue dots) on the applied magnetic field in aluminium (nominal chemical valence $Z_v = 3$). The two physical parameters are represented by the dimensionless quantity $-1/n_e e R_H^*$ and $\omega_c \tau_e$, respectively. Data are taken from [8].

usually done by reporting on the horizontal axis the dimensionless quantity $\omega_c \tau_e$, where $\omega_c = eB/m_e$ is the cyclotron frequency. In figure 7.4 we illustrate the paradigmatic case of aluminium, showing that $-1/n_e e R_H^*$ is not at all a constant and, therefore, that R_H^* is experimentally found to depend on the magnetic field, contrary to the prediction of the Drude theory; furthermore, at high magnetic field it assumes positive values. This latter feature is quite intriguing: first of all, we observe that $-1/n_e e R_H^* = n_e^*/n_e$, where n_e^* is the experimental density of the electron gas which turns out to differ from the one calculated by equation (7.1) within the free electron theory (in the case of aluminium shown in figure 7.4 the high field value of the Hall coefficient suggests that there is just one carrier per atom, instead of three); next, the change in sign suggests that the carriers have positive charge. These results, similarly found in other metals like for instance zinc and cadmium, do not find an explanation within the classical free electron theory.

7.3 The quantum (Sommerfeld) theory of the conduction gas

The main source of failure of the Drude theory lies in fact that the electrons have been treated classically. The next obvious step in improving our theory, is, therefore, to apply quantum mechanics to the conduction gas.

7.3.1 The ground-state

Let us consider a metal specimen at zero temperature. Since its bulk properties do not depend on the shape, for mathematical convenience we will consider a cubic sample with side L and faces normal to the x, y, and z Cartesian axes. The single-particle wavefunction for any (free and independent) electron of the conduction gas is obtained by solving the Schrödinger equation

$$-\frac{\hbar^2}{2m_{\rm e}}\nabla^2\psi(\mathbf{r}) = E\psi(\mathbf{r}), \qquad (7.22)$$

where E is the electron energy. By imposing the Born–von Karman condition stated in equation (1.4), we easily get the normalised wavefunction

$$\psi_{\mathbf{k}}(\mathbf{r}) = \frac{1}{L^{3/2}}\exp(i\mathbf{k}\cdot\mathbf{r}) = \frac{1}{\sqrt{V}}\exp(i\mathbf{k}\cdot\mathbf{r}), \qquad (7.23)$$

where $V = L^3$ is the system volume and the *electron wavevector* $\mathbf{k}$ has the following Cartesian components[11]

$$k_x = \frac{2\pi}{L}\xi_x \quad k_x = \frac{2\pi}{L}\xi_y \quad k_x = \frac{2\pi}{L}\xi_z, \qquad (7.24)$$

with ξ_x, ξ_y, $\xi_z = 0, \pm1, \pm2, \pm3, \dots$. We stress that (i) the free electron wavefunction given in equation (7.23) has been labelled by $\mathbf{k}$ which *plays the role of a quantum number* for the crystalline states[12], (ii) the wavefunction given in equation (7.23) does *obey the Bloch theorem* discussed in section 6.3: in this specific case we simply have $u_{\mathbf{k}}(\mathbf{r}) = 1$. The electron energy is

$$E = \frac{\hbar^2 k^2}{2m_{\rm e}} = \frac{\hbar^2}{2m_{\rm e}}\left(k_x^2 + k_y^2 + k_z^2\right), \qquad (7.25)$$

a result which makes quite evident the function of quantum numbers associated with k_x, k_y, and k_z. By using the quantum mechanical operator $\hat{\mathbf{p}} = -i\hbar\nabla$, we easily obtain the *electron momentum*

$$\mathbf{p} = \hbar\mathbf{k}, \qquad (7.26)$$

and the corresponding *electron velocity* $\mathbf{v} = \hbar\mathbf{k}/m_{\rm e}$.

Because of equation (7.24), the allowed wavevectors are spaced by $2\pi/L$ along any direction and, therefore, their number density in the reciprocal space is $L^3/(2\pi)^3 = V/(2\pi)^3$. Furthermore, since the electron energies only depend on the magnitude of $\mathbf{k}$ (and not on its direction in space), the number of allowed wavevectors corresponding to quantum levels with energy in between E and $E + dE$ is given by the product between their number density and the infinitesimal volume $4\pi k^2 dk$ of the reciprocal space[13]. While the discussion has so far proceeded in analogy with the vibrational case discussed in section 3.7, now we must duly take into account that we can place two electrons (with opposite spin) on each allowed quantum level: as a matter of fact, each $\mathbf{k}$ actually corresponds to two different electron states with the same energy[14]. Therefore, by calculating

---

[11] We remark that in this chapter the Latin indices i and j will be used to label particles, while the Cartesian components will be indicated explicitly by x, y, and z.

[12] The wavevector $\mathbf{k}$ is the counterpart of the principal, angular, and magnetic quantum numbers found in atomic physics [9] and outlined in section 1.2.

[13] This volume corresponds to the spherical shell contained in between the two surfaces E and $E + dE$.

[14] By assumption, there is no external magnetic field acting on the system.

$$G(E)dE = 2\frac{V}{(2\pi)^3}4\pi k^2 dk \tag{7.27}$$

$$= \frac{V}{2\pi^2\hbar^3}\,(2m_{\mathrm{e}})^{3/2}\,E^{1/2}\,dE, \tag{7.28}$$

we either define the *electronic density of states* (eDOS) $G(E)$ and provide a direct way to calculate *the number of electron states with energy in the interval* $[E,\ E + dE]$ given by $G(E)dE$.

The *ground-state configuration of the conduction gas in a metal* is found at temperature $T = 0$ K by placing the N free electrons on the various quantum levels provided by equation (7.25), starting from the lowest one. This is a subtle task: we know that each $\mathbf{k}$ vector corresponds to two different states and that electrons obey the Pauli principle; accordingly, we can place just one electron on each spin-resolved state (which, in practice, corresponds to populate pairs of electron states degenerate in energy). This way of proceeding is perfectly consistent with the statistical arguments developed in section 6.2, provided that we take the results there obtained in the limit $T \to 0$. Since the total number N of electrons is finite (although very large, as discussed in section 7.1), this necessarily implies that *all quantum states are totally filled up to a maximum energy* E_{F}, while for any $E > E_{\mathrm{F}}$ we find empty states. Hence

$$N = \int_0^{E_{\mathrm{F}}} G(E)\,dE \tag{7.29}$$

$$= \frac{V}{3\pi^2\hbar^3}\,(2m_{\mathrm{e}})^{3/2}\,E_{\mathrm{F}}^{3/2} \tag{7.30}$$

$$= \frac{2}{3}\,G(E_{\mathrm{F}})\,E_{\mathrm{F}}, \tag{7.31}$$

from which we calculate the *Fermi energy* of the free electron gas

$$E_{\mathrm{F}} = \frac{\hbar^2}{2m_e}\,(3\pi^2)^{2/3}\,n_{\mathrm{e}}^{2/3}, \tag{7.32}$$

where $n_{\mathrm{e}} = N/V$ is the electron density: E_{F} *represents the energy of the highest occupied state at zero temperature*. The corresponding *ground-state energy* E_{GS} of the electron gas can be calculated either using the eDOS or summing over the single-particles energies

$$E_{\mathrm{GS}} = \int_0^{E_{\mathrm{F}}} E\,G(E)\,dE = \frac{V}{5\pi^2\hbar^3}\,(2m_{\mathrm{e}})^{3/2}E_{\mathrm{F}}^{5/2} \tag{7.33}$$

$$= 2\sum_{k<k_{\mathrm{F}}}\frac{\hbar^2 k^2}{2m_{\mathrm{e}}}, \tag{7.34}$$

where $k_F = \sqrt{2m_e E_F/\hbar^2}$ is the *Fermi wavevector*

$$k_F = (3\pi^2 n_e)^{1/3}, \tag{7.35}$$

which corresponds to the radius of the sphere in the reciprocal space containing the $N/2$ states able to host N electrons in pairs with opposite spin[15]. Its surface marks the boundary between filled and empty states at $T = 0$ K: it is commonly referred to as the *Fermi surface* of the metal. Typical values of the Fermi energy and wavevector are reported in table 7.2. It is interesting to observe that k_F is of the order of ~ 1 Å^{-1} and, therefore, the de Broglie wavelength of the matter waves [9] associated with the most energetic electrons is comparable to the typical interatomic distances. This suggests that the free electron picture is conceptually wrong: conduction electrons are not free because their interactions with ions are negligibly small but, rather, because matter waves freely propagate within a periodic lattice of ions.

The *Fermi velocity* v_F

$$\mathbf{v}_F = \frac{\hbar \mathbf{k}_F}{m_e}, \tag{7.36}$$

represents the quantum expectation value for the maximum electron speed in the conduction gas. As reported in table 7.2, its values are typically quite large, contrary to what was predicted by the classical kinetic theory according to which they should be zero at $T = 0$ K; we also observe that they even exceed the room-temperature thermal velocity of a classical particle with mass m_e. This is indeed an interesting feature, providing evidence that, consistently with the conclusion of the previous section, the conduction gas can hardly be treated as a classical one. This conclusion is naturally explained by introducing the *Fermi temperature* $T_F = E_F/k_B$, defining the inherent temperature scale of the conduction gas in a metal: such a gas can be treated classically only at temperatures comparable to or larger than T_F, but given the typical values reported in table 7.2, this situation is never reached because T_F

Table 7.2. Fermi energy E_F (units eV), Fermi wavevector k_F (units 10^8 cm^{-1}), Fermi velocity v_F (units 10^8 cm s^{-1}), and Fermi temperature T_F (units 10^4 K) for some metallic elements. Values are approximated to the first digit.

	Li	Na	Cu	Ag	Au	Mg	Ca	Fe	Zn	Al	Sn	Pb
E_F	4.7	3.2	7.0	5.4	5.5	7.1	4.7	11.1	9.5	11.7	10.2	9.5
k_F	1.1	0.9	1.4	1.2	1.2	1.4	1.1	1.7	1.6	1.7	1.6	1.6
v_F	1.3	1.1	1.6	1.4	1.4	1.6	1.3	2.0	1.8	2.0	1.9	1.8
T_F	5.5	3.8	8.2	6.4	6.4	8.2	5.4	13.0	11.0	13.6	11.8	11.0

---

[15] This calculation is simple: the product between the volume of the sphere and the number density of allowed wavevectors $(4\pi k_F^3/3)[V/(2\pi)^3] = N/2$ immediately drives to equation (7.35).

exceeds the melting temperature of the metal. We draw the conclusion that *the electron conduction gas in a metal is genuinely quantum at any temperature at which the solid state exists.*

7.3.2 Finite temperature properties

Let us now consider a metal in equilibrium at temperature $T > 0$ K. In this case the eDOS is written as

$$G(E, T) = G(E) \, n_{\text{FD}}(E, T) \tag{7.37}$$

$$= \frac{V}{2\pi^2 \hbar^3} \, (2m_{\text{e}})^{3/2} \, \frac{1}{1 + \exp[(E - \mu_c)]/k_{\text{B}}T} \, E^{1/2}, \tag{7.38}$$

where we have combined the expression given in equation (7.28), which is a mere counting of states, with the finite-temperature probability $n_{\text{FD}}(E, T)$ that the quantum level E is occupied, a correction entering our theory through equation (6.7). The $G(E, T)$ function is plotted in figure 7.5 (thick blue line), together with its zero-temperature counterpart (thin black line). We remark that *in plotting this figure we have neglected the temperature dependence of the chemical potential* and, accordingly, we have set $\mu_c = E_{\text{F}}$ at any $T \geqslant 0$. We will very soon critically re-address this assumption, proving that it is valid to a very good extent.

The number N of electrons is obviously unaffected by temperature and we can therefore cast the normalisation condition (previously expressed as in equation (7.29)) in a new form

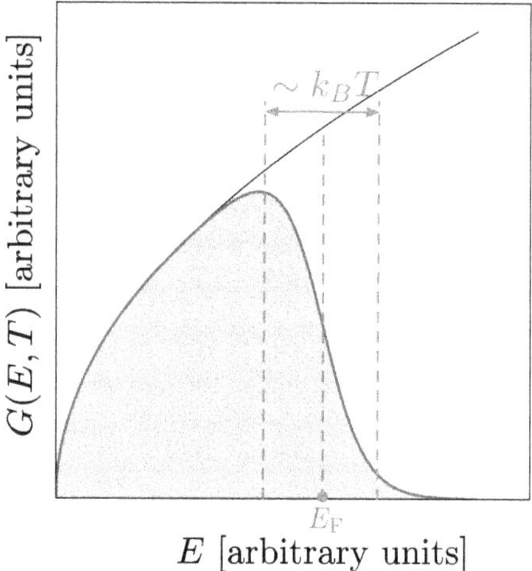

Figure 7.5. The electronic density of states $G(E, T)$ (thick blue line) of a metallic free electron gas in equilibrium at temperature $T > 0$ K. The thin black line represents the plot of its zero-temperature counterpart (see equation (7.28)). The energy range $\sim k_{\text{B}}T$ marked by a red double arrow corresponds to the interval over which the Fermi–Dirac distribution is non-constant: it is centred at the zero-temperature value of the chemical potential, namely at the Fermi energy E_{F}.

$$N = \int_0^{+\infty} G(E, T)\, dE = \int_0^{+\infty} G(E)\, n_{\text{FD}}(E, T)\, dE, \tag{7.39}$$

which allows us to interpret the shaded area of figure 7.5 as the conserved number of electrons. Since this notion is valid for any selected range of energy, we can develop a new interesting concept.

At first we remark that the area under the $G(E, T)$ function corresponding to the energy interval $E_F \leqslant E \leqslant +\infty$ represents the number of those few electrons that, upon increasing temperature from zero to T, have been promoted to energies above the Fermi energy. Obviously, this promotion has emptied an equal number of states just below E_F: this number corresponds to the area in between the function $G(E, T)$ and its zero-temperature counterpart and calculated for $0 \leqslant E \leqslant E_F$. This phenomenon is usually referred to as *thermal excitation* of electrons and it is summarised by stating that *at any finite temperature a number of electrons of the conduction gas are promoted from quantum states just below the Fermi energy to states just above it*. The thermal excitation phenomenon mostly involves electrons with energy in the interval $\sim k_B T$ centred at E_F.

The time has come to calculate once and for all the actual $\mu_c(T)$ dependence, thus checking the accuracy of the assumption which has replaced it with E_F at any temperature. To this aim we are going to follow a two-step procedure: first, we calculate N as a function of μ_c and T; next, we invert the expression so obtained to eventually get $\mu_c(T)$. The integral appearing in equation (7.39) can only be calculated by the asymptotic Sommerfeld expansion [3, 10], as outlined in appendix F. Given the key role of such an expansion, the resulting quantum theory is commonly referred to as the *Sommerfeld theory of the metallic state*. By using equation (F.10) we can write

$$N = \int_0^{\mu_c} G(E)\, dE + \frac{\pi^2}{6}\, G^{(1)}(\mu_c)\, (k_B T)^2 + \frac{7\pi^4}{360}\, G^{(3)}(\mu_c)\, (k_B T)^4 + \cdots, \tag{7.40}$$

where $G^{(m)}(\mu_c)$ indicates the mth order derivative of the zero-temperature eDOS calculated at energy μ_c. Even if we are now acknowledging that the chemical potential is a function of temperature, we can nevertheless say that, as shown in figure 6.2, $\mu_c \sim E_F$ whenever $T \ll T_F$; this allows us to approximate the integral appearing in the above equation as

$$\int_0^{\mu_c} G(E)\, dE \sim \int_0^{E_F} G(E)\, dE - (E_F - \mu_c)\, G(E_F) = \tag{7.41}$$
$$= N - (E_F - \mu_c)\, G(E_F),$$

where we used equation (7.29). The graphical rendering of this calculations is reported in figure 7.6. By inserting this result into equation (7.40) and retaining just the first leading term of the Sommerfeld expansion we easily get

$$E_F - \mu_c = \frac{\pi^2}{6}\, (k_B T)^2\, \frac{G^{(1)}(\mu_c)}{G(E_F)}, \tag{7.42}$$

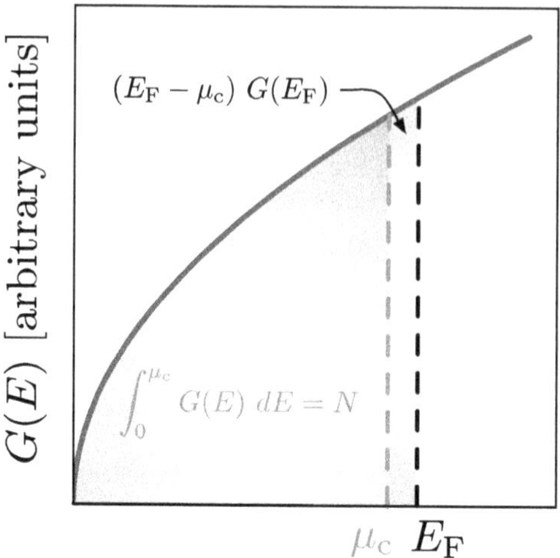

Figure 7.6. Graphical rendering of the calculation of the integral given in equation (7.41). The chemical potential and Fermi energy are indicated by μ_c and E_F, respectively. The function $G(E)$ is the zero-temperature eDOS given in equation (7.28).

which represents *the formal justification of the adopted approximation* $\mu_c \sim E_F$: the difference between the Fermi energy and the chemical potential is proportional to T^2 and, therefore, to a large extent it is negligible in the limit of small temperatures[16]. Another consequence of this result is that

$$\mu_c(T) = E_F \left[1 - \frac{\pi^2}{12} \left(\frac{T}{T_F} \right)^2 \right],\tag{7.43}$$

which further confirms that for any $T \ll T_F$ (see table 7.2) we can safely set $\mu_c(T) = E_F$. While this is a common practice, the reader is warned that *identifying the Fermi energy with the chemical potential is just a (motivated and convenient) approximation*, but the two concepts should not be confused.

The method of the Sommerfeld expansion is also useful to calculate the *internal energy of the free electron gas* which is obtained by generalising equation (7.33) as

$$U = \int_0^{+\infty} E\, G(E, T)\, dE = \int_0^{+\infty} E\, G(E)\, n_{FD}(E, T)\, dE,\tag{7.44}$$

By defining the zero-temperature internal energy as

$$U_0 = \int_0^{E_F} E\, G(E)\, dE,\tag{7.45}$$

and making use of equation (7.42) we proceed with the Sommerfeld expansion of equation (7.44) as follows (see appendix F)

---

[16] This result suggests that figure 6.2 has been obtained for very high $T_{2,3,4}$ temperatures, that is, much higher than typical temperatures of physical interest. As a matter of fact, the calculations performed to obtain figure 6.2 were executed for $T_{2,3,4}$ of the order of $\sim 10^4$ K.

$$\mathcal{U} = \mathcal{U}_0 + (\mu_c - E_F)\, E_F\, G(E_F) + \frac{\pi^2}{6}\, [G(E_F) + E_F\, G^{(1)}(E_F)]\, (k_B T)^2 + \cdots$$

$$= \mathcal{U}_0 - \frac{\pi^2}{6}\, (k_B T)^2\, E_F\, G^{(1)}(\mu_c) + \frac{\pi^2}{6}\, [G(E_F) + E_F\, G^{(1)}(E_F)]\, (k_B T)^2 + \cdots \quad (7.46)$$

$$= \mathcal{U}_0 + \frac{\pi^2}{6}\, G(E_F)\, (k_B T)^2 + \cdots,$$

which eventually leads to a *full quantum expression of the specific heat $c_V^e(T)$ of the free electron gas*

$$c_V^e(T) = \frac{1}{V}\frac{\partial \mathcal{U}}{\partial T}\bigg|_V = \frac{1}{V}\frac{\pi^2}{3}\, G(E_F)\, k_B^2\, T + \cdots$$

$$= \frac{\pi^2}{2} n_e k_B \frac{T}{T_F} + \cdots \quad (7.47)$$

$$= \frac{3}{2} n_e k_B \left(\frac{1}{3}\pi^2 \frac{T}{T_F}\right) + \cdots.$$

By comparing this expression with the classical result $c_V^e(T) = 3n_e k_B/2$ provided by the Drude theory developed in section 7.2, we understand that quantum mechanics predicts a lower specific heat than classical physics, in full agreement with experimental observations. Furthermore, the quantum correction term put in parenthesis in the above equation introduces a linear dependence on T which, once again, corrects a major failure of the Drude theory (according to which $c_V^e(T)$ should be constant). Anyway, we stress that the electron contribution to the total specific heat of a metallic crystal is proportional to T/T_F which, at temperatures of physical interest, makes it really marginal as compared to the lattice contribution c_V^l discussed in sections 4.1.2 and 4.1.3.

It is instructive to consider in more detail what happens at low-temperature: to this aim, we will make use of equations (7.47) and (4.11), respectively[17]. Basically, they dictate that

$$c_V^e(T) = \xi_1\, T \quad \text{and} \quad c_V^{\text{Debye}}(T) = \xi_2\, T^3, \quad (7.48)$$

where $\xi_1 = (\pi^2 n_e k_B/2T_F)$ and $\xi_2 = 12R\pi^4/5VT_D$. In figure 7.7 the variation of the quantity

$$\frac{1}{T}c_V^{\text{tot}} = \frac{1}{T}(c_V^e + c_V^l) = \xi_1 + \xi_2\, T^2, \quad (7.49)$$

upon the square temperature is shown in the case of copper (red dots), where we have expressed the lattice contribution to the specific heat by the Debye result setting $c_V^l = c_V^{\text{Debye}}$. The observed linear trend (blue dashed line) in the reported low

---

[17] We know that the Debye model does not provide the most fundamental description of the lattice specific heat c_V^l, but for the purpose of the present discussion either it is accurate enough and it predicts the needed analytic trend of c_V^l.

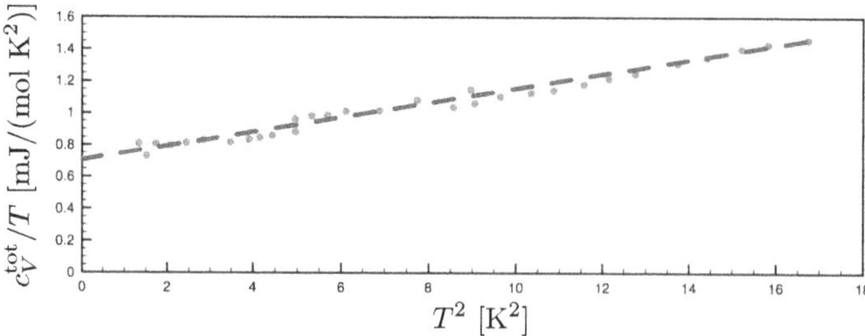

Figure 7.7. The quantity c_V^{tot}/T for copper (red dots) measured at low-temperature (data taken from [11]). The blue dashed line is the trend predicted by equation (7.49).

temperature range provides evidence that the electronic contribution is here dominant.

This way of representing the laboratory data has become a widely adopted practice, since the zero-temperature value of the quantity c_V^{tot}/T directly provides an experimental determination of ξ_1, that is, of the Fermi temperature[18]. However, a non negligible discrepancy is often found between the experimental and the theoretical[19] values; for instance, this discrepancy is as large as 48% in the case of aluminium. In order to reconcile the free electron theory with experimental evidence, it is customary to replace the bare electron mass m_e by an *effective mass* m_e^* which basically corrects the fact that the electrons of the conducting gas are not really free, but rather diffuse within a crystalline environment and, therefore, their inertia to motion is different than in empty space. The Fermi temperature is corrected by a factor m_e^*/m_e and, therefore, by setting a suitable value of effective mass it is possible to fit the theoretical prediction onto experimental data[20].

Finally, at higher temperatures the ionic degrees of freedom dominate the specific heat or, equivalently, $c_V^{\text{tot}}/T \sim c_V^{\text{l}}/T$. This is qualitatively shown in figure 7.8: *in metals the electronic and lattice contributions to the specific heat are comparable only at very low temperatures*, that is, up to just few degrees Kelvin, for most metallic systems.

The Sommerfeld theory of the free electron gas also provides *a more correct estimation of the electron contribution to the thermal conductivity* κ_e than the Drude theory. While we maintain the kinetic theory as our reference theoretical framework, we now replace in equation (7.15) the correct expression for the specific heat provided in equation (7.47) and estimate the electron square velocity by means of equation (7.36) so that

$$\kappa_e^{\text{Sommerfeld}} = \frac{1}{3}\,\tau_e\,v_{\text{F}}^2\,\frac{3}{2}n_e k_{\text{B}}\left(\frac{1}{3}\pi^2\frac{T}{T_{\text{F}}}\right), \tag{7.50}$$

---

[18] We remark that a similar way of plotting the specific heat of an insulating material would provide a straight line with the zero-temperature intercept found right at the origin of the axes: in these materials the electronic contribution is absent. This is consistent with the experimental evidence that electrons do not provide any measurable contribution to the specific heat of a non metallic system at room temperature.

[19] We recall that we have defined $T_{\text{F}} = E_{\text{F}}/k_{\text{B}}$, where E_{F} is given in equation (7.32).

[20] The effective mass concept is widely used in solid state physics. Therefore, one should always pay attention to the genuine meaning of effective mass which is used in the various contexts.

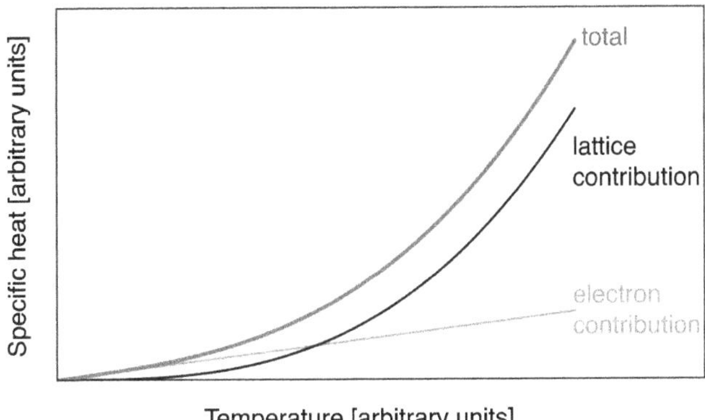

Figure 7.8. A qualitative comparison between the electron (red line) and lattice (black line) contributions to the total specific heat (blue line) of a metallic system in the range of very small temperatures (just a few degrees Kelvin above zero temperature).

which leads to

$$\frac{\kappa_e^{\text{Sommerfeld}}}{\sigma_e T} = \frac{\pi^2}{3}\left(\frac{k_B}{e}\right)^2 = 2.44 \times 10^{-8} \text{ W } \Omega \text{ K}^{-2} \tag{7.51}$$

a prediction in fairly good agreement with the experimental values of the Lorenz number which, for the metal elements listed in table 7.2, has an average value of 2.41×10^{-8} W Ω K^{-2}. This remarkable result derives from the rectification of the double error contained in the classical theory, namely: the Sommerfeld expression for $c_V^e(T)$ now includes a corrective term which decreases the Drude estimation by a factor proportional $k_B T/E_F$, while the Fermi velocity is larger than the classical average thermal velocity by an inverse factor $E_F/k_B T$. The two errors cancel out in the Drude theory, thus providing the correct order of magnitude for the Lorenz number, but still remain in its separate evaluation of the specific heat and electron speed.

Finally, the Sommerfeld theory also overcomes other failures of the classical theory, remarkably in describing thermoelectric effects. For instance, Drude theory wrongly predicts the ratio $Q = |\mathbf{E}|/|\nabla T|$ where ∇T is the gradient of temperature applied across a metal sample and $\mathbf{E}$ is the observed electric field established within it. Q is a material-specific property named *thermopower*: its determination does require the use of the Fermi–Dirac statistics for the conduction electrons [3, 10].

7.3.3 More on relaxation times

In our discussion on transport coefficients σ_e and κ_e we have twice introduced the notion of relaxation time which, although conceptually different in the two cases, was considered the same for charge and heat currents. It is now necessary to reconsider this aspect in greater detail.

Let us start by readdressing the direct-current conductivity. Electrons, during their drift motion under the action of an external electric field $\mathbf{E}$, undergo scattering with lattice defects and ionic oscillations[21]. The former provide a constant contribution τ_d to the electron relaxation time, while the effect of the ionic oscillation can be described as electron–phonon scattering events: their contribution $\tau_{ph}(T)$ is inherently dependent on temperature since the phonon population of each mode is so. If we assume that the two mechanisms are independent (that is, if the number of defects is small enough to leave unaffected the vibrational spectrum of the system), then we can apply the same Matthiessen rule already introduced in section 4.3 to understand thermal transport and write

$$\frac{1}{\tau_e} = \frac{1}{\tau_d} + \frac{1}{\tau_{ph}(T)}. \tag{7.52}$$

By now inserting this expression for the electron relaxation time into equation (7.7), we immediately obtain the *resistivity ρ_e of a metal* in the form

$$\rho_e = \frac{m_e}{n_e e^2} \frac{1}{\tau_e} = \frac{m_e}{n_e e^2} \frac{1}{\tau_d} + \frac{m_e}{n_e e^2} \frac{1}{\tau_{ph}(T)} = \rho_d + \rho_{ph}(T), \tag{7.53}$$

where the two contributions are referred to as the *residual resistivity* and *ideal resistivity*, respectively, since ρ_d is the only one active even at zero temperature, while $\rho_{ph}(T)$ is the only one found even in a totally defect-free system. The electron–phonon scattering largely affects the relaxation time, which is typically decreased from 10^{-11} s at $T = 0$ K down to 10^{-14} s at room temperatures. By multiplying the Fermi velocity by τ_e we can easily estimate the order of magnitude of the electron mean free path λ_e to be as large as dozens of nm at room temperature or dozens of μm at zero temperature. This is indeed a much more accurate estimation of λ_e than provided by the Drude theory and, more importantly, it better proves that the average distance covered between two successive collisions is much larger than the lattice interatomic spacing: as far as charge current phenomena are concerned, the electrons in a metal can be really considered as free, that is not colliding with lattice ions.

The discussion so far has not yet considered the collisions between electrons which represent another source of scattering affecting the relaxation time: however, *neglecting electron-electron scattering is largely justified by the occurrence of (i) screening effects and (ii) exclusion principle.* Screening is something we can understand only by going beyond the independent electron approximation and, therefore, it falls beyond the level of treatment we are developing: in fact, screening is an advanced topic of solid state theory which is presented elsewhere [3, 10, 12]. Here we limit ourselves to observing that the motion of the electrons in the conducing gas is highly correlated and, therefore, they can adjust in such a way as to largely cancel the single-particle long-range Coulomb field. The resulting screened potential turns out to have the form of a Yukawa potential: a Coulomb term multiplied by an exponentially decaying term. The interaction strength is accordingly reduced to a negligibly small value above a characteristic distance named *screening length,*

---

[21] The effect of the regular ions of the ideal lattice can be included in the electron effective mass, as discussed in the previous section.

typically calculated of the order of ~1Å. Such a dramatic reduction of the interaction range makes the mean free path between two electron–electron scattering events much larger than in the case of impurity or phonon scattering which, therefore, can be considered as the by far dominant ones at room temperature. As for the Pauli principle, the scattering between two electrons can only occur if there are vacant states in which to accommodate the electrons after the collision. In practice, *the exclusion principle reduces the frequency of electron–electron scattering events* because it makes fewer scattering channels available to electrons [2].

Now that we have better characterised the different scattering mechanisms affecting electrons during their drift motion, let us consider how an externally applied electric field affects their quantum states, that is their wavevector. In the absence of field, the Fermi surface is a sphere centred at $\mathbf{k} = 0$ (see equation (7.35) and related comments) with a boundary which is slightly blurred by the thermal excitation mechanism. Under the action of the field $\mathbf{E}$ the sphere is shifted by an amount that corresponds to a wavevector variation $\Delta \mathbf{k} = -(e\tau_e/\hbar)\mathbf{E}$ for each electron[22]. The situation is pictorially represented in figure 7.9. The relaxation time entering this equation is the same discussed above, but it is now naturally interpreted as *the time needed to the Fermi sphere to relax back to its unshifted position, after that the electric field is turned off.* The key point is that the microscopic events driving this displacement are precisely the collisions suffered by the electrons. More specifically, the relaxation is ruled over by those scattering processes removing one electron from the right side of the shifted Fermi sphere and adding it to its left side. As shown in figure 7.9 we estimate a wavevector change as large as $\sim 2k_F$ for these processes.

It is now time to address the thermal current case and its corresponding relaxation time. A similar analysis of the scattering events as above can be elaborated when a

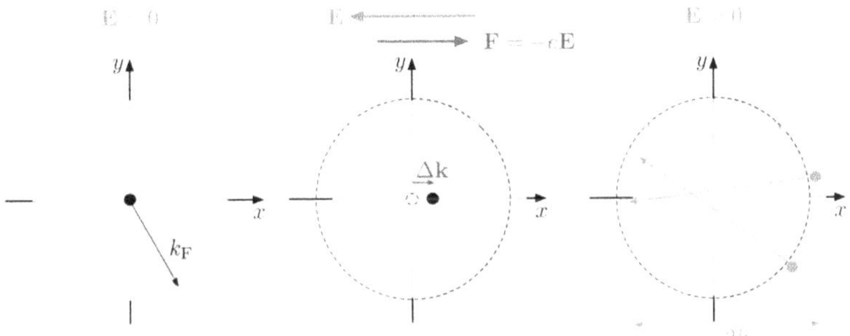

Figure 7.9. Left: the Fermi sphere centred at $\mathbf{k} = 0$ in absence of electric field (k_F is the Fermi wavevector, that is, the radius of the Fermi sphere whose blurred boundary is not shown explicitly for sake of clarity). Middle: an electric field $\mathbf{E}$ is applied so that a constant force $\mathbf{F} = -e\mathbf{E}$ acts on each electron causing a $\Delta \mathbf{k}$ shift of the centre of the Fermi sphere. Right: by turning off the applied electric field, the Fermi sphere relaxes back to its equilibrium position because of electron scattering events pictorially shown by purple arrows (the filled and empty dots represent the initial and final electron state before and after the scattering event, respectively).

---

[22] This result is obtained by integrating the electron equation of motion under the action of the field $\mathbf{E}$.

metal is under the action of a thermal gradient ∇T. In this case, the blurring of the Fermi sphere is different on its opposite sides taken along the direction of ∇T because of the different local temperature. We can equivalently state that *there are more thermal excitations on the hotter side of the Fermi sphere than on the colder side.* Following the same reasoning as in the charge current case, the relaxation time entering equation (7.50) is now interpreted as *the time needed to the Fermi sphere to even its left and right blurring, after that the temperature gradient is turned off.* Once again, this evolution is regulated by microscopic events that mainly consist in electron–phonon scattering processes. In this case, however, they occur by means of either large or small momentum changes, as pictorially indicated in figure 7.10.

In summary, just one leading scattering process has been identified in the case of a charge current (associated with a large $\sim 2k_{\rm F}$ wavevector change), while two unalike processes (characterised by a large and small wavevector change, respectively) are at work in the case of a heat current. Since in this latter case more scattering occurs, *we expect the relaxation time associated with a heat current is shorter than the one associated with a charge current.* This qualitatively explains why, contrary to what is predicted by the Drude and Sommerfeld free electron theories (see equations (7.17) and (7.51)), the Lorenz number does depend on temperature[23], as found experimentally at low temperatures.

In concluding this section we observe that *for both a charge and a heat current we have understood that the overall relaxation time actually depends on temperature,* since phonon populations enter into their evaluation. The separate calculation of $\tau_{\rm e}$ for both cases leads to the accurate prediction of the temperature dependence of the electrical resistivity and thermal conductivity of a metal as reported elsewhere [3, 4].

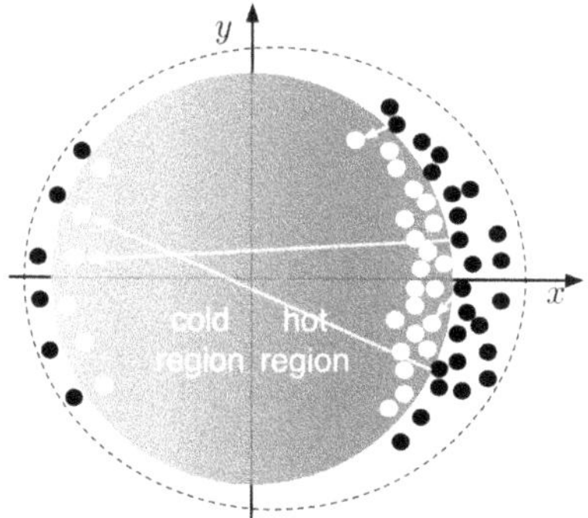

Figure 7.10. The Fermi sphere of a metal under the effect of a temperature gradient. The cold (cyan) and hot (red) boundaries are blurred (dashed line) to a different extent since they feel a different local temperature. For illustration purposes the blurring effect is much magnified. Yellow lines pictorially represent electron–phonon scattering events which can imply a small as well as a large momentum change. They drive the Fermi sphere back to its equilibrium state as soon as the temperature gradient is removed. This occurs by displacing electrons from thermally-excited states (filled dots) to empty ones (empty dots).

---

[23] More precisely, we have eventually understood that it is incorrect to assume that the relaxation time for a charge and a heat current is the same. In other words, the simplification of the $\tau_{\rm e}$ factor operated in deriving equations (7.17) and (7.51) is only valid as a first approximation.

7.3.4 Failures of the Sommerfeld theory

The Sommerfeld theory outclasses the Drude one by more accurately predicting many physical properties of metals; it also enlightens some important concepts like the difference between the chemical potential and the Fermi energy or the real need to treat the electron conduction gas as a fermion gas obeying the Fermi–Dirac statistics. However, it cannot yet be regarded as the most complete and predictive quantum theory of electron states in a crystal, since its predictions are still not in good agreement with experiments in some important cases.

First of all, we remark that the Sommerfeld theory for the charge current is basically the same as the Drude one and, therefore, it suffers the same limitation, in particular as regards the wrong predictions about the Hall coefficient[24]. This is mainly due to the fact that in deriving such a coefficient the fermionic nature of the charge carriers is not explicitly taken into account[25]. Even the alternate-current conductivity provided by the two free electron models is only grossly adequate in describing metal reflectivity, but it falls short with other optical properties of metals like, notably, their colour. Finally, the Fermi surface of real metals is not a simple sphere with radius k_F [3, 4].

A part for these phenomenological failures, the Sommerfeld theory is unable to explain a very fundamental fact: *why in Nature do insulators exist?* Our basic assumption was to treat the system of valence electrons as a gas of delocalised charge carriers. Why is this a reasonably good approximation in some materials (metals) and not in many others (insulators)? The Sommerfeld theory does not provide an answer to this question. We need a more refined approach to the electronic structure of a crystalline solid.

References

[1] Reif F 1987 *Fundamentals of Statistical and Thermal Physics* (New York: McGraw-Hill)

[2] Kittel C 1996 *Introduction to Solid State Physics* 7th edn (Hoboken, NJ: Wiley)

[3] Ashcroft N W and Mermin N D 1976 *Solid State Physics* (London: Holt-Saunders)

[4] Hook J R and Hall H E 2010 *Solid State Physics* (Hoboken, NJ: Wiley)

[5] Fox M 2001 *Optical Properties of Solids* (Oxford: Oxford University Press)

[6] Ehrenreich H, Phillipp H R and Segall B 1962 *Phys. Rev.* **132** 1918

[7] Ziman J M 1960 *Electrons and Phonons* (Oxford: Oxford University Press)

[8] Lück R 1966 *Phys. Status Solidi* **18** 49

[9] Colombo L 2019 *Atomic and Molecular Physics: A Primer* (Bristol: IOP Publishing)

[10] Grosso G and Pastori Parravicini G 2014 *Solid State Physics* 2nd edn (Oxford: Academic)

[11] Corak W S, Garfunkel M S, Satterthwaite C B and Wexler A 1955 *Phys. Rev.* **98** 1699

[12] Jones W and March N H 1973 *Theoretical Solid State Physics* vol 1 (New York: Dover)

---

[24] This also reflects in a wrong prediction about the magnetoresistance: the free electron theory, in both formulations, estimates that the resistance of a metal sample measured normally to an externally applied magnetic field does not depend on the field. This is not true, as experimentally found in many real metals [3, 4].

[25] See section 7.2.1 where σ_e was calculated without any need to use statistical laws.

IOP Publishing

Solid State Physics
A primer
Luciano Colombo

Chapter 8

The band theory

Syllabus—The band theory for the crystalline electron states is developed through a hierarchy of increasingly accurate models. At first we discuss the nearly free electron approach, where the crystal field potential is treated just as a perturbation on the free electron states. Despite its qualitative character, this model is nevertheless able to support the key concept that energy states are grouped into allowed bands, separated by forbidden gaps. The corresponding band filling will allow us to distinguish among metals, insulators, and semiconductors. Next we will elaborate the tight-binding theory which leads to the accurate prediction of the dispersion relations for the crystalline electrons. A number of general features about the band structure will be eventually discussed, with focus on electron dynamics. In particular, we will discuss the concepts of hole and effective mass, indeed two key ingredients of the microscopic theory of the charge transport in solids.

8.1 The general picture

8.1.1 Bands and gaps

As explained in section 7.3.4, *the Sommerfeld theory is unable to explain why some materials are metals and others not,* although the corresponding atomic elements have a similar ground state electronic configuration. Ultimately, the limitations of this theory are related to the fact that valence electrons are treated as free and independent particles. In order to improve our description of electron states in crystalline systems, we should therefore re-address these assumptions and accordingly develop a more refined theory.

Actually, it is not necessary to question the entire theoretical apparatus set up in chapter 1 since we already decided once and for all to treat electrons within the single-particle approximation: according to what is explained in section 1.4.1, each electron is assumed to independently move under the action of a local crystal field potential $V_{\text{cfp}}(\mathbf{r})$. The electron quantum many-body problem is therefore disentangled into a set of similar single-particle Schrödinger equations (1.22), which

doi:10.1088/978-0-7503-2265-2ch8

provide the energy and the wavefunction for each electron. The implication of this approximation is twofold: while electrons will be similarly treated as independent particles as in the Sommerfeld theory of metals, nevertheless *they are subject to a potential $V_{\mathrm{cfp}}(\mathbf{r}) = V_{\mathrm{cfp}}(\mathbf{r} + \mathbf{R}_l)$ which has the same periodicity of the crystal*. This immediately calls into question two general quantum mechanical results, namely: (i) the energies of the crystalline electron states are grouped into bands, separated by forbidden gaps (see section 6.4) and (ii) the wavefunction must have the Bloch form (see section 6.3).

Let us consider in more detail the structure of allowed bands and forbidden gaps obtained within the Kronig–Penney model and reported in figure 6.4. We remark that, although this result has been obtained for a simplified form of $V_{\mathrm{cfp}}(\mathbf{r})$, it has a general validity: the energy spectrum of an electron subject to *any* periodic potential consists in bands and gaps. The parameter $\xi = \alpha a$ appearing in equation (6.35) is nothing other than the normalised single-electron energy E and, therefore, we can draw *a conceptual diagram representing the band structure of a crystal* such as the one shown figure 8.1: here we report the case of both a free electron (left) and an electron subject to a periodic potential (middle); for sake of comparison, the energy spectrum of an electron confined with an infinite potential well is also reported (right): ideally it represents the extreme case of an electron so tightly bound to its nucleus that it is not affected by the crystalline environment.

We can easily read the physical situation as follows. If we consider the case of a one-dimensional infinite chain, we understand that each potential well corresponding to a core region is just equivalent to any other and, therefore, the electron quantum states there confined have the same discrete energies as in any other similar well.

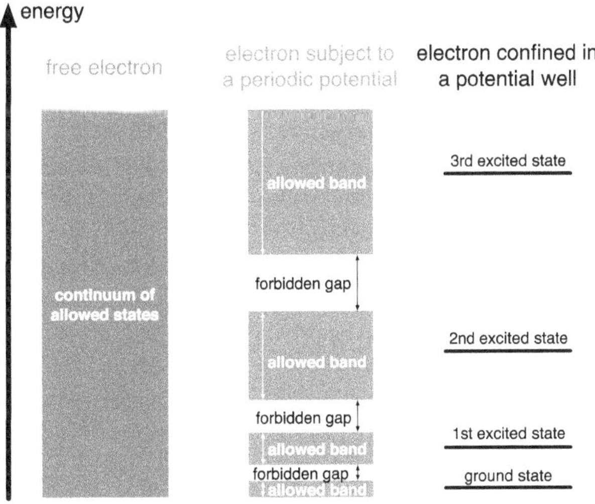

Figure 8.1. Comparison of the energy levels allowed to a free electron (left), to an electron subject to periodic crystalline potential (middle) and to an electron confined within an infinite potential well (right). This diagram represents the graphical conceptualisation of what is obtained within the Kronig–Penney model and reported in figure 6.4.

Since, however, the wells are not actually infinitely deep, there is a non-zero tunnelling probability for one electron to diffuse from a given core region to a neighbouring one. This, in turn, makes the set of discrete confined levels (each of which would be degenerate in the idealised situation of infinite wells) to open into a band of allowed crystalline states. It is convenient to label bands by a *band index n* which we must duly append to any symbol identifying electron energies E_n and wavefunction $\psi_n(r)$. On the other hand, so as not to complicate formalism too much, we will hereafter omit the apex $(\mathbf{R})$ identifying the specific ionic configuration for which the electron problem is solved (see equation (1.22)). It will be understood that, according to the adiabatic approximation, if ions are displaced then the electron problem will have to be rewritten and solved again.

Let us now consider the implications of the Bloch theorem. We know that the wavefunction of an electron moving in a periodic potential has the form of a product between a plane wave $\exp(i\mathbf{k} \cdot \mathbf{r})$ and a periodic function $u_\mathbf{k}(\mathbf{r}) = u_\mathbf{k}(\mathbf{r} + \mathbf{R}_l)$ (see equation (6.24)). This implies that *the wavevector* $\mathbf{k}$ *is the real quantum number identifying the crystalline state* and, therefore, we agree to index the wave function as $\psi_{n\mathbf{k}}(\mathbf{r})$ and the corresponding energy as $E_{n\mathbf{k}}$. However, a new problem now arises: how does the wavevector $\mathbf{k}$ enter a conceptual diagram like the one shown in figure 8.1? The answer is: it doesn't! In other words, *this way to represent the allowed states is unable to provide any information about the relationship between the wavevector and the corresponding energy.* This is enough to understand that we cannot settle to represent the band structure of a crystal only this way: we do need *dispersion relations* $E_{n\mathbf{k}} = E_n(\mathbf{k})$.

8.1.2 The weak potential approximation

A first qualitative attempt to calculate the dispersion relations consists in considering *the periodic potential as a small perturbation* with respect to the free electron situation. While this approach, usually referred to as *nearly free electron theory* [1–3], is unable to provide quantitative $E_{n\mathbf{k}} = E_n(\mathbf{k})$ relations for real crystals, it nevertheless provides very useful qualitative information about their band structure. Following a pedagogical approach already adopted several times in this textbook, we will develop this model just in one dimension, so as to keep the mathematics simple and focus on the physical implications.

Let us consider a monoatomic linear chain with lattice spacing a; assuming that a valence electron is subject to a *weak periodic potential*, we indicate by $\hat{V}_{\text{pert}}(x) = \hat{V}_{\text{pert}}(x + sa)$ (with s any positive or negative integer) the operator describing the *perturbation with respect to the free electron situation*. Because of its periodicity, we can represent the perturbation as a Fourier series

$$\hat{V}_{\text{pert}}(x) = -\sum_{m=1}^{+\infty} V_m \cos\left(\frac{2\pi m}{a}x\right), \tag{8.1}$$

where, in order to fulfil the convention adopted in figure 6.3(top), we assume that $V_m > 0$. To the first order in the perturbation, the correction ΔE_{pert} to the free electron energy is calculated as

$$\Delta E_{\text{pert}} = \frac{\int \phi^*(x)\,\hat{V}_{\text{pert}}(x)\phi(x)\;dx}{\int \phi^*(x)\phi(x)\;dx},$$ (8.2)

where $\phi(x)$ is the unperturbed wavefunction: it is the one-dimensional counterpart of the general solution given in equation (7.23) and corresponds to a plane wave[1]. This result of standard quantum mechanical perturbation theory must be handled with some care: the unperturbed energy levels $E^{(0)} = \hbar^2 k^2/2m_e$ are twofold degenerate since the progressive $\phi_p(x) = \exp(ikx)$ and regressive $\phi_r(x) = \exp(-ikx)$ waves correspond to the same electron energy. Therefore, the more rigorous way to calculate the integrals appearing in equation (8.2) is to replace the unperturbed wavefunctions with two orthogonal linear combinations of $\phi_p(x)$ and $\phi_r(x)$—which we will indicate by $\phi_1(x)$ and $\phi_2(x)$—and to impose the constraint $\int \phi_1^*(x)\hat{V}_{\text{pert}}(x)\phi_2(x)\,dx = 0$ [4, 5]. It is easy to prove that this constraint is fulfilled for any value of k only by the real functions $\phi_1(x) = \cos(kx)$ and $\phi_2(x) = \sin(kx)$. Let us proceed by inserting $\phi_1(x)$ into equation (8.2)

$$\Delta E_{\text{pert}} = \frac{\int \phi_1^*(x)\left[-\sum_{m=1}^{+\infty} V_m \cos\left(\frac{2\pi m}{a}x\right)\right]\phi_1(x)\;dx}{\int \phi_1^*(x)\phi_1(x)\;dx}$$

$$= -\frac{\sum_{m=1}^{+\infty} V_m \int \cos^2(kx) \cos\left(\frac{2\pi m}{a}x\right) dx}{\int \cos^2(kx)\,dx}$$ (8.3)

$$= -\frac{\sum_{m=1}^{+\infty} V_m \int [1 + \cos(2kx)] \cos\left(\frac{2\pi m}{a}x\right) dx}{\int [1 + \cos(2kx)]\,dx}.$$

The numerator is nonzero only if the periodicity of $\cos(2kx)$ and $\cos(2\pi mx/a)$ are the same or, equivalently, only if $k = m\pi/a$. For these values of the wavevector we straightforwardly calculate $\Delta E_{\text{pert}} = V_m/2$. On the other hand, by inserting $\phi_2(x)$ into equation (8.2) we obtain the same constraint on k and similarly calculate $\Delta E_{\text{pert}} = -V_m/2$. The conclusion is remarkable: *the perturbation operated by the weak potential on the free electron energies affects only those pairs of states lying at the Brillouin zone boundaries* (which correspond to $m = \pm 1, \pm 2, \pm 3, \ldots$) by shifting the two degenerate states by a similar amount $V_m/2$ but in opposite directions.

---

[1] We remark that in this calculation we are taking an infinite system and, therefore, unperturbed wavefunctions are not normalised as in equation (7.23). Accordingly, in equation (8.2) the expectation value of the perturbation operator is properly normalised [4, 5].

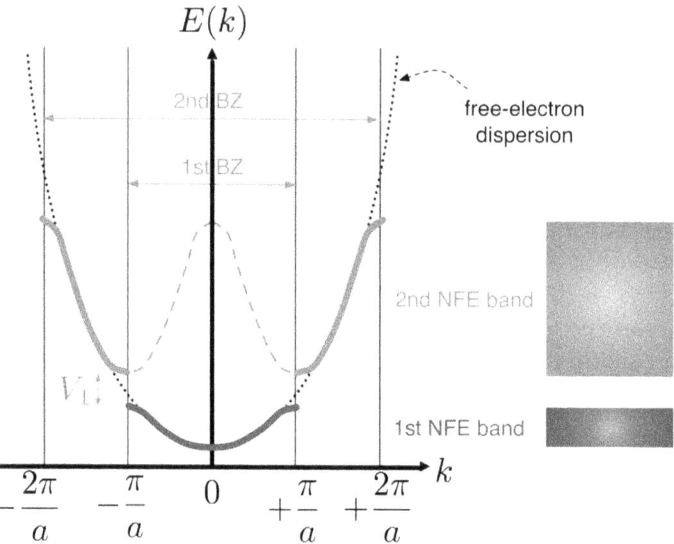

Figure 8.2. Right: full lines (blue and red) represent the dispersion relations $E = E(k)$ of a crystal treated within the nearly free electron (NFE) model; the dashed line (black) corresponds to the dispersion relation $E = \hbar^2 k^2/2m_{\mathrm{e}}$ of a free electron. The weak NFE potential opens gaps at the zone boundaries; the first gap has width V_1. Right: the corresponding band structure diagram. The reduced zone scheme consists in folding the NFE dispersions within the first Brillouin zone (1BZ).

In figure 8.2 it is shown the dispersion relation of the one-dimensional crystal treated in the weak potential approximation and, for comparison, the dispersion predicted by the free electron model. The first important feature to observe is that *the true existence of a periodic potential acting on the valence electrons opens gaps in the energy spectrum*, in full agreement with the Kronig–Penney model. Furthermore, by only considering the vertical ordering of the allowed energies (that is, by neglecting the role played by the wavevector) we recover the band diagram shown in figure 8.1. These two ways of representing the energies of crystalline electrons are fully equivalent and equally useful, but only through the dispersion relations can we graphically render the role of the wavevector. Next, we rigorously observe that the perturbative calculation only indicates that gaps are opened at $k = m\pi/a$; on the other hand, in figure 8.2 the dispersion relation in proximity of the Brillouin zone boundaries is shown to deviate from the purely parabolic trend predicted by the free electron model. In order to calculate the exact $E_n(k)$ values near such boundaries a more refined calculation is indeed required. Let us for instance consider the boundary of the first Brillouin zone at $m = 1$. The arguments developed after the derivation of equation (8.3) indicate that here the only non vanishing term of the perturbation potential is $V_1 \cos(2\pi x/a)$ so that the Schrödinger equation can be written in the form

$$\left[-\frac{\hbar^2}{2m_{\mathrm{e}}}\nabla^2 + V_1 \cos\left(\frac{2\pi x}{a}\right) \right]\psi(x) = E\psi(x). \tag{8.4}$$

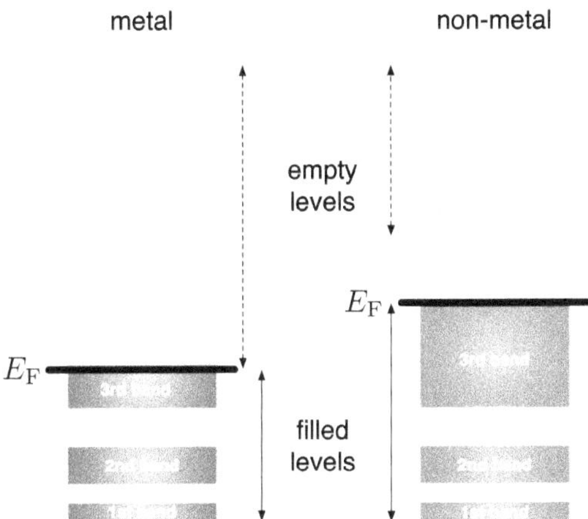

Figure 8.3. Band filling at $T = 0$ K. The allowed energy levels occupied by electrons are shown in blue or red, while empty ones are shown in grey. For metals (left) the Fermi energy E_F falls within the upmost partly occupied band. For non-metals (right) E_F coincides with the highest level of the last fully occupied band.

Since we understand that the potential term appearing in this quantum problem is weak, in the spirit of the perturbation theory we attempt a trial solution of the form

$$\psi(x) = \xi_1 \exp(ikx) + \xi_2 \exp[i(k - 2\pi/a)x], \tag{8.5}$$

where the first term corresponds to the free electron case, while the second term describes the wavefunction distortion generated by the perturbation. The constants ξ_1 and ξ_2 are determined by a boring, but straightforward calculation[2] which leads to

$$E(k) = \frac{\hbar^2 k^2}{2m_e} + \frac{\hbar^2 \pi}{m_e a}\left[k' \pm \sqrt{k'^2 + (m_e a V_1/2\pi\hbar^2)^2} \,\right], \tag{8.6}$$

with $k' = \pi/a - k$. This result reduces to the free electron one for any value of k well within the Brillouin zone, while it provides the bending of the dispersion relation nearby its boundaries, as shown in figure 8.2. The same calculation is performed for any $m > 1$, leading to similar results.

8.1.3 Band filling: metals, insulators, semiconductors

Before developing a theory more rigorous than the nearly free electron one, we address another important qualitative question which is inherently linked to the actual existence of a band structure, namely: *how are the allowed states populated by electrons?* To answer, it is useful to adopt the kind of band representation shown in figure 8.3.

---

[2] The trial solution must be at first inserted into equation (8.4); the resulting equation is then multiplied by $\exp(ikx)$ and eventually integrated over the space. These two last steps are repeated a second time but now multiplying by $\exp[i(k - 2\pi/a)x$. The resulting two equations form an algebraic system where ξ_1 and ξ_2 are the unknown variables. By imposing the secular determinant to be zero in order to have nontrivial solutions, we get a second order equation in $E(k)$ whose solutions are given by equation (8.6).

Let us suppose that the crystal is in thermal equilibrium at temperature $T = 0$ K and that a total number N_{val} of valence electrons are available: we want to elaborate a procedure for filling the allowed bands. In order to do so, we must operate under the Pauli principle: the maximum number of electrons that we can accommodate on each level is equal to double the degree of degeneracy of that level[3], since only up to two electrons with opposite spin can correspond to the same quantum state labelled by the given pair of band index n and wavevector $\mathbf{k}$. Accordingly, we start by placing the maximum possible number of electrons on the lowest level of the lowest allowed band. Having filled the first level, we pass to the second one of that same band, arranging the electrons with similar criteria as before. Then we proceed in the same way for all the other levels of the first band until it is completely filled. At this point, in order to place the remaining electrons, we have to pass directly to the second band, since no states are available in the forbidden gap separating the first band from the second one. By repeating the same procedure as in the previous case, the second band will also be filled; next, the third one and so on $\cdots$ until all the N_{val} valence electrons are eventually used. As in the case of the free electron gas, the energy E_{F} of the highest occupied level in this $T = 0$ K condition will be referred to as Fermi level.

The intriguing point is that, if we consider the position of the Fermi level, *only two possible cases are given*: either (i) E_{F} falls within the last (that is: highest in energy) band or (ii) E_{F} coincides with the upmost level of the last band. The situation is pictorially summarised in figure 8.3. The two cases are inherently different since the last band is either partially or completely occupied and, therefore, *according to the position of the Fermi level we distinguish between metals (first case) and non-metallic materials (second case)*. While this distinction could appear as mostly conventional, we can motivate it by a simple qualitative argument addressing the charge transport properties in the two opposite cases. In metallic systems an applied electric field shifts the electron distribution similarly to the case of a free electron gas (see figure 7.9) simply because there are vacant states just above E_{F}: this generates a crystalline quantum state able to carry electric current. On the other hand, in the case of a totally filled last band an applied electric field is unable to promote electrons above E_{F} since the energy it provides is not large enough to bypass the energy gap: in this condition the material is an insulator. There are, however, solids whose energy gap between the last fully occupied band and the first empty one is sufficiently small that at any finite temperature some electrons are thermally excited to the upper band. The graphical rendering of the situation is reported in figure 8.4. These materials are, therefore, insulators at zero temperature and (poor) conductors at $T > 0$ K: they are named *semiconductors*.

A more fundamental explanation of the charge transport characteristic will be developed in section 8.3.3, but as of now we can already understand why in Nature do materials exist with metallic or insulating or semiconducting behaviour

---

[3] The qualitative discussion so far developed has not yet highlighted the fact that states belonging to different bands, but with the same wavevector, can be degenerate in energy. This will be fully exploited in the next section. Here it is sufficient to admit that, in general, energy levels can be degenerate.

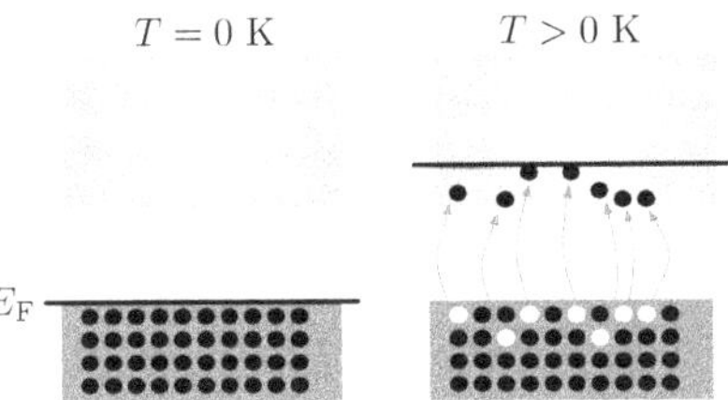

Figure 8.4. Graphical rendering of the thermal excitation (red arrows) of electrons (black dots) in a small-gap insulator. White dots: emptied energy levels. The topmost occupied level is shifted from the Fermi energy E_F position to an intermediate position of the upper band, thus reproducing a situation qualitatively similar to that of a metal (see figure 8.3).

$$T = 0\ \text{K}$$

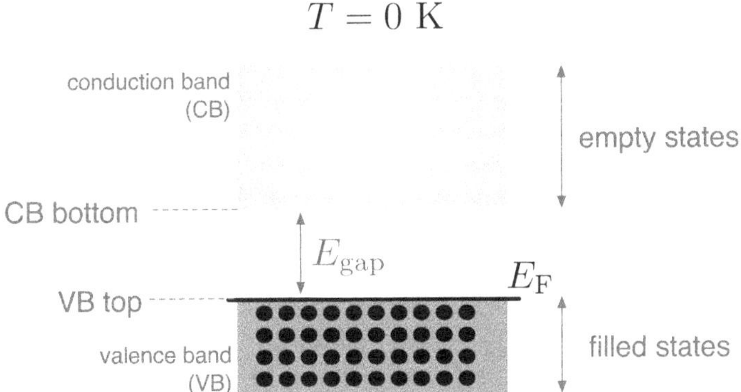

Figure 8.5. Standard nomenclature of the band theory (only the topmost occupied band and the first empty one are shown).

(something that the free electron theory was unable even to predict): ultimately, *the ability of a material to carry an electric current is determined by its band structure and by the relative filling with valence electrons.* We also understand that this is a truly quantum mechanical property since it is derived from general properties of electrons subject to a periodic potential[4].

We complete the nomenclature commonly used within the band theory by introducing some further conventions. First of all, we agree to refer to *the highest energy level of a band as the 'band top'* and, likewise, to *the lowest energy level as the 'band bottom'*. Furthermore, in non-metallic systems *the last occupied band is called the valence band,* while *the first empty band is called the conduction band.* The two bands will be hereafter indicated by the acronyms VB and CB, respectively. The energy difference between the VB top and the CB bottom will be indicated with the explicit name of *forbidden gap*: it represents the interval of forbidden energies that separates the last totally full band from the first completely empty one. We will always indicate with the symbol E_{gap} the energy amplitude of such a forbidden gap. These conventions are summarised in figure 8.5.

---

[4] Historically, this achievement was welcome as one of the most convincing successes in support of quantum mechanics [6].

8.2 The tight-binding method

The time has come to proceed with a rigorous calculation of the band structure of a crystalline solid. As in the case of lattice dynamics, we will first consider a model one-dimensional crystal where the formal development can be exploited in full analytical detail and the most relevant physical features emerge in a simple way. Next, we will outline the calculation of the band structure of real solids, using the formal results reported in appendix G.

8.2.1 Bands in a one-dimensional crystal

Let us consider a monoatomic linear chain of atoms with lattice spacing a so that ion positions are given by $x_s = sa$ with $s = 0, \pm 1, \pm 2, \pm 3, \dots$. Let us also suppose that there is just one valence electron for each atom in the chain. The Born–von Karman boundary condition given in equation (1.4) is applied to a crystal portion containing a suitably large number N of atoms (and, therefore an equal number $N_{\text{val}} = N$ of valence electrons).

The preliminary step in our approach is to consider the case of a single isolated atom of the same chemical species present in the chain. Let $\hat{V}_{\text{a}}$ be the quantum operator describing the potential V_{a} felt by the valence electron and let us suppose that the corresponding Schrödinger problem

$$\left[-\frac{\hbar^2}{2m} \nabla^2 + \hat{V}_{\text{a}} \right] \phi_{\text{a}} = E_{\text{a}} \phi_{\text{a}}, \tag{8.7}$$

has been solved by means of the standard methods of atomic physics [7–9]. In our formalism ϕ_{a} e E_{a} are the atomic wavefunction and energy of the atomic states, respectively.

Once the atom is placed in some lattice position along the chain, we can assume to a very good approximation that its valence electron is now subject to a potential $V_{\text{c}}(x)$ written as[5]

$$V_{\text{c}}(x) = V_{\text{a}} + \Delta V(x), \tag{8.8}$$

where $\Delta V(x)$ *describes the difference between the crystalline environment and the isolated atom situation.* Our physical intuition suggests that $\Delta V(x)$ is vanishingly small in the core regions, while it significantly differs from zero in the interstitial ones, as qualitatively reported in figure 6.3. Obviously $\Delta V(x) = \Delta V(x + sa)$. The crystalline Schrödinger problem is therefore written as

$$\left[-\frac{\hbar^2}{2m} \frac{d^2}{dx^2} + \hat{V}_{\text{c}}(x) \right] \psi_{\text{c}}(x) = E_{\text{c}} \psi_{\text{c}}(x), \tag{8.9}$$

where $\hat{V}_{\text{c}}(x)$ is the quantum operator corresponding to the potential given in equation (8.8), while $\psi_{\text{c}}(x)$ and E_{c} are the wavefunction and energy of the crystalline

---

[5] We choose to align the monoatomic chain along the x direction.

states, respectively. For further convenience, we recast this equation in a more compact form

$$\hat{H}\psi_c(x) = E_c\psi_c(x) \quad \text{where} \quad \hat{H} = -\frac{\hbar^2}{2m}\frac{d^2}{dx^2} + \hat{V}_c(x). \tag{8.10}$$

By multiplying on the left by $\psi_c^*(x)$ and by taking the integral over the crystal volume containing the selected N atoms we get

$$\psi_c^*(x)\hat{H}\psi_c(x) = E_c\psi_c^*(x)\psi_c(x) \quad \rightarrow \quad \int\psi_c^*(x)\hat{H}\psi_c(x)dx = E_c\int\psi_c^*(x)\psi(x)dx. \tag{8.11}$$

Since it is understood that $\psi_c(x)$ is (or can straightforwardly be) normalised we have

$$\int \psi_c^*(x)\hat{H}\psi_c(x)dx = E_c. \tag{8.12}$$

The calculation of the energies of the valence electrons is therefore traced back to evaluate integrals like the one appearing on the left-hand side of this equation. To this aim we should know $\psi_c(x)$.

The key idea is to represent $\psi_c(x)$ as a *linear combination of atomic wavefunctions* (LCAO approximation), each centred on a given lattice site. Since we are now addressing VB states, our physical intuition leads us to use a basis set made of atomic wavefunctions $\phi_a(x)$ corresponding to valence states which are filled when the atom is in its ground configuration. This is tantamount to writing

$$\psi_c(x) = \frac{1}{\sqrt{N}} \sum_{s=0}^{N} \exp(iksa) \, \phi_a(x - sa), \tag{8.13}$$

where k is the wavefunction of the electron matter wave. Thanks to the prefactor $1/\sqrt{N}$ such a wavefunction is properly normalised[6]. Since the acting potential $V_c(x)$ is periodic, we must verify that this guessed form of $\psi_c(x)$ *has the due mathematical property defining a Bloch function*. To this aim we proceed with the following calculation

$$\psi(x) = \frac{1}{\sqrt{N}} \sum_{s=0}^{N} \exp(iksa) \, \phi_a(x - sa)$$

$$= \exp(ikx) \, \frac{1}{\sqrt{N}} \sum_{s=0}^{N} \exp[-ik(x - sa)] \, \phi_a(x - sa)$$

$$= \exp(ikx) \, u(x), \tag{8.14}$$

---

[6] This holds true under the plain assumption that the atomic wavefunctions are also normalised.

where it is easy to realise that $u(x) = u(x + a)$ as requested (see section 6.3). The wavefunctions given by equation (8.14) are known as *Bloch sums*.

Let us then proceed with the calculation of the integral appearing in equation (8.12): by there inserting the Bloch sum given by equation (8.13) we get

$$
\begin{aligned}
E_c &= \frac{1}{\sqrt{N}} \frac{1}{\sqrt{N}} \sum_{s'=0}^{N} \sum_{s=0}^{N} \exp(-iks'a)\exp(iksa) \int \phi_a^*(x - s'a)\hat{H}\phi_a(x - sa)dx \\
&= \frac{1}{N} \sum_{s'=0}^{N} \sum_{s=0}^{N} \exp[ik(s - s')a] \int \phi_a^*(x - s'a)\hat{H}\phi_a(x - sa)dx \\
&= \frac{1}{N} \sum_{s'=0}^{N} \sum_{s=0}^{N} \exp[ik(s - s')a] \int \phi_a^*(x')\hat{H}\phi_a(x' - (s - s')a)dx',
\end{aligned}
\tag{8.15}
$$

where in the last step we performed the change of variable $x' = x - s'a$. This is mostly convenient since, for any value of s selected by the corresponding sum, the variable $s'' = s - s'$ assumes the values 0, 1, 2, 3, ... upon summing over s'. Therefore, we can write

$$
\begin{aligned}
E_c &= \frac{1}{N} \sum_{s'=0}^{N} \sum_{s=0}^{N} \exp[ik(s - s')a] \int \phi_a^*(x')\hat{H}\phi_a(x' - (s - s')a)dx' \\
&= \frac{1}{N} \sum_{s=0}^{N} \underbrace{\sum_{s''=0}^{N} \exp(iks''a) \int \phi_a^*(x')\hat{H}\phi_a(x' - s''a)dx'}_{\text{these values are all equal for any } s} \\
&= \frac{1}{N} N \sum_{s''=0}^{N} \exp(iks''a) \int \phi_a^*(x')\hat{H}\phi_a(x' - s''a)dx'.
\end{aligned}
\tag{8.16}
$$

It is now useful to split the sum over s'' by isolating the $s'' = 0$ term

$$
E_c = \int \phi_a^*(x')\hat{H}\phi_a(x')dx' + \sum_{s''=1}^{N} \exp(iks''a) \int \phi_a^*(x')\hat{H}\phi_a(x' - s''a)dx',
\tag{8.17}
$$

so as to separately treat the two terms that appear on the right-hand side of this equation.

In the first term both the wavefunction ϕ_a and its complex conjugate are calculated in the same position x'. Since they are atomic wavefunctions, they are pretty much localised in the neighbourhood of the ion core they belong to. Accordingly, if x' lies in an interstitial region, we can to a good approximation assume $\phi_a(x') = 0 = \phi_a^*(x')$. On the other hand, if x' is instead close to a core region, then we can locally assume $V_c(x') = V_a$, according to equation (8.8). This reasoning leads to

$$\int \phi_a^*(x')\hat{H}\phi_a(x')dx' \simeq \int \phi_a^*(x')\left[-\frac{\hbar^2}{2m}\nabla^2 + \hat{V}_a\right]\phi_a(x')dx'$$
$$= E_a \int \phi_a^*(x')\phi_a(x')dx' \qquad (8.18)$$
$$= E_a,$$

because of the already mentioned normalised character of the atomic wavefunction.

In the second term on the right-hand side of equation (8.17) the ϕ_a wavefunction and its complex conjugate are calculated in positions $(x' - s''a)$ and x', respectively. Even in this case we can take profit from their localised character: if the distance $s''a$ between such positions is large enough, then one out of the two wavefunctions turns out to be negligibly small. Or, equivalently: *the localised character of the basis set functions used in the linear combination allows us to assume that the integral contained in this term is nonzero only provided that x' and $(x' - s''a)$ fall near to two first next-neighbouring lattice positions, that is, only when $s'' = \pm1$.* Under this assumption, we are left to calculate just two integrals for $(x' - s''a) = x' \pm a$, something that can be done straightforwardly once the atomic wavefunctions are known. We remark that two such integrals are equal by symmetry and that even now we can set $V_c(x') = V_a$, thus handling them as in the previous case. More explicitly we have

$$\int \phi_a^*(x')\hat{H}\phi_a(x' \pm a)dx' \simeq \int \phi_a^*(x')\left[-\frac{\hbar^2}{2m}\nabla^2 + \hat{V}_a\right]\phi_a(x' \pm a)dx' = \gamma, \quad (8.19)$$

where γ is in principle calculated once and for all, since all ingredients are known[7]. This remarkable results leads to the following *dispersion relation $E_{VB}(k)$ for the valence band*

$$E_{VB}(k) = E_a + \gamma \exp(ika) + \gamma \exp(-ika) = E_a + 2\gamma \cos(ka). \qquad (8.20)$$

Typically it turns out that $\gamma > 0$.

A similar calculation can be executed for the conduction band, provided that a properly different set of atomic wavefunctions is used; more specifically, it is intuitive to assume that we must represent crystalline CB states (that are empty at $T = 0$ K) by atomic wavefunctions $\varphi_a(x)$ describing excited states. The corresponding CB integrals of the same kind as in equation (8.19) are typically negative and, therefore, we conveniently write the *dispersion relation $E_{CB}(k)$ for the conduction band* as

$$E_{CB}(k) = E_a' - 2\delta \cos(ka), \qquad (8.21)$$

---

[7] In practice, it is possible to adopt two different strategies: the most fundamental approach consists in the numerical calculation of the γ integral by using some explicit (exact or approximated) representation of the atomic wavefunctions [7–9]; alternatively, integrals like this one are fitted on a suitable set of available experimental data [1, 10]. The first approach is more fundamental and superior, but it could result in a heavy numerical effort; the second approach is much more computationally light, but it is limited in accuracy and transferability. In modern solid state physics the two approaches are, respectively, referred to as an *ab initio* or a semi-empirical theory.

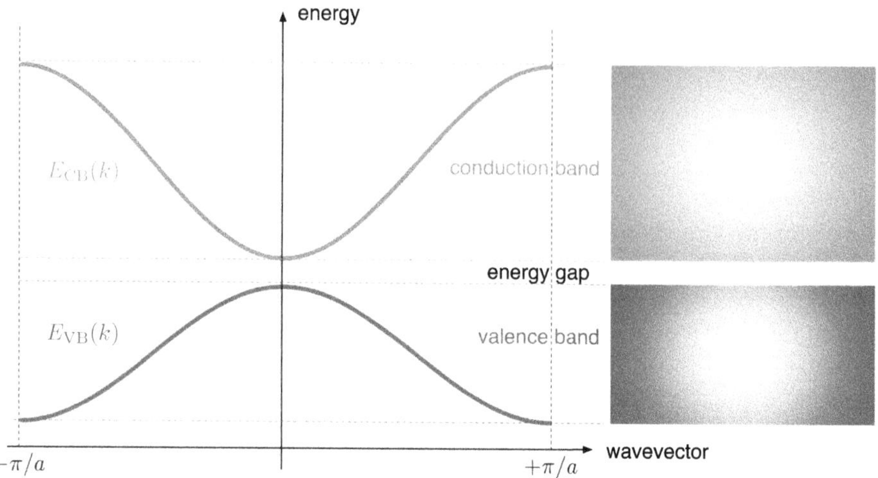

Figure 8.6. Valence (VB, blue) and conduction (CB, red) band of a monoatomic linear chain with lattice spacing a. Left: dispersion relations calculated by the tight-binding method and represented in the reduced zone scheme. Right: corresponding band diagram.

where $E_a' \neq E_a$ is the energy of the atomic excited states described by $\varphi_a(x)$ and it is understood that δ is the absolute value of the integral $\int \varphi_a^*(x') \hat{H} \varphi_a(x' \pm a) dx'$. It is interesting to observe that both VB and CB dispersion relations can be obtained directly from the exact result provided by the Kronig–Penney model through equation (6.33) by exploiting the LCAO approximation [11].

The theoretical procedure we have developed takes the name of *tight binding theory* (or method), where the locution 'tight-binding' underlines the fact that the crystalline wave function is built out of atomic wave functions. In this respect, the tight binding theory is dual to the nearly free electron one: the actual potential felt by electrons is considered pretty much atom-like (at least in the core regions). As in the case of phonons (see section 3.2), it is customary to adopt the *reduced zone scheme* for plotting the electron dispersion relations; in other words, the $E_{VB}(k)$ and $E_{CB}(k)$ curves are plotted for wavevector values within the first Brillouin zone, that is, for $k \in [-\pi/a, +\pi/a]$ as reported in figure 8.6.

8.2.2 Bands in real solids

The tight-binding theory can also be applied to three-dimensional solids [1, 10, 12, 13] in any possible crystal structure or chemical composition, as well as containing an arbitrary number of valence electrons. Although the theory is developed in the same way as described in the previous section, the resulting mathematics is definitely more complicated, as shown in full detail in appendix G: here we simply outline the procedure from a conceptual point of view and discuss a few paradigmatic applications.

The starting point is to write the crystalline wavefunction in a LCAO form by using a set of suitable localised orbitals $\{\varphi_{\alpha l b}(\mathbf{r}) = \varphi_\alpha(\mathbf{r} - \mathbf{R}_l - \mathbf{R}_b)\}$ centred on the different ion positions[8]; the label α stands for the full set of quantum numbers defining the corresponding state. In principle, such orbitals can be true atomic wavefunctions which, however, form a non-orthogonal basis set since orbitals centred on different lattice positions are not so; alternatively, an orthogonalisation procedure can be operated, as detailed in appendix G, still preserving the s-, p-, d-, $\cdots$ character of the atomic orbitals.

In order to set up a formalism naturally obeying the Bloch theorem, the following Bloch sums are defined

$$\varphi_{\alpha bk}^{\text{Bloch}}(\mathbf{r}) = \frac{1}{\sqrt{N}} \sum_{l} e^{i\mathbf{k}\cdot\mathbf{R}_l}\varphi_{\alpha}(\mathbf{r} - \mathbf{R}_l - \mathbf{R}_b), \tag{8.22}$$

where N is the number of unit cells contained in the crystal portion subject to the periodic Born–von Karman boundary condition. The electron wavefunction for the nth band is accordingly cast in the following LCAO form

$$\begin{aligned}
\psi_{n\mathbf{k}}(\mathbf{r}) &= \frac{1}{\sqrt{N_b}} \sum_{\alpha b} \tilde{B}_{n\alpha b}\varphi_{\alpha bk}^{\text{Bloch}}(\mathbf{r}) \\
&= \frac{1}{\sqrt{NN_b}} \sum_{\alpha lb} e^{i\mathbf{k}\cdot\mathbf{R}_l}\tilde{B}_{n\alpha b}\varphi_{\alpha}(\mathbf{r} - \mathbf{R}_l - \mathbf{R}_b) \\
&= \frac{1}{\sqrt{NN_b}} \sum_{\alpha lb} B_{n\alpha lb}(\mathbf{k})\varphi_{\alpha}(\mathbf{r} - \mathbf{R}_l - \mathbf{R}_b),
\end{aligned} \tag{8.23}$$

where N_b is the number of atoms in the lattice basis, $\tilde{B}_{n\alpha b}$ are the LCAO expansion coefficients, and for brevity we have set $B_{n\alpha lb}(\mathbf{k}) = \exp(i\mathbf{k} \cdot \mathbf{R}_l)\tilde{B}_{n\alpha b}$.

By inserting the trial wavefunction given in equation (8.23) into the eigenvalue equation[9]

$$\left(-\frac{\hbar^2}{2m_e}\nabla^2 + \hat{V}_{\text{cfp}}(\mathbf{r})\right)\psi_{n\mathbf{k}}(\mathbf{r}) = E_n(\mathbf{k})\psi_{n\mathbf{k}}(\mathbf{r}), \tag{8.24}$$

we are reduced to solving the secular problem

$$\sum_{\alpha lb}[H_{\alpha'l'b',\alpha lb} - E_n(\mathbf{k})S_{\alpha'l'b',\alpha lb}]B_{n\alpha lb}(\mathbf{k}) = 0, \tag{8.25}$$

where

$$H_{\alpha'l'b',\alpha lb} = \langle\varphi_{\alpha'}(\mathbf{r} - \mathbf{R}_{l'} - \mathbf{R}_{b'})| - \frac{\hbar^2}{2m_e}\nabla^2 + \hat{V}_{\text{cfp}}(\mathbf{r}) \mid \varphi_{\alpha}(\mathbf{r} - \mathbf{R}_l - \mathbf{R}_b)\rangle, \tag{8.26}$$

---

[8] We recall that, consistently with the notation used elsewhere in this Primer, the vectors $\mathbf{R}_l$ and $\mathbf{R}_b$, respectively, indicate the unit cell position and the ion position within the lattice basis.

[9] We are working in the single-particle approximation developed in section 1.4.1 and, therefore, each electron is considered under the action of a crystal field potential $V_{\text{cfp}}(\mathbf{r})$.

are the elements of the *tight-binding matrix* while

$$S_{\alpha'l'b',\alpha lb} = \langle \varphi_{\alpha'}(\mathbf{r} - \mathbf{R}_{l'} - \mathbf{R}_{b'}) \mid \varphi_{\alpha}(\mathbf{r} - \mathbf{R}_l - \mathbf{R}_b) \rangle \qquad (8.27)$$

are the elements of the *overlap matrix*: in the case where the φ_{α} orbitals are orthogonal, it reduces to the unit matrix.

When implementing this theory in practice, we must cope with a diagonalisation problem (see equation (8.25)) which must be repeated as many times as the number of $\mathbf{k}$ wavevectors we like to use for plotting the dispersion relations. Since a diagonalisation task corresponds to an overall computational effort showing a cubic scaling with the matrix rank, we need strategies to reduce as much as possible the number of tight-binding and overlap matrix elements to be evaluated (which is itself a very demanding task). To this aim, we first of all observe that the number of integrals entering the theory through equations (8.26) and (8.27) can be really huge, depending on the completeness of the basis set which has been chosen, as well as on the space extension of the orbitals there included[10]. Many different approaches can be followed and approximations can be adopted in order to reduce this number as much as possible, still keeping a meaningful description of the true physics. For instance we can choose a minimal basis set consisting of just a few orbitals for each ion, and/or we can restrict to close-neighbours interactions only, and/or we can agree to determine the surviving matrix entries through an empirical fitting procedure, instead of calculating them explicitly by some first principles method. A detailed discussion of these issues is reported in appendix G. All in all, the tight-binding method—in whatever flavour—has been extensively applied in solid state physics and crystalline bands have been accordingly calculated for most materials [10, 13, 14].

In order to illustrate a practical application of the tight-binding method, we will illustrate the case of silicon and gallium arsenide, which represent two prototypical materials of paramount importance for applications in the information technology. Silicon is a group-IV elemental semiconductor with diamond structure, while gallium arsenide is a III–V compound semiconductor with zincblend structure (see section 2.3.2). In both cases we have eight valence electrons per unit cell (four for each silicon atom and three or five for the gallium or arsenic atom, respectively). The dispersion relations will be calculated with a minimal tight-binding basis set consisting in just a few atomic orbitals whose energy is close to the energy of the highest VB states and the lowest CB ones: accordingly, we will focus our attention just on the crystalline states which fall near the forbidden gap. This is justified by the fact that only electrons in this energy window are affected by thermal excitation and, therefore, mainly contribute to transport and optical properties. Nevertheless, it is in principle conceivable the use of a larger tight-binding basis set which would enable the calculation of a wider spectrum of bands [14]. We will follow the standard convention to represent the $E_n(\mathbf{k})$ dispersions by choosing $\mathbf{k}$ to

---

[10] It is clear from equations (8.26) and (8.27) that the larger the delocalisation of the φ_{α} orbitals, the larger is the number of integrals to be evaluated.

vary along the high-symmetry directions of the irreducible part of the 1BZ (see figure 2.16).

In figure 8.7 the band structure of silicon is reported, where the energy scale has been chosen by setting its zero at the VB top: the four valence or conduction bands correspond to negative or positive energies, respectively. This is obviously purely conventional: the physics does not change if the band structure is rigidly shifted upwards or downwards by an arbitrary amount; for example, the optical properties of a crystal solely depend on energy differences between different electronic states which, of course, are not altered by a rigid displacement of the entire band structure. At $T = 0$ K each valence band is fully occupied[11] while all conduction bands are empty. This band structure suggests a number of meaningful physical features.

First of all, we observe that along some direction bands can be degenerate in energy. Let us for instance consider the valence bands along Λ and Δ directions: it seems we have three bands. On the other hand, the same valence bands appear in the number of four if we look at the Σ direction. Obviously, the number of bands cannot change by varying the direction of the wavevector in the reciprocal space, as confirmed by the careful analysis of what happens at the X point as far as the two valence bands higher in energy are concerned: coming from the Σ direction they end up overlapping, merging into two different but degenerate sequences of quantum states along the Δ direction. There are also cases in which degenerate states are found only at a specific $\mathbf{k}$-point, as at Γ (three degenerate valence bands) or X (two different pairs of degenerate bands). This adds a very practical motivation to use the combination of the band index and the wavevector to label the electron crystalline states.

Next, we can estimate $E_{\mathrm{gap}}^{\mathrm{Si}} = 1.1\,\mathrm{eV}$. This allows us to elaborate a more quantitative distinction between insulators and semiconductors than provided in

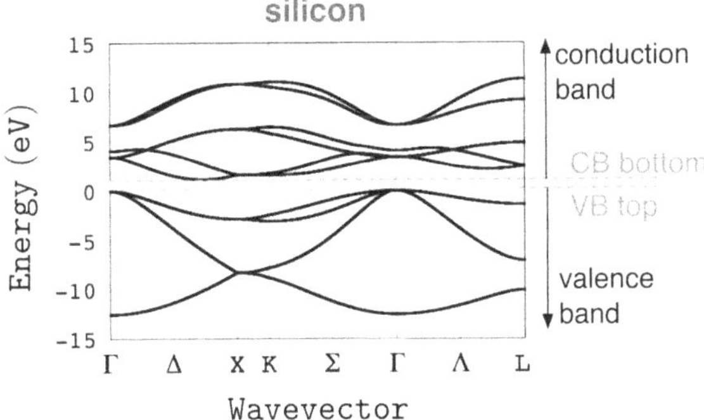

Figure 8.7. Band structure of silicon, as obtained by a tight-binding calculation (see appendix G for technical details). The energy gap is in yellow colour. Tight-binding parameters are taken from [15].

---

[11] As well as all other lower-energy bands not represented in figure 8.7. We remark once again that we are not representing the full band structure, but only its part close to the energy gap. In other words, figure 8.7 is the three-dimensional counterpart of figure 8.6.

section 8.1.3. More precisely, *it is customary to define semiconductor any non-metallic crystalline system* (i.e. any solid whose last occupied band at $T = 0$ K is fully filled) *with an energy gap of the order of ~1 eV*. According to this criterion, we classify silicon as a semiconductor[12].

Another interesting feature is that VB top and the CB bottom correspond to different wavevectors, that is, they are not vertically aligned. Therefore, silicon is said to be *an indirect gap semiconductor*. This intrinsic characteristic determines much of its application potential in technology; for example, this is the reason why crystalline silicon is not a good material for optoelectronics, as more extensively discussed in section 9.4.

Let us now turn to consider figure 8.8 where the band structure of gallium arsenide is reported, following the same conventions and notation as in the previous case. Even this material can be classified as a semiconductor since we have $E_{\text{gap}}^{\text{GaAs}} = 1.43$ eV. The most important difference with silicon is that gallium arsenide is a *direct gap semiconductor*: its BV top and CB bottom are both found at the Γ point of the Brillouin zone. On the other hand, the general topological features and degenerate character of the bands are quite similar to the case of silicon, to report the fact that the two materials have identical crystalline structure and very similar electronic structure.

We conclude this overview with an observation, full of consequences, about the two dispersions at the VB top and CB bottom. It is better to focus on the case of GaAs (neglecting, for simplicity, the degeneracy of its valence states), because it is more immediate to read, and to consider the band topology just near the Γ point. The similarity with the bands calculated through the simple one-dimensional model of section 8.2.1 (see figure 8.6) is really remarkable: this motivates us to continue the study of the bands in one dimension, confident that their main physical properties will be (at least qualitatively) similarly valid even in the case of real three-dimensional crystals.

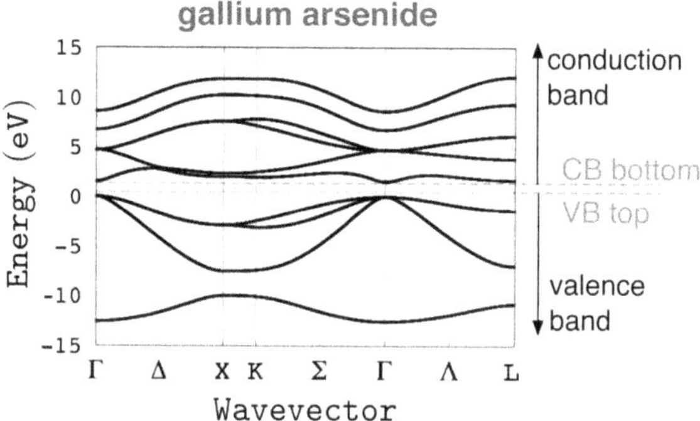

Figure 8.8. Band structure of gallium arsenide, as obtained by a tight-binding calculation (see appendix G for technical details). The energy gap is in yellow colour. Tight-binding parameters are taken from [15].

---

[12] To give a more direct feeling about the difference between a true insulator and a semiconductor, we observe that the native oxide of silicon, that is SiO_2, has an energy gap as large as $E_{\text{gap}}^{\text{SiO}_2} = 8.9$ eV, which corresponds to an experimentally observed excellent insulating behaviour at any temperature of physical interest.

In concluding this discussion on the band structure of real solids we remark that *the width of the energy gap is an intrinsic property of an insulating material* which, however, is affected by temperature and by any applied stress, as ultimately due to the variation of the interatomic distances with respect to the zero temperature equilibrium configuration. Some details on this issue are reported in appendix H.

8.3 General features of the band structure

Interestingly enough, the general features of the band structure discussed so far are similarly found both in the nearly free electron and tight-binding models, although they are based on somewhat opposite approaches to the problem. Therefore, *it is quite reasonable to expect that most of the physical implications deduced from the band structure predicted by them apply to real solids*, where an intermediate situation is likely expected to occur.

In order to explore the most relevant of such implications, we will investigate a one-dimensional model solid whose electrons are accommodated on band states described by dispersion relations $E = E_n(k)$. The corresponding matter waves will be described by $\psi_{nk}(x)$ wavefunctions. This choice of convenience will make the formalism simple and the underlying physics transparent, but is should be noted that the band theory can be developed at a more mathematically rigorous level, as found in [1, 2, 12].

8.3.1 Parabolic bands approximation

If we concentrate on considering only the band portions close to the VB top and the CB bottom, we recognise that in these regions the bands have a trend that, in very good approximation, can be described as parabolic (see figures 8.7 and 8.8 at the Γ point). Therefore, whenever we are interested in studying the physics of electrons in the proximity of the forbidden gap, we can meaningfully use the so called *parabolic bands approximation*.

This approximation is straightforwardly applied in the one-dimensional case by taking the limit of $k \to 0$ of equations (8.20) and (8.21)

$$
\begin{aligned}
\lim_{k \to 0} E_{\mathrm{VB}}(k) &= E_{\mathrm{a}} + 2\gamma\left[1 - \frac{(ka)^2}{2}\right] \\
\lim_{k \to 0} E_{\mathrm{CB}}(k) &= E_{\mathrm{a}}' - 2\delta\left[1 - \frac{(ka)^2}{2}\right].
\end{aligned}
\tag{8.28}
$$

Taking this limit is justified by the fact that at $k = 0$ we found both the VB top and the CB bottom, as shown in figure 8.6. A general remarkable feature is drawn: *the thermal excitation of electrons basically occurs within a parabolic band scheme.*

8.3.2 Electron dynamics

It is very convenient to address electron dynamics within a semi-classical scheme according to which: (i) *electron energy states are described quantum mechanically,* but (ii) *their equations of motion are classical.* This approximated scheme is trustworthy when one wants to study the motion of the electrons over a length scale much larger than the interatomic distances. This is, for instance, the relevant case of motion under the action of an externally applied and slowly varying electric field, that is an electric field which is practically constant over the length scale of interatomic lattice distances. On the other hand, the results of this approximation can hardly be extended to the case of nanostructures [16, 17], that is to solid state systems whose structural features display on the 10^{-9} m scale: here a full quantum theory of electron transport is needed, as detailed elsewhere [12, 17–19].

Let us at first consider the dynamics of an electron in the absence of an external electric field. If it is accommodated on the nth band, then its velocity $v_n(k)$ is given by the group velocity of the corresponding matter wave

$$v_n(k) = \frac{d\omega_n(k)}{dk} = \frac{1}{\hbar}\frac{dE_n(k)}{dk}, \tag{8.29}$$

where of course we used the $E_n(k) = \hbar\omega_n(k)$ matter–wave duality relation. This result is as much mathematically straightforward as it is dense in physical implications: *the velocity depends on electron wavevector and the kind of band on which the electron is accommodated.* In other words, electrons with identical wavevector, but belonging to different bands, have different velocity; equation (8.29) indicates that their velocity is calculated as the angular coefficient of the band tangent at the selected k-point. Similarly, an electron accommodated on a given band will have different velocity depending on its wavevector. Incidentally, this is one of the most important reasons why it is so necessary to know in detail the band structure of a solid: our ability to correctly describe the electron dynamics ultimately depends on it[13]. In any case, this result indeed represents a major conceptual difference with respect to a pure free electron theory, where just the same (average) drift or thermal velocity was attached to each electron. Another intriguing feature is that a VB electron and a CB electron with the same wavevector move in opposite directions, as clearly understood by inspection of figure 8.6: indeed an unexpected result which, as soon to be explained, has many implications in the theory of charge transport mediated by electrons subject to a periodic crystal field potential.

In the one-dimensional case the calculation of the electron velocity is easy: simply, we must combine equation (8.29) with equations (8.20) and (8.21) to obtain the results summarised in figure 8.9. Interestingly enough, this simple case illustrates quite an important new feature, namely: once a band is selected, *it is possible that an electron placed there has zero velocity* (this corresponds to a zone-centre or zone-boundary wavevector) *and it also moves in both possible directions, depending on its wavevector* (in a one-dimensional system, it can move both to the right and to the left).

---

[13] Soon we will meet a second very important reason of importance of bands, related to the concept of electron effective mass.

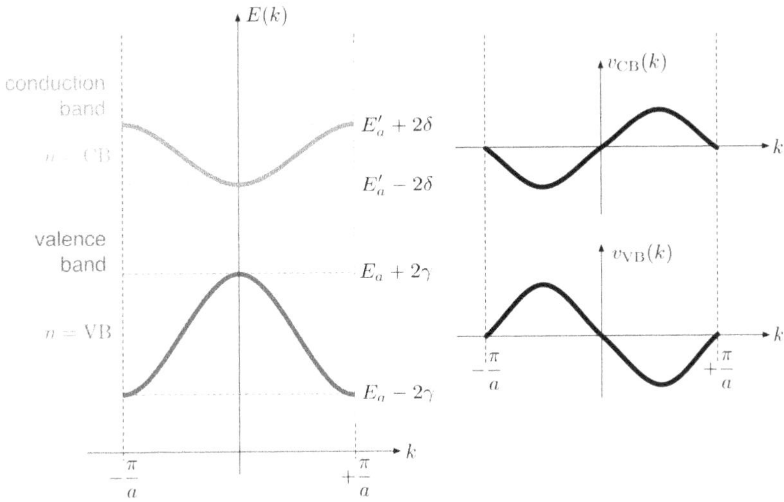

Figure 8.9. Left: the band structure of a one-dimensional crystal. Left: the velocity $v_{VB}(k)$ and $v_{CB}(k)$ for the valence and conduction band, respectively.

These results are naturally extended to the three-dimensional case by setting

$$\mathbf{v}_n(\mathbf{k}) = \frac{1}{\hbar}\, \nabla_{\mathbf{k}} E_n(\mathbf{k}), \tag{8.30}$$

where we understand that $\nabla_{\mathbf{k}} = (\partial/\partial k_x)\hat{\mathbf{i}} + (\partial/\partial k_y)\hat{\mathbf{j}} + (\partial/\partial k_z)\hat{\mathbf{k}}$. Once again, the important result is that *the velocity of a crystalline electron has magnitude, orientation and direction depending on its actual quantum state*, which is in turn characterised by the pair of band index and wavevector.

8.3.3 Electric field effects

Let us now apply a constant and uniform electric field $\mathbf{E}$ along a one-dimensional crystal. Within the semi-classical scheme a driving force

$$F = -e|\mathbf{E}| = \hbar\frac{dk}{dt}, \tag{8.31}$$

is calculated, governing the drift motion of the electron. The solution of this equation of motion is

$$k(t) = k_0 - \frac{e|\mathbf{E}|}{\hbar}t, \tag{8.32}$$

where k_0 is the electron wavevector at time $t = 0$, that is, when the electric field is turned on. By making use of equation (8.29) this result reflects in *a time-dependent electron velocity*

$$v_n(k, t) = \frac{1}{\hbar}\frac{dE_n}{dk(t)}, \tag{8.33}$$

suggesting the practical rule that *under the action of an electric field, the electron velocity at time t is calculated by evaluating the slope of band tangent at the point $k(t)$ given in equation (8.32)*. This result has a quite interesting implication, as we easily understand by considering the case of an electron in the valence band: under the action of the electric field, which we consider oriented to the left with no loss of generality, the wavevector varies linearly with time, assuming gradually increasing values and, therefore, it will sooner or later end up reaching the right edge of the 1BZ. However, given the crystalline periodicity, the $k = +\pi/a$ value defines a quantum state equivalent to the one described by $k' = k + G$ with $G = -2\pi/a$ a reciprocal lattice vector. This is tantamount to saying that the electron, once it reaches the right edge, is flipped back to a state corresponding the left one. Next, as time goes by, the electron will again assume increasing wavevector values, as before eventually reaching the right edge of 1BZ: here its wavevector will be flipped back once more. And so on ... This periodic back-and-forth variation of $k(t)$ in the Brillouin zone will continue as long as the electric field is present. This phenomenon is described by saying that *under the action of an electric field a band electron is subjected to Bloch oscillations*: their graphical rendering is reported in figure 8.10.

We remark that this result has been obtained by guessing the equation of motion (8.31) where no scattering phenomena appear, contrarily to what we discussed in section 7.1. This is of course a very crude approximation: in practice, it is very difficult to experimentally observe Bloch oscillations in real materials just because ionic motions and defects disturb the electron motion. Such oscillations are only detected at low temperature and in chemically pure systems, since the occurrence of such circumstances makes the periodic variation of $k(t)$ only marginally affected by electron–phonon and electron-defect scattering events or, equivalently, the friction term appearing in equation (7.3) to play a marginal role.

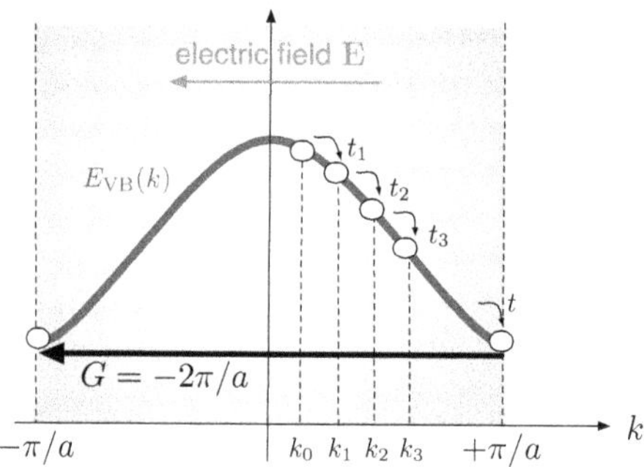

Figure 8.10. Graphical rendering of a Bloch oscillation. A valence band electron is initially located on the state with wavevector k_0. By applying a constant and uniform electric field **E** oriented to the left, the wavevector changes through the sequence $k_1 < k_2 < k_3 < \cdots$ which correspond to the values at time $t_1 < t_2 < t_3 < \cdots$ predicted by equation (8.32). Eventually the electron will occupy a state described by a wavevector falling just at the right edge of the first Brillouin zone, which is represented by the shaded area. Because of the crystalline periodicity, this $k = +\pi/a$ value defines a quantum state equivalent to the one described by $k' = k + G$ with $G = -2\pi/a$ and, therefore, we can state that the electron is effectively flipped back to the left edge of the zone. This oscillation lasts until the electric field is turned off.

Let us now derive a new result of major conceptual importance by considering *a completely filled band*: in this case we have

$$\sum_{k \in 1\mathrm{BZ}} v_n(k) = 0,\tag{8.34}$$

because for any electron with velocity $v_n(k)$ we have another one with velocity $-v_n(k)$, as easily understood by applying equation (8.29) to the band structure reported in figure 8.9. This implies that the current density vector[14] is zero, since it is proportional to the sum of the velocities of all available charge carriers (in the present case: all the electrons accommodated on the band). *The striking feature of the full-filling situation is that this conclusion applies even if the crystal is under the action of an external electric field*: although electrons are accelerated by the field as predicted by equation (8.33), their new velocities nevertheless correspond to values already found for that band, as summarised in figure 8.10, and therefore corresponding to occupied quantum states. In other words, equation (8.34) still holds under the action of an electric field. The conclusion is remarkable: *a completely filled band cannot conduct electricity*, since the current density vector associated with its electrons is in any case zero.

This outcome is really very important since it quantitatively justifies the classification presented in section 8.1.3 where we discussed the possible band filling situations and correspondingly associated the metallic or insulating behaviour. On the one hand: metals can transport electricity simply because they have the topmost band only partially occupied and, therefore, equation (8.34) does not hold or, equivalently, they have an inherently non-zero current density vector. On the other hand, the same argument applies to semiconductors to explain their (poor) conducting behaviour at finite temperature: because of the thermal excitation of electrons from the VB to the CB (see figure 8.4), a semiconductor material results as having *two partially filled bands* which provide a non zero density current vector. However, since $\mathbf{J}_q$ is proportional to the carrier density, we easily understand that such a material can only be a poor conductor: the Fermi–Dirac distribution function is significantly blurred by temperature only in the range $\sim k_{\mathrm{B}}T$ around the Fermi level and, therefore, just very few electrons are in fact promoted to the CB, while only an equally small number of empty states are created in the VB[15]. Finally, an insulator has such a large energy gap that the number of electrons actually promoted to the CB by thermal excitation is so small that their contribution to the current density vector is totally negligible.

---

[14] The charge current density vector $\mathbf{J}_q$ appearing in the Ohm law $\mathbf{J}_q = \sigma\mathbf{E}$ is defined as $\mathbf{J}_q = (\sum_i q_i \mathbf{v}_i)/V$, where the sum is over all the carriers with charge q_i and moving with velocity $\mathbf{v}_i$, while V is the system volume. In the case where all carriers have the same charge q and move with the same drift velocity $\mathbf{v}_d$, we recover the elementary definition $\mathbf{J}_q = nq\mathbf{v}_d$.

[15] This argument rigorously applies to intrinsic semiconductors only, that is, to undoped ones. Doping significantly alters this picture, as discussed in section 9.1.1.

8.3.4 Electrons and holes

The case of semiconductors, characterised by two partially filled topmost bands, is especially intriguing and deserves more attention.

Let us consider a semiconductor with N_{val} valence electrons and suppose that N_{CB} ones have been promoted to the conduction band by thermal excitation, while $N_{VB} = N_{val} - N_{CB}$ electrons are left in the valence band. If an electric field is applied, the N_{CB} electrons in CB provide a contribution $J_{q,CB}$ to the total charge current density as large as

$$J_{q,CB} \sim \sum_{k_{occ} \in CB} v_{CB}(k_{occ}), \tag{8.35}$$

where the sum runs over the wavevectors corresponding to the occupied CB states. Similarly, the VB electrons provide a contribution

$$J_{q,VB} \sim \sum_{k_{occ} \in VB} v_{VB}(k_{occ}), \tag{8.36}$$

where in this case the sum runs over the wavevectors of the VB states still occupied by the unexcited electrons. It is useful to rewrite this current density contribution in a different way

$$\begin{aligned} J_{q,VB} &\sim \sum_{k_{occ} \in VB} v_{VB}(k_{occ}) + \sum_{k_{empty} \in VB} v_{VB}(k_{empty}) - \sum_{k_{empty} \in VB} v_{VB}(k_{empty}) \\ &\sim \sum_{\forall k \in VB} v_{VB}(k) - \sum_{k_{empty} \in VB} v_{VB}(k_{empty}), \end{aligned} \tag{8.37}$$

where the $\sum_{k_{empty} \in VB} v_{VB}(k_{empty})$ term is obtained by summing over the wavevectors corresponding to the emptied states in the valence band. By applying the result 8.34, we understand that the first term on the right-hand side of the above equation is zero since it contains a sum over all the states of a totally filled band. This leads to the remarkable result

$$J_{q,VB} \sim - \sum_{k_{empty} \in VB} v_{VB}(k_{empty}), \tag{8.38}$$

which in physical terms is interpreted as if *the $J_{q,VB}$ contribution is given by a new type of charge carriers which occupy the VB states left empty by excited electrons.* These new carriers have opposite charge with respect to electrons, as proved by the minus sign appearing in equation (8.38), and they equal in number the thermally promoted electrons. The new carriers are referred to as *holes* and, because of their positive charge, their acceleration has the same direction as the applied electric field. The picture emerging from this analysis is really compelling: *the thermal excitation phenomena deeply affect the zero temperature insulating behaviour of semiconductors in that a double population of carriers is generated, respectively, electrons in CB and* *holes in VB.* Their number is just the same since temperature cannot of course generate or annihilate charge carriers.

The advantage of building a theory of electric transport based on two types of carriers is mainly practical and can be explained in a basic way by using a simple hydrodynamic analogue. Let us consider the case of a tube only partially filled with a fluid: by placing the tube in vertical position, we will observe an air bubble at the top. If we now overturn the tube, we observe a flow of fluid matter downwards that we could in principle describe by calculating the equations of motion of all the running molecules. An alternative, but much simpler, way is to describe the observed situation by calculating the trajectory of just the air bubble; in doing so we introduce a fictitious 'air bubble' particle (which is a 'hole' of molecules) with an interesting characteristic: subject to the action of the gravitational force, it always moves upwards as if it has a negative mass. The parallel with the case of electrons and holes in a semiconductor is really transparent: the valence band conduction is more easily described in terms of the motion of just a few holes, instead of considering all the unexcited electrons still accommodated there.

Thermally generated electrons (in CB) and holes (in VB) are named *intrinsic carriers* since their number is only determined by the combination of intrinsic materials properties (mainly, the energy gap) and the environment temperature. Their separate contributions to the total current density vector add up: as a matter of fact, electrons and holes do carry opposite charges but they also have opposite velocity at equal wavevector (see figure 8.9) and the current density is given by products (charge) × (velocity). We nevertheless remark that the magnitudes of $J_{q,VB}$ and $J_{q,CB}$ are different since electrons and holes have unalike absolute velocity at equal wavevector (the slope of the VB and CD bands are inherently different).

8.3.5 Effective mass

We now readdress the electron dynamics under the action of an external electric field **E**, preliminarily treated in section 8.3.3 where, however, we only considered the effect of the external force $-e|\mathbf{E}|$. Even if we still assume for the moment an idealised picture where no defects are present and ions are clamped at their ideal positions, we know that *an electron travelling within a crystal experiences as well the interactions with the lattice and with all the remaining electrons.* In short, we are faced with the problem of inserting the fundamental notion that *we are dealing with band electrons, rather than free electrons.*

By considering an infinitesimal time interval dt, the work done by external force $-e|\mathbf{E}|$ changes the electron energy by the amount

$$dE_n(k) = -e|\mathbf{E}|\, v_n(k)\, dt$$
$$= \frac{-e|\mathbf{E}|}{\hbar}\, \frac{dE_n(k)}{dk}\, dt, \tag{8.39}$$

leading to

$$\frac{dk}{dt} = \frac{-e|\mathbf{E}|}{\hbar}.$$

(8.40)

The rate of change of the electron velocity is

$$\frac{dv_n}{dt} = \frac{dv_n}{dk}\frac{dk}{dt} = \frac{1}{\hbar}\frac{d^2E_n(k)}{dk^2}\frac{-e|\mathbf{E}|}{\hbar},$$

(8.41)

so that its equation of motion is conveniently written in the form

$$-e|\mathbf{E}| = \left(\frac{1}{\hbar^2}\frac{d^2E_n(k)}{dk^2}\right)^{-1}\frac{dv_n}{dt},$$

(8.42)

which indeed represents a very suggestive way of putting things: the quantity

$$m_e^* = \left(\frac{1}{\hbar^2}\frac{d^2E_n(k)}{dk^2}\right)^{-1},$$

(8.43)

plays the role of an *electron effective mass*, fully incorporating any band feature through the second derivative of the $E_n(k)$ dispersion.

The conclusion is striking: *whenever we aim at investigating the dynamics of a band electron under the action of an external force, we can treat it as a free particle, provided that its rest mass is replaced by the effective mass m_e^*.* It is hard to underestimate the importance of this unexpected result: we have traced the overwhelmingly complicated dynamical problem of an electron simultaneously subject to an external applied field and to the crystalline potential to that of a free particle just subject to the external force, at the only price of replacing m_e with m_e^*. This is yet another issue supporting the fundamental role played by the band structure in the theory of the crystalline solid state.

Since the value of m_e^* depends on the dispersion $E_n(k)$ we understand that the effective mass of an electron is not unique, but it depends on the specific quantum state on which it is accommodated. In other words, while all electrons have the same rest mass, they have different m_e^* values, depending (i) on the actual band they occupy and (ii) on the wavevector describing their crystalline state, as reported in figure 8.11 in the case of a one-dimensional crystal. Therefore, within the same system we will observe that *electrons on different quantum states will be differently accelerated by the applied electric field.* Accurate calculations of the bands in real semiconductors [20–23] show that the values of m_e^* in typical semiconductors are smaller than the rest mass value by a factor in between 10 and 100. The effective mass can also assume negative values, whenever the band assumes a downward curvature: this implies that an electron is accelerated along the direction of the electric field (contrary to what happens for a free electron). There are also states for which the effective mass is infinitely large: they correspond to the flex points of the $E = E_n(k)$ dispersion curve. When an electron is accommodated on such states, an electric field—regardless of its strength—cannot accelerate it. All features so far discussed also apply to holes and, therefore we introduce the more refined formalism

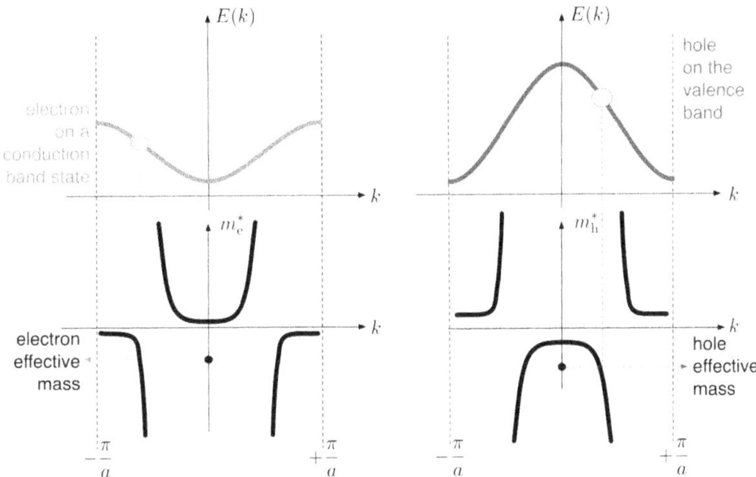

Figure 8.11. Left: dispersion relation (top) and effective mass (bottom) for a conduction band electron (filled dot). Right: dispersion relation (top) and effective mass (bottom) for a valence band hole (empty circle).

$$m_e^* = \left(\frac{1}{\hbar^2}\frac{d^2 E_{CB}(k)}{dk^2}\right)^{-1} \qquad m_h^* = \left(\frac{1}{\hbar^2}\frac{d^2 E_{VB}(k)}{dk^2}\right)^{-1}, \tag{8.44}$$

for the effective mass of a CB electron and a VB hole, respectively.

The extension to the case of three-dimensional crystals is straightforward, but it still requires some attention to mathematical details because of the term containing the second derivative of the band energy. More precisely, in real materials we calculate a *tensorial effective mass for both electrons and holes* as

$$m_{e,ij}^* = \left(\frac{1}{\hbar^2}\frac{d^2 E_{CB}(k)}{dk_i dk_j}\right)^{-1} \qquad m_{h,ij}^* = \left(\frac{1}{\hbar^2}\frac{d^2 E_{VB}(k)}{dk_i dk_j}\right)^{-1}, \tag{8.45}$$

where i and j are the Cartesian indices. The tensorial character has an important physical consequence: by applying an electric field $\mathbf{E} = (E_x, E_y, E_z)$ the equation of motion of, say, an electron must be written in the following matrix form

$$-e\begin{pmatrix} E_x \\ E_y \\ E_z \end{pmatrix} = \begin{pmatrix} m_{e,xx}^* & m_{e,xy}^* & m_{e,xz}^* \\ m_{e,yx}^* & m_{e,yy}^* & m_{e,yz}^* \\ m_{e,zx}^* & m_{e,zy}^* & m_{e,zz}^* \end{pmatrix}\begin{pmatrix} a_x \\ a_y \\ a_z \end{pmatrix}, \tag{8.46}$$

where $\mathbf{a} = (a_x, a_y, a_z)$ is of course its acceleration. This implies that the net force along the ith direction is

$$-eE_i = m_{e,ix}a_x + m_{e,iy}a_y + m_{e,iz}a_z, \tag{8.47}$$

or, equivalently, the acceleration is not necessarily along the direction of the external field, as the case of a free electron. It is quite intuitive to understand that this property, closely related to the underlying band structure, largely affects the charge transport properties of a real crystal. The very same conclusions are drawn for holes.

The practical calculation of m_e^* and m_h^* is made easy by the parabolic band approximation introduced in section 8.3.1. To be concrete, let us consider the two paradigmatic cases of silicon and gallium arsenide: figures 8.7 and 8.8 suggest that within the energy window limited to just above the CB bottom and just below VB top their bands can be parabolically approximated. Once again: these are the only crystalline states which are interested by the thermal excitation phenomena and, therefore, this is the actual energy interval where electrons and holes are, respectively, promoted to and generated in conducting states. Two important details need nevertheless to be duly taken into account: (i) we must include in our model the existence of two almost degenerate valence bands; and (ii) the silicon CB is actually not symmetric around its minimum. Including these features defines the minimum level of sophistication to elaborate a meaningful picture. The true occurrence of two valence bands in this parabolic model implies that we should expect the existence of *two different kinds of holes*, accommodated on bands with different curvature or, equivalently, *with different effective mass*: the higher (smaller) curvature corresponds to the smaller (higher) value of m_h^*. It is therefore customary to distinguish between *light holes* and *heavy holes*, respectively.

The model is implemented in practice as shown in figure 8.12 where the bands are represented as

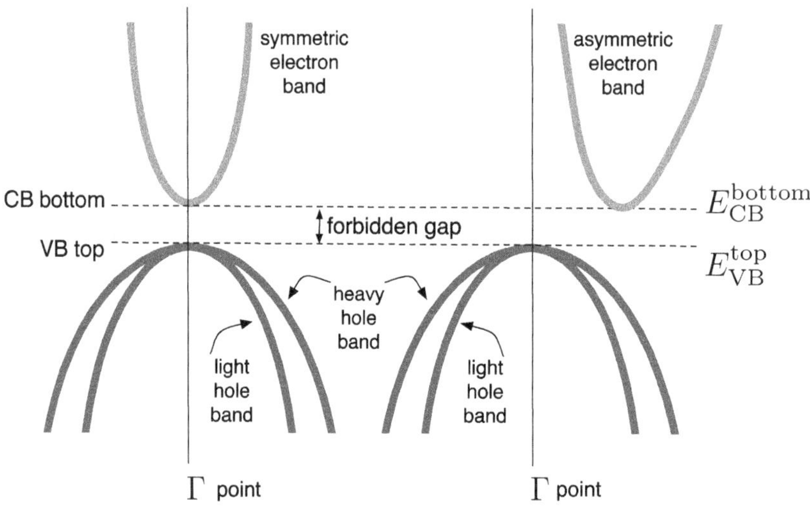

Figure 8.12. Parabolic band approximation applied to the case of gallium arsenide (left), a direct gap semiconductor, and silicon (right), an indirect gap semiconductor. The Γ point lies at the centre of the 1BZ.

Table 8.1. The energy gap E_{gap} (eV units) and the electron m_{e}^* and hole m_{h}^* effective mass of some semiconductors. In a few cases distinction is made between the longitudinal $m_{\text{e},\parallel}^*$ and transverse $m_{\text{e},\perp}^*$ effective mass of electrons, as well as between the light hole m_{lh}^* and heavy hole m_{hh}^* effective mass. All effective masses are in units of the electron rest mass m_{e}.

	Si	Ge	AlP	AlAs	GaP	GaAs	InP	InAs	CdTe	CdSe	GaN
E_{gap}	1.11	0.66	2.43	2.16	2.26	1.43	1.35	0.36	1.50	1.73	3.40
m_{e}^*			0.13	0.50	0.13	0.067	0.07	0.028	0.11	0.13	0.20
$m_{\text{e},\parallel}^*$	0.98	1.58									
$m_{\text{e},\perp}^*$	0.19	0.08									
m_{h}^*					0.67		0.40	0.33	0.35	0.40	0.80
m_{lh}^*	0.16	0.04		0.49		0.12					
m_{hh}^*	0.50	0.30		1.06		0.50					

$$E_{\text{CB}}(k) = E_{\text{CB}}^{\text{bottom}} + \frac{\hbar^2 k^2}{2m_{\text{e}}^*} + f(k)$$

$$E_{\text{VB}}^{\text{lh}}(k) = E_{\text{VB}}^{\text{top}} - \frac{\hbar^2 k^2}{2m_{\text{lh}}^*} \tag{8.48}$$

$$E_{\text{VB}}^{\text{hh}}(k) = E_{\text{VB}}^{\text{top}} - \frac{\hbar^2 k^2}{2m_{\text{hh}}^*},$$

where $f(k)$ is a suitable function correcting for the possibly non-perfect parabolic curvature of the conduction band (in gallium arsenide this term is missing), while m_{lh}^* and m_{hh}^* are the light hole and heavy hole effective mass, respectively.

The non perfect symmetry of the conduction band at its minimum (typical of semiconductors with indirect gap) is properly taken into account by calculating two different values of the electron effective mass, according to the actual reciprocal space direction chosen to measure the band curvature. Let us consider the silicon case reported in figure 8.7: the minimum of the CB is found along the Δ direction, linking the Γ and X points of the 1BZ. The standard notation is to name 'longitudinal' the effective mass value obtained by considering the band curvature along Δ and to use the symbol $m_{\text{e},\parallel}^*$ to indicate it; on the other hand, the value calculated through the curvature found along a direction normal to Δ is in turn named the 'transverse' effective mass with symbol $m_{\text{e},\perp}^*$. In table 8.1 we report the effective mass values for electrons and holes as calculated for some semiconductor materials.

8.4 Experimental determination of the band structure

Many experimental techniques used to measure the band structures in solids use magnetic fields to enable the Landau quantisation of electronic orbits. Other methodologies are instead optical: they measure either photon absorption or reflection phenomena occurring during the interaction between a solid specimen and some electromagnetic probe. Since the detailed description of such magnetic or optical techniques can hardly be exploited by means of such an elementary theory of the solid state as presented in this Primer, we limit ourselves to outlining just the *photoelectron spectroscopy* (PES) technique, which is (at least conceptually) very simple, while directing the interested reader to other textbooks [1, 2, 22, 24] for a more thorough presentation.

PES is the solid state counterpart of the photoelectric effect [7, 8] in that an electron initially located on an occupied crystalline band state with energy E_n is promoted, upon absorption of a photon with energy $\hbar\omega$, to an empty state with energy E_{empty} above the vacuum level E_{vacuum} of the investigated material[16]. The excited electron eventually escapes from the solid, moving as a free particle with kinetic energy E_{out} simply given by the balance

$$
\begin{aligned}
E_{\text{out}} &= \underbrace{E_{\text{empty}}} - E_{\text{vacuum}} \\
&= (E_n + \hbar\omega) - E_{\text{vacuum}}.
\end{aligned}
\tag{8.49}
$$

While the photon energy $\hbar\omega$ is controlled by the experimental setup, a direct measurement of E_{out} allows us to determine the energy E_n of the crystalline band state (the vacuum level is known). The intensity of the PES signal is proportional to the number of electrons occupying the initial state or, equivalently, to the density of electronic states at energy E_n.

From this simplified explanation it is deduced that PES just provides the band diagram reported in figure 8.2 left, but not the dispersion relations $E = E_n(\mathbf{k})$ which indeed require the knowledge of both the wavevectors and the corresponding energies. This limitation is overcome by a more advanced version of this experimental technique, known as angle-resolved photoelectron spectroscopy (ARPES). We observe that, in general, the photoemission process must conserve both energy, as reported in equation (8.49), and momentum. In particular, it is proved [1] that parallel to the crystal surface the electron momentum obeys the following conservation law

$$
\hbar\mathbf{k}_n^{(\text{surf})} = \hbar\mathbf{k}_{\text{empty}}^{(\text{surf})} + \hbar\mathbf{G}^{(\text{surf})},
\tag{8.50}
$$

where $\mathbf{G}^{(\text{surf})}$ is a reciprocal lattice vector of the surface[17], while $\hbar\mathbf{k}_n^{(\text{surf})}$ and $\hbar\mathbf{k}_{\text{empty}}^{(\text{surf})}$ are the surface projections of the initial and final electron momenta. Therefore, if we measure the angle θ at which the electron is emitted[18], then we can write [1]

$$
\hbar|\mathbf{k}_{\text{empty}}^{(\text{surf})}| = (2m_e E_{\text{out}})^{1/2} \sin\theta,
\tag{8.51}
$$

---

[16] The vacuum level is the energy of a free electron outside a solid material. The minimum work that must be spent on promoting a crystalline electron to the vacuum level is commonly referred to as the work function of that material.

[17] Crystalline surfaces have their own two-dimensional crystallography which, similarly to the bulk one, is described in terms of direct and reciprocal lattice vectors [25–27].

[18] More precisely, θ is the angle formed by the crystal surface normal and the emission direction of the electron.

which in turn provides $\hbar\mathbf{k}_n$ through equation (8.50). Excluding the details of the experimental apparatus and protocol of measurement, this is enough to understand that equations (8.49) and (8.51) are enough to reconstruct the E_n *versus* $\mathbf{k}$ relationship.

8.5 Other methods to calculate the band structure

The accurate determination of the band structure of crystalline solids is an art extensively developed in the second half of the XXth century, in parallel with the development of increasingly powerful digital computers: advances in theoretical methods and numerical techniques have been tightly interlaced and mutually beneficial. A number of different methods have been set up, the tight-binding approach—here privileged for pedagogical reasons—being just one among many others. The mathematics of such methods represents a very technical issue of the solid state theory, which is fully exploited elsewhere [2, 12, 13]. Here we limit ourselves to outlining some general features.

The conceptual framework is that defined by the adiabatic, frozen core, non-magnetic, and single-particle approximations. The Schrödinger problem to be solved is provided by equation (1.22), where the local potential $V_{\text{cfp}}(\mathbf{r})$ acting on the electron is typically determined by a self-consistent procedure. Also, the single-particle wave-function must have the form of a Bloch wavefunction. Finally, core and valence wavefunctions have a remarkably different space dependence: they both display strong atomic-like oscillations near each ion, while in the interstitial regions core wave-functions are vanishingly small and valence ones are instead slowly-varying plane-wave like. This ultimately dictates that core and valence wavefunctions are orthogonal.

A first class of band structure methods is based on the idea of representing the crystalline states as Bloch wavefunctions independent of the energy of the valence state of interest. This is the case of the tight-binding method where atomic orbitals are used to create a Bloch state; the same concept is also adopted by using *orthogonalised plane waves* (OPWs), where the orthogonality between core and valence states is enforced by constructing the valence Bloch state by means of plane waves suitably orthogonalised to core states.

A different choice is operated in the so called *cellular methods*, where just a single Wigner–Seitz cell is considered, where the single-electron potential is approximated within a sphere centred on each lattice site so as to describe an isolated ion (and, therefore, is spherically symmetric), while outside the sphere is taken to be zero. The radial Schrödinger equation for such a *muffin tin potential* is then solved and these solutions are used as a basis set for the crystalline wavefunctions. More specifically, a set of *augmented plane waves* (APWs) are generated, consisting in a combination of a spherical wave (describing the electron state within the Wigner–Seitz cell) and a plane wave (describing the electron state in the interstitial regions). Suitable boundary conditions are applied at the borders of the Wigner–Seitz cell.

Finally, a completely different strategy is to eliminate the core states from our electronic structure problem. This can be done by replacing the real crystalline potential with a suitable *pseudo-potential* providing the accurate determination of both valence and conduction band states. The key idea is to replace the true strong potential with a much softer one within the core regions. This replacement is operated under the constraint that the corresponding pseudo-wavefunctions do represent the genuine crystalline wavefunctions outside the core region, while no care is played within it. Also, the pseudo-wavefunctions must be norm-conserving. The pseudo-potential calculations are typically performed by OPWs expansions.

References

[1] Singleton J 2001 *Band Theory and Electronic Properties of Solids* (Oxford: Oxford University Press)

[2] Ashcroft N W and Mermin N D 1976 *Solid State Physics* (London: Holt-Saunders)

[3] Kittel C 1996 *Introduction to Solid State Physics* 7th edn (Hoboken, NJ: Wiley)

[4] Miller D A B 2008 *Quantum Mechanics for Scientists and Engineers* (New York: Cambridge University Press)

[5] Griffiths D J and Schroeter D F 2018 *Introduction to Quantum Mechanics* 3rd edn (Cambridge: Cambridge University Press)

[6] Eisberg R and Resnick R 1985 *Quantum Physics of Atoms, Molecules, Solids, Nuclei, and Particles* 2nd edn (Hoboken, NJ: Wiley)

[7] Colombo L 2019 *Atomic and Molecular Physics: A Primer* (Bristol: IOP Publishing)

[8] Demtröder W 2010 *Atoms, Molecules and Photons* (Berlin: Springer)

[9] Bransden B H and Joachain C J 1983 *Physics of Atoms and Molecules* (Harlow: Addison-Wesley)

[10] Harrison W A 1980 *Electronic Structure and the Properties of Solids* (New York: Dover)

[11] Morrison M A, Estle T L and Lane N F 1976 *Quantum States of Atoms, Molecules, and Solids* (Upper Saddle River, NJ: Prentice-Hall)

[12] Grosso G and Pastori Parravicini G 2014 *Solid State Physics* 2nd edn (Oxford: Academic)

[13] Bassani F and Pastori-Parravicini G 1975 *Electronic States and Optical Transitions in Solids* (Oxford: Pergamon)

[14] Papacostantopoulos D A 2015 *Handbook of the Band Structure of Elemental Solids* 2nd edn (New York: Springer)

[15] Vogl P, Hjilmarsson H H and Dow J D J 1983 *Phys. Chem. Solids* **44** 365

[16] Davies J H 1998 *The Physics of Low-dimensional Semiconductors* (Cambridge: Cambridge University Press)

[17] Di Ventra M 2008 *Electrical Transport in Nanoscale Systems* (Cambridge: Cambridge University Press)

[18] Ridley B K 1988 *Quantum Processes in Semiconductors* (Oxford: Oxford University Press)

[19] Heikkilä T T 2013 *The Physics of Nanoelectronics* (Oxford: Oxford University Press)

[20] Grundmann M 2010 *The Physics of Semiconductors* (Heidelberg: Springer)

[21] Cardona M and Yu P Y 2010 *Fundamentals of Semiconductors* (Heidelberg: Springer)

[22] Seeger K 1989 *Semiconductor Physics* (Heidelberg: Springer)

[23] Balkanski M and Wallis R F 1989 *Semiconductor Physics and Applications* (Oxford: Oxford University Press)

[24] Fox M 2001 *Optical Properties of Solids* (Oxford: Oxford University Press)

[25] Prutton M 1983 *Surface Physics* (Oxford: Oxford University Press)

[26] Zangwill A 1988 *Physics at Surfaces* (Cambridge: Cambridge University Press)

[27] Bechstedt F 2003 *Principles of Surface Physics* (Berlin: Springer)

IOP Publishing

Solid State Physics
A primer
Luciano Colombo

Chapter 9

Semiconductors

Syllabus—The physics of charge carriers in a semiconductor and related transport phenomena is developed considering both the case of a drift current generated by electric fields and the case of a diffusion current generated by concentration gradients. In this context we introduce the concept of mobility, relaxation time and conductivity, which have paramount importance in semiconductor physics and engineering. We then move on to study the carrier statistics in a semiconductor at thermal equilibrium, eventually leading to the general laws that predict the concentration of charge carriers in intrinsic and extrinsic semiconductors. Next, we outline the non-equilibrium processes involving the generation and recombination of carriers and we obtain the most general form for the continuity equation for the charge current. This chapter ends with a phenomenological description of the optical properties of semiconductors, mainly developed through the concept of inter-band transitions.

9.1 Some preliminary concepts

In our presentation of elementary solid state physics, semiconductors play a twofold pedagogical and applicative role: on the one hand, they represent a paradigmatic playground for developing a microscopic theory of charge transport in non-metallic systems (that is in systems where the free electron model cannot be used as effectively as in metals); on the other hand, they have a fundamental role in modern information and communication (nano)technologies. For these reasons they deserve the careful treatment here developed where both issues are addressed, together with a statistical treatment of charge carrier populations either in equilibrium condition or out of equilibrium.

Throughout this chapter we assume that the band structure of a semiconductor is known, as obtained by applying anyone of the methods developed in chapter 8. Therefore, concepts like the valence (VB) and conduction (CB) band, the energy gap and the effective mass of charge carriers will be extensively used.

9.1.1 Doping

We will distinguish between *intrinsic and extrinsic semiconductors*, provided that a pristine sample or a doped one (see section 2.5.1) is, respectively, considered. Basically, they differ in that only thermally excited carriers from the VB to the CB are available in pristine materials (hereafter referred to as intrinsic carriers), while additional electrons in CB and holes in VB are added in doped semiconductors (hereafter referred to as extrinsic carriers). This is of course the result of doping, that is, the alteration of the chemistry of the pristine material by insertion of *donor or acceptor impurities*: in the first case, the impurity atoms carry an excess of electrons with respect to the pristine material, while in the second case they lack of electrons or, equivalently, they carry an excess of holes with respect to the intrinsic population. In engineering applications doped semiconductors are mainly used.

Given the fundamental role played in semiconductor physics by doping, a preliminary question should be addressed, namely: *what is the effect of doping on the underlying band structure of the material?* The answer to this question is not at all trivial, since it requires the extensive use of the quantum mechanical perturbation theory: the presence of a dopant defect is described as a perturbation to the ideal case of perfect crystal and the way such a perturbation affects the band structure is accordingly calculated[1]. The graphical rendering of this concept is reported in figure 9.1, which will soon be explained in full detail.

In principle, by using perturbative methods it is possible to rigorously calculate the effects of dopant species both on the electron energy levels (and, therefore, the

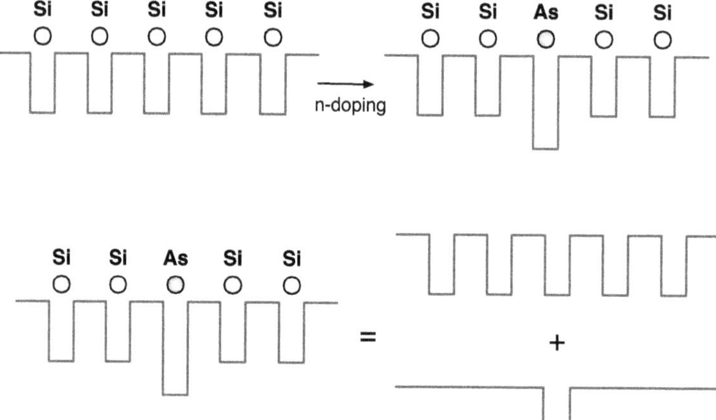

Figure 9.1. Graphical rendering of the effect of n-doping on the crystal field potential (Kronig–Penney model) in silicon.

---

[1] The tacit assumption is that the density of dopants is not too high, so we can still think in terms of a pristine material—with its own band structure—and a perturbation due to the dirty chemistry. There are cases of extremely high doping levels for which this assumption is actually not valid and, therefore, a different theoretical approach should be followed. This topic is treated in specialised textbooks [1–4].

new band structure) and on the wavefunctions describing the corresponding quantum states (and, therefore, their localised or extended character). This is a heavy indeed approach and we rather prefer a semi-quantitative argument based on the Kronig–Penney model (see section 6.4) which, despite its simplified character, nevertheless leads to the correct framing of the problem.

Let us consider the case of an ideal one-dimensional semiconductor: the crystal field potential felt by an electron in conduction band is represented in figure 9.1, top left. To make things concrete, we suppose that the semiconductor is silicon. Let us now suppose that, following a doping process, a single lattice site has been occupied by an arsenic atom: this situation is usually referred to as *n-doping*, where n stands for 'negative' as justified by the observation that arsenic injects an excess electron (which is a negative charge) into the pristine material. To a very good approximation, we can describe the variation of potential due to the Si → As substitution as a variation of the depth of the potential well at the lattice site where this substitution has occurred. Since the atomic number of arsenic is greater than that of silicon [5], we can qualitatively predict that the variation consists in a depth increase of the well: the As nucleus more strongly attracts an electron than a silicon nucleus does. This is graphically rendered in figure 9.1, top right.

After doping, the resulting state of affairs can be described as the sum of an 'ideal situation' and a 'perturbation': the 'ideal situation' corresponds to a potential profile typical of a non-doped crystal, while the 'perturbation' only consists in a depth variation of the potential well at the lattice site interested by the Si → As substitution. Graphically, the situation is described in figure 9.1, bottom right. The energy levels of the doped crystal are ultimately obtained by adding to the normal band structure of the ideal crystal the discrete levels confined in the additional potential well generated by doping. In practice, it is assumed that the additional well is so shallow that only one state can be there confined. Moreover, the explicit perturbative calculation provides evidence that the new energy level falls within the forbidden gap of the intrinsic semiconductor. In particular, in the case of n-doping this level always falls near the CB bottom. In summary, this simple model allows us to state that *the effect of a single Si → As replacement on the silicon band structure is equivalent to the addition of a single energy level in the forbidden gap, just below the conduction band.* As befits electron states of a potential well, also the new state generated by the doping is confined at the site interested by the Si → As replacement: for this reason, it is called *impurity level*. Equivalently, we can state that the matter wave describing an electron accommodated on the impurity level is mostly confined nearby the dopant site.

To complete the picture, we must clarify an important issue: so far it was supposed to insert just one dopant atom into the host lattice. In fact our model is still valid even if many dopant impurities are placed at as many lattice sites, *as long as their density is not too high*. If the dopants are sufficiently diluted, they are on average at a fairly large distance from each other. Because of their confined character, the new impurity levels do not interact[2]: each will be confined in its

---

[2] Equivalently: the tails of the matter waves describing the impurity states have a negligibly small overlap.

own well. Since all the wells are just the same (all dopants are of the same chemical species), each of them contributes to the band structure with the same impurity level as all the others. In brief: if they have been inserted N_D doping elements and a low-dilution situation is in fact verified, then the band structure will have only one N_D-fold degenerate impurity level.

The model we have developed can be straightforwardly extended to the case of a dopant lacking a valence electron with respect to the replaced silicon atom: a situation that is reported as *p-doping*. Let us consider, a Si → B substitution: since boron has a smaller atomic number than silicon [5], the corresponding potential well is shallower. The situation is summarised in figure 9.2. There should be noted an important difference with respect to the previous case: the term describing the 'perturbation' seems to represent a potential barrier, rather than a well. However, we remark that in this case we are discussing the motion of the carriers in the valence band: therefore, *we are allowed to look at the barrier as a potential well for the holes accommodated in the VB*. All the previous results also apply to p-doping: the impurity levels are degenerate in energy (in the low dilution limit) and they lie in the forbidden gap, but this time just above the top of the VB (as proved by an explicit perturbative calculation).

Finally, in figure 9.3 we sketch the band structure of a semiconductor in the intrinsic case as well as in the two different cases of n- and p-doping. The impurity levels for n-doping are called *donor levels* because at $T = 0$ K they are fully occupied by electrons that, upon heating, can be easily excited to the CB. In other words, since

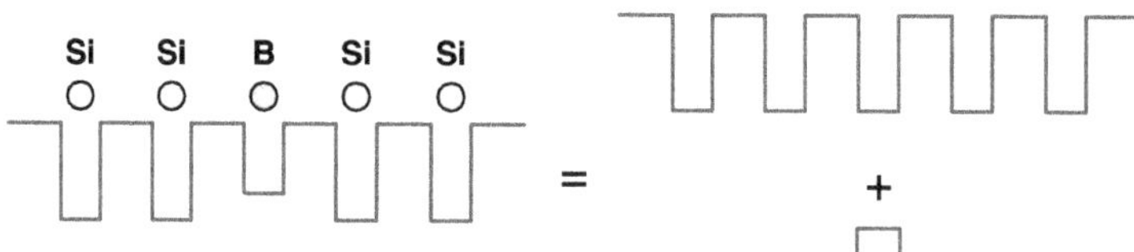

Figure 9.2. Graphical rendering of the effect of p-doping on the crystal field potential (Kronig–Penney model) in silicon.

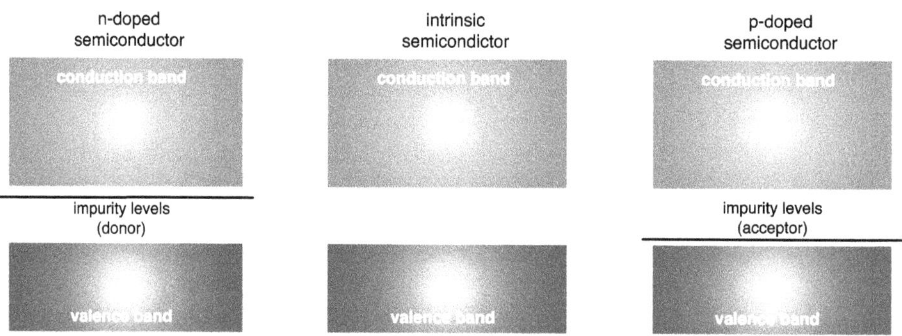

Figure 9.3. The band structure of an n-doped (left), intrinsic (middle), and p-doped (right) semiconductor. Impurity levels are marked by a thick black line: they lie just below the CB minimum and just above the VB maximum in the case of n-doping and p-doping, respectively.

the donor levels lie just below the CB minimum, the thermal excitation mechanism is much more efficient for them, than for the intrinsic electrons in the VB. The levels introduced by p-doping are, instead, called *acceptor levels* because at $T = 0$ K they are empty and become populated by thermal excitation of electrons (eventually trapped there) from the VB.

Nowadays, the exact position of donor and acceptor levels in the gap of any semiconductor is precisely known [3] and, therefore, it is possible to engineer the band structure according the specific application of interest. In practice, by selecting the kind and density of dopant impurities it is possible to have a (doped) semi-conductor with any density of holes in VB and/or electrons in CB as required by the target application.

9.1.2 Density of states for the conduction and valence bands

For further convenience we elaborate an explicit expression of the eDOS (previously introduced in section 7.3.1 only for metals) for a specific semiconductor band n. To this aim, we will combine the parabolic bands approximation outlined in section 8.3.1 with the concept of effective mass introduced in section 8.3.5.

Let us consider a generic band $E_n(\mathbf{k})$ and indicate by $G_n(E)dE$ the corresponding number of electron states with energy in the interval $[E, E + dE]$. If we know the single-electron energies, then we can calculate the eDOS $G_n(E)$ as

$$G_n(E) = \sum_{\mathbf{k} \in \mathrm{1BZ}} \delta(E - E_n(\mathbf{k})) = \frac{V}{(2\pi)^3} \int_{\mathbf{k} \in \mathrm{1BZ}} \delta(E - E_n(\mathbf{k}))\, d\mathbf{k}, \qquad (9.1)$$

where the discrete sum takes into account that only a set of discrete $\mathbf{k}$-points are rigorously allowed in the 1BZ, while the integral expression is a good approximation valid in the limit of a very large crystal when we can treat the wavevector as a continuous variable[3]. It is convenient to normalise the eDOS with respect to the crystal volume, so as to obtain an expression independent of geometrical factors. To this aim, we introduce the *density of states per unit volume $g_n(E)$ for the nth band* as

$$g_n(E) = \frac{G_n(E)}{V} = \frac{1}{(2\pi)^3} \int_{\mathbf{k} \in \mathrm{1BZ}} \delta(E - E_n(\mathbf{k}))\, d\mathbf{k}, \qquad (9.2)$$

which allows us to directly calculate the *total density of states* (per unit volume) as $g_{\mathrm{tot}}(E) = \sum_n g_n(E)$.

While equation (9.2) represents the most accurate way to calculate the eDOS for each band, in the following we can take profit from the twofold fact that: (i) most of the semiconductor physics is ruled over by carriers in the proximity of the forbidden gap; and (ii) the valence and the conduction bands are to a very good approximation

---

[3] We have similarly found this twofold way of writing a density of states in section 4.1.3, where the phonon density of states was used in both forms to evaluate the heat capacity. In general, each allowed $\mathbf{k}$ wavevector has a volume $\Delta \mathbf{k} = (2\pi)^3/V$ in the reciprocal space, so that for any function $f(\mathbf{k})$ it holds $\sum_{\mathbf{k}} f(\mathbf{k}) = [V/(2\pi)^3]\sum_{\mathbf{k}} f(\mathbf{k})\Delta \mathbf{k}$. In the limit $\Delta \mathbf{k} \to 0$ (that is in the limit of a very large crystal) we can therefore replace any sum $\sum_{\mathbf{k}} (\cdots)$ with the integral expression $[V/(2\pi)^3] \int (\cdots)\, d\mathbf{k}$.

parabolic near such a gap. This twofold observation greatly simplifies the mathematics since we can straightforwardly extend to semiconductors the result given in equation (7.28), which was derived under a similar parabolic k-dependence of the energy in the free electron gas. Therefore, we define the *CB density of states per unit volume* $\bar{g}_{CB}(E)$ *in parabolic bands approximation* as

$$\bar{g}_{CB}(E) = \frac{4\pi}{h^3} (2m_e^*)^{3/2} (E - E_c)^{1/2}, \tag{9.3}$$

where E_c is the energy at the bottom of the CB. Similarly, we define the *VB density of states per unit volume* $\bar{g}_{VB}(E)$ *in parabolic bands approximation* as

$$\bar{g}_{VB}(E) = \frac{4\pi}{h^3} (2m_h^*)^{3/2} (E_v - E)^{1/2}, \tag{9.4}$$

where E_v is the energy at the top of the VB.

Equations (9.3) and (9.4) are deceptively simple at first sight, but they hide a subtlety: the effective mass is used for both electrons and holes, but it is left ambiguous which actual value should be used. More specifically, we must be more accurate in defining whether we should use the m_{lh}^* or the m_{hh}^* mass for holes, or the $m_{e,\parallel}^*$ or the $m_{e,\perp}^*$ mass for electrons, as explained in section 8.3.5. In order to work out a general theory, it is customary to define effective masses suitably averaged for the density of states calculations: they are usually referred to as *density of states effective masses*.

Let us consider first the case of three-dimensional non-perfectly parabolic conduction band. In this case the electron tensorial mass defined in equation (8.45) is replaced by an average scalar quantity defined as

$$m_e^* = \left(m_{e,xx}^* \, m_{e,yy}^* \, m_{e,zz}^* \right)^{1/3}. \tag{9.5}$$

If we further consider the case of a non-symmetric CB shown in figure 8.12, typically found in indirect gap semiconductors with diamond structure, we conveniently define a density of states effective mass which properly takes into account the CB anisotropicity by setting

$$m_e^* = N^{2/3} \left[m_{e,\parallel}^* \left(m_{e,\perp}^* \right)^2 \right]^{1/3}, \tag{9.6}$$

where the numerical pre-factor N appears since the CB has N equivalent minima (for silicon $N = 6$, while for germanium $N = 4$).

Finally, if we consider the case of degenerate valence bands (see figure 8.12), we define a density of states effective mass which properly takes into account the existence of light and heavy holes by setting

$$m_h^* = [(m_{lh}^*)^{3/2} + (m_{hh}^*)^{3/2}]^{2/3}, \tag{9.7}$$

which holds for semiconductors with both diamond and zincblend structure.

We conclude this part by pointing out that, in order not to overcharge the formalism, we have used the same symbol for the effective mass and the density of states effective mass. This will not generate any ambiguity in the following.

9.2 Microscopic theory of charge transport

Charge carriers can be accelerated, so as to give rise to a conduction regime, either: (i) by an external electric field (which generates a *drift current*); or (ii) by the existence of a spatial gradient of their concentration (which generates a *diffusion current*). In both cases we have to explicitly consider the contributions to the current density $\mathbf{J_q}$ of both electrons and holes.

9.2.1 Drift current in a weak field regime

Let us consider the case in which an external electric field $\mathbf{E}$ is applied to a semiconductor. For simplicity, we assume a constant, spatially uniform and not too intense field. This last condition is normally referred to as the *weak field regime* and it corresponds to a wide class of situations really encountered in engineering applications. From a physical point of view, we can consider 'weak' an electric field provided that it is able to accelerate electrons only up to a velocity not larger than their classical thermal velocity $v_e^{th} = \sqrt{3k_B T/m_e^*}$.

The semi-classical calculation that solves the CB electron dynamics proceeds similarly to section 7.2 and the key result provided by equation (7.6) only needs to be corrected by replacing the electron rest mass by its effective counterpart, which introduces full information on the underlying band structure, thus obtaining[4]

$$\mathbf{J}_{q,CB} = -n_e e \mathbf{v_d} = \frac{n_e e^2 \tau_e}{m_e^*} \mathbf{E}, \qquad (9.8)$$

from which we understand that the *CB electron transport is ruled over by three intrinsic material parameters*, namely: (i) the electron effective mass m_e^*; (ii) the relaxation time τ_e for the scattering events; and (iii) the density n_e of the electrons in the CB. While we have already developed robust knowledge on m_e^*, we still need to deepen the analysis of τ_e and n_e. This will be done soon, respectively, in sections 9.2.2 and 9.2.3, but we can anticipate that the important issue for technological applications is that *all these quantities are not assigned once and for all, but instead can be engineered according to specific needs*. In fact: the number of electrons in the CB can be controlled by the doping level; the effective mass of the carriers is determined by the band structure, that is by the selected semiconductor; the relaxation time is governed by the kind and by the amount of imperfections present in the lattice (both features affect the frequency of electron-defect scattering events), as well as by the temperature (which determines the phonon population and, therefore, the occurrence of electron–phonon scattering events). Given the

---

[4] We are using the same notation introduced in section 8.3.4.

fundamental role played in semiconductor transport theory by the combination of these three parameters, it is quite useful to define the *electron mobility*

$$\mu_e = \frac{e\tau_e}{m_e^*},\tag{9.9}$$

which is typically expressed in units [cm^2 V^{-1} s^{-1}]. This quantity provides a direct measure of the ease with which electrons can be accelerated, since the mobility increases with increasing relaxation time (i.e. with decreasing frequency of scattering events) and decreasing effective mass (which measures inertia to motion). The *direct-current electrical conductivity* is written as

$$\sigma_e = \frac{e^2\tau_e}{m_e^*}\,n_e = n_e\,e\,\mu_e,\tag{9.10}$$

which is the counterpart of equation (7.7) valid for metals.

The above analysis is not complete since at $T > 0$ K in any semiconductor (regardless if doped or intrinsic) both electrons and holes are in principle found, respectively, in CB and VB. In other words, we must duly take into account the existence of two kinds of charge carriers. The hole transport in VB is basically described by the same equations (9.8)–(9.10) as above, provided that suitable values for the effective mass m_h^*, relaxation time τ_h and carrier density n_d are used. In brief, the *total current density* $\mathbf{J}_{q,tot}$ in a semiconductor is written as

$$\mathbf{J}_{q,tot} = \mathbf{J}_{q,CB} + \mathbf{J}_{q,VB} = (\sigma_e + \sigma_h)\,\mathbf{E} = \sigma_{tot}\,\mathbf{E},\tag{9.11}$$

where we have introduced the *total direct-current conductivity*

$$\sigma_{tot} = e(n_e\mu_e + n_h\mu_h),\tag{9.12}$$

as the sum of the electron and hole ones. This equation makes it clear that the electron and hole contributions to the total current are always additive (that is, they have the same sign) since electrons and holes carry an opposite charge, but also move with opposite drift velocity under the action of the external electric field $\mathbf{E}$. The electron and hole mobility values in some semiconductors are reported in table 9.1.

9.2.2 Scattering

As anticipated, we now address the relaxation time topic. In order to make the physical picture clear, we are going to discuss it in the specific case of electrons.

Table 9.1. Room temperature electron μ_e and hole μ_h mobility of some semiconductors in units cm^2 V^{-1} s^{-1}.

	Si	Ge	AlP	AlAs	GaP	GaAs	InP	InAs	CdTe	CdSe	GaN
μ_e	1350	3900	80	1000	300	8500	4000	22600	1050	650	300
μ_h	480	1900		180	150	400	600	200	100		

The approach to the scattering problem will be phenomenological, a more advanced and rigorous discussion is reported in [6].

In a real semiconductor there is a great variety of electron scattering mechanisms, basically due to two different classes of phenomena: (i) scattering by lattice vibrations; and (ii) scattering by defects (native, contaminant or dopant), either charged or electrically neutral. They differently affect electron mobility and we assume to be valid, as in the case of metals discussed in section 7.3.3, the *Matthiessen rule*, according to which *each scattering mechanism acts individually*. Therefore, the resulting total relaxation time $\tau_{e,tot}$ is given by

$$\frac{1}{\tau_{e,tot}} = \sum_i \frac{1}{\tau_{e,i}}, \tag{9.13}$$

where the sum runs over all the different scattering mechanisms labelled by the i index. This assumption is useful to develop a very detailed analysis by raising the question: *what would be the electron mobility $\mu_{e,i}$ if only the ith scattering mechanism was present?* It is possible to answer to this question since the direct consequence of equation (9.13) is that we can write

$$\frac{1}{\mu_{e,tot}} = \sum_i \frac{1}{\mu_{e,i}}, \tag{9.14}$$

where $\mu_{e,tot}$ is the real electron mobility[5]. Since any mechanism has a very specific temperature dependence, the Matthiessen rule allows for studying them separately, thus determining which one dominates in different temperature regimes.

In order to address the scattering by lattice vibrations it is very convenient to adopt the phonon picture, which allows for a very effective corpuscular language. The scattering phenomena are therefore described as impacts between particles (electrons/holes and phonons, respectively) which, in a weak field regime, are treated as elastic: phonon scattering events modify electron trajectories, leaving their energies unaffected.

Long wavelength acoustic phonons give rise to a periodic compression/dilatation of the lattice: it is as if a spatially modulated pressure was applied. Pressure, in turn, alters the band structure[6] and it is precisely in this phenomenon that the effect of scattering from acoustic phonons resides: due to the band structure modification, the effective mass of electrons is affected, and so is their mobility. Since the amplitude of acoustic oscillations depends on temperature, electron mobility will also depend on temperature. It is experimentally found that for this mechanism $\mu_e \sim 1/T^{3/2}$, that is: *the scattering by acoustic phonons dominates* (or, equivalently, it is maximally effective in decreasing the electron mobility) *at high temperatures*. In the case of elemental semiconductors this is the leading mechanism due to lattice vibrations.

In compound (that is, polar) semiconductors long wavelength optical phonons generate an internal electric field that, obviously, interacts with the charge carriers

---

[5] Real in this framework means the experimentally measured value.
[6] In appendix H we outline the variation of the energy gap, as a simple example of pressure effects.

altering their dynamics: also in this case the scattering by optical phonons results in a decreasing mobility as the temperature increases. This dependence is very complex, but if the temperature is not too high it is experimentally found that for this mechanism $\mu_e \sim 1/T$.

Finally, *piezoelectric scattering* is observed in those semiconductors that show polarisation under deformation. This is for instance the notable case of GaAs and II–VI semiconductors: acoustic phonons induce polarisation and the resulting electric field affects electron mobility. For this mechanism, experimental data provide a $\mu_e \sim 1/T^{1/2}$ dependence.

We now move to consider the scattering by defects and, to keep things simple, we limit our discussion to point defects only. Let us start by considering the case of a charged defect, which is commonly associated with the presence of a contaminating impurity or an ionised dopant: an electron approaching the defect site will be affected by Coulomb attraction or repulsion which, of course, will affect its dynamics and ultimately its mobility. Also for the Coulomb scattering there is observed a dependence on temperature that, this time, is found in the form $\mu_e \sim T^{3/2}$. The result is qualitatively explained by observing that upon increasing temperature, the carrier kinematics becomes faster and, therefore, electrons on average remain near charged defects for a shorter time: ultimately, the scattering mechanism becomes less effective. Limited to this case, we can conclude that the mobility increases with increasing temperature.

Finally, there is also observed the scattering by neutral (not electrically charged) defects. These are non-ionised dopants or contaminants, as well as isovalent substitutional defects (for example: a Ge atom replacing a Si atom in a silicon crystal). In this case the scattering is purely mechanical, basically due to the local variation of the crystal potential in the proximity of the defect. This scattering phenomenon modifies the dynamics of the carriers, but it is practically independent of temperature; in any case it is measured to be comparatively much less important than the other mechanisms.

In figure 9.4 we report the temperature dependence of the electron mobility in an n-doped GaAs sample: the different $\mu_{e,i}$ values (corresponding to an ideal mobility only limited by just the single ith scattering mechanism) are shown by coloured lines, while the experimental (total) mobility is shown by dots. In order to qualitatively justify the observed T-dependence, we can focus on the two main scattering mechanisms, namely: the optical phonon and Coulomb ones. Accordingly, we approximate equation (9.14) as follows

$$\frac{1}{\mu_{e,tot}} \sim \frac{1}{\mu_{e,opt.phonon}} + \frac{1}{\mu_{e,Coulomb}}, \tag{9.15}$$

and elaborate the following rationale: at low temperatures the Coulomb scattering mechanism dominates in limiting the electron mobility and, therefore, we can set $\mu_{e,tot} \sim \mu_{e,Coulomb}$; in the opposite regime of high temperatures, the optical phonon scattering mechanism dominates and therefore we approximate $\mu_{e,tot} \sim \mu_{e,opt.phonon}$;

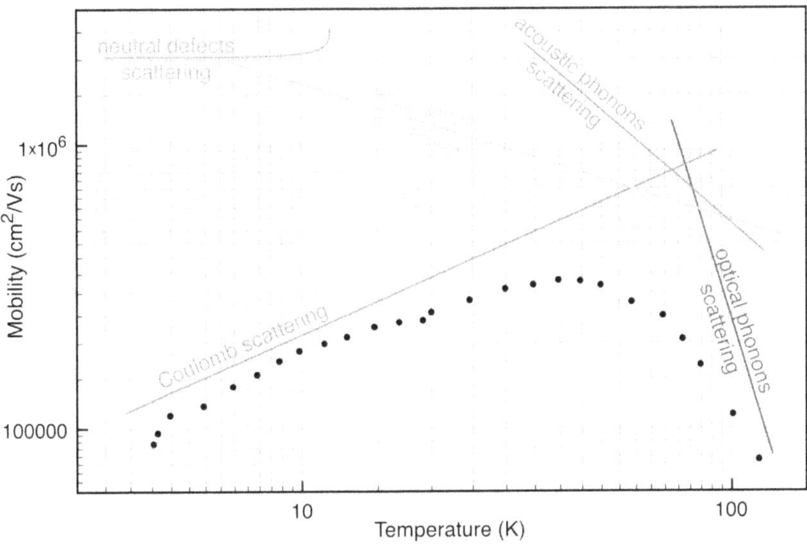

Figure 9.4. Black dots: electron mobility as function of temperature (bi-log scale) in an n-doped GaAs sample (data taken from [7]). Colour lines: contribution of each single scattering mechanism.

at intermediate temperatures, experimental data show the transition between the two limiting situations.

We remark that (i) all the arguments developed here apply both in the case of electrons and in the case of holes and (ii) all the conclusions drawn in this section are actually generally valid for any semiconductor. In particular, this holds for the Coulomb scattering which, in doped semiconductors, is governed by the actual concentration of donors or acceptors. This gives rise to a new question: *how does the carrier mobility depend on the dopant concentration?* The experimental data show that at low doping levels the carrier mobility is basically unrelated to the actual number of impurities per unit volume, while in heavily doped samples the mobility monotonically decreases with increasing defect concentration. This is shown in figure 9.5 in the case of silicon, for both electrons and holes. The experimental trends are explained by these simple arguments: at low dopant concentrations, the carrier scattering is dominated by phonon-like mechanisms, while the Coulomb mechanism plays a marginal role; therefore, the mobility in this regime is basically constant (we are keeping constant the temperature of the sample). In the opposite case of heavy doping, the Coulomb scattering dominates in increasing reason with the number of impurities present: more dopants → increased scattering efficiency → smaller carrier mobility.

The result shown in figure 9.5 deserves a further comment. Depending on the specific technological application, we may be interested in maximising either the density of carriers or their absolute mobility. In this respect, figure 9.5 clearly shows that the two conditions of 'high doping' and 'high mobility' are in contrast as regards their overall effect on the ability to conduct electric current. The mutual influence of temperature and doping level is revealed in a complex dependence of

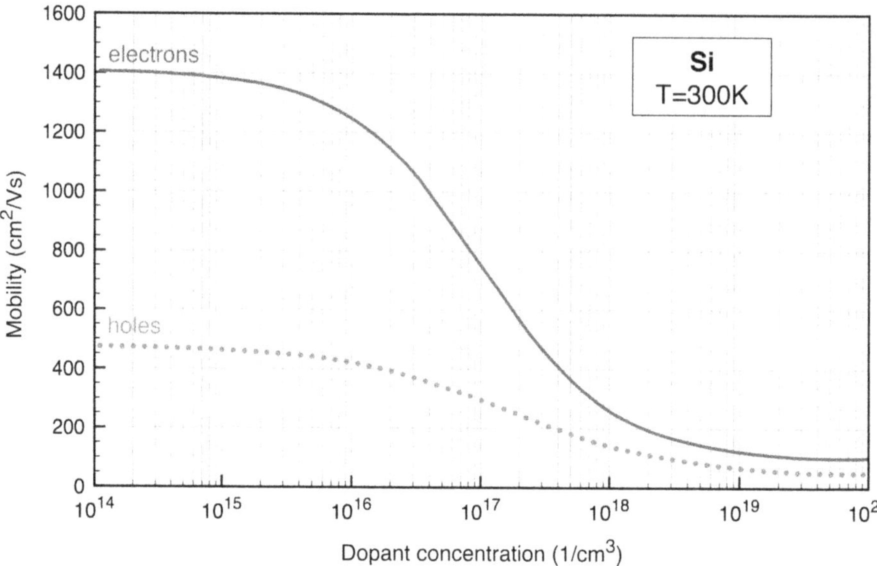

Figure 9.5. Electron and hole mobility in silicon at room temperature as function of the dopant concentration (semi-log scale). Data taken from [8].

mobility on these two factors, as shown in figure 9.6 in the specific case of electrons in an n-doped silicon sample[7]. So, as far as the ability to support high intensity electric currents is concerned, the 'best' semiconductor is not necessarily the one with maximum carrier mobility (a condition that necessarily corresponds to a low concentration of carriers, given the fact that their number is obviously determined by the number of dopant atoms), nor the one with very high concentration of carriers (a condition that necessarily implies a low mobility). In other words, in practical applications it will always be necessary to determine the optimal balance between dopant concentration and carrier mobility: this is an illuminating example of 'materials engineering'.

9.2.3 Carriers concentration

The last parameter to be discussed affecting charge transport is the concentrations of carriers, that is, the number of electrons and holes per unit volume in CB and VB, respectively. Like mobility, these concentrations vary with temperature. We discuss the paradigmatic case of an n-doped semiconductor, whose band structure is sketched in figure 9.3.

At zero temperature all impurity levels are occupied by electrons confined in the potential well associated with the dopant atom, while the CB is empty. Upon increasing temperature, a number of such electrons are thermally excited to the

---

[7] We remark that in the high doping regime (that is, for concentrations larger than 10^{17} dopants cm^{-3}) the T-dependence of electron mobility is similar to the trend reported in figure 9.4: this provides full consistency with the discussion developed in section 9.2.2.

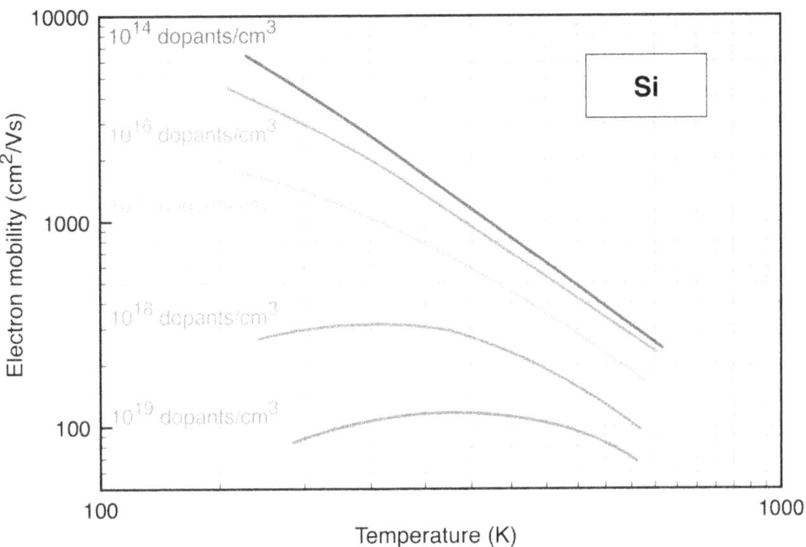

Figure 9.6. Electron mobility in a n-doped silicon sample as function of temperature and dopant concentration (bi-log scale). Data taken from [9].

conduction band: this process is possible even at moderately low temperatures due to the proximity of the impurity level to the CB bottom. By progressively increasing the temperature, the number of excited electrons increases correspondingly, until all the impurity levels have lost their electrons[8]: therefore, a further temperature increase does not cause any increase of the electron population in CB. This regime lasts until such a high temperature is eventually reached to make possible the direct VB → CB excitation. Starting from this condition, any further temperature rise will accordingly increase the carrier concentration in CB, directly taking more and more electrons from the VB. This promotion becomes exponentially efficient with increasing temperature.

In figure 9.7 the variation of the electron concentration in the conduction band is reported as a function of the inverse temperature for the case under discussion. We can identify three regions with different physical features:

- in the *high temperature range* the concentration of carriers is very high: in fact, not only have all impurities been ionised, but also the thermal excitation of electrons from the valence band to the conduction band is particularly efficient. In this temperature range the conductivity is dominated by the electrons coming from the VB, that is, by the *intrinsic carriers*[9]: accordingly, *this regime is called 'intrinsic region'*;

- in the *intermediate temperature range* the concentration of carriers remains basically constant because (i) all impurities have been ionised, but (ii) the temperature is no longer so high as to generate a sizeable number of electron

---

[8] It is equivalently said that all dopant atoms have been ionised.

[9] This is the only mechanism in fact at work in the absence of any doping.

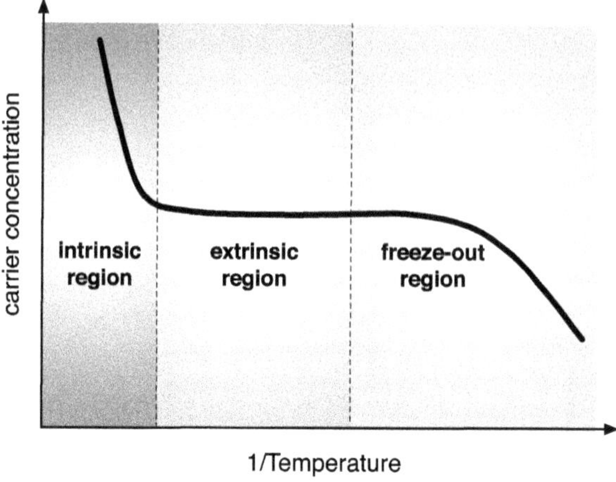

Figure 9.7. Qualitative trend of the electron concentration in conduction band (log scale) in an n-doped semiconductor as function of inverse temperature.

> excitations from the VB to the CB. We accordingly refer to this intermediate situation as the *'extrinsic region'*, since the conductivity is dominated by electrons mainly provided by an extrinsic cause, such as doping;
> - in the *low temperature range* no VB carrier is directly excited to CB, while the number of those made available by doping decreases as the temperature decreases. This temperature regime is referred to as the *'freeze-out region'*.

It is important to point out that for all semiconductors used in modern information technologies, the room temperature falls within the intrinsic region: this implies that at normal operating temperatures *the number of carriers in a semiconductor device is typically regulated by the doping level.* Finally, we remark that the above qualitative discussion and conclusions are of general validity, that is, they also apply to holes in p-doped materials.

9.2.4 Conductivity

Equation (9.12) makes it clear that the total direct-current conductivity depends on the concentration of both carrier types, as well as on their mobilities. Since we learned that such quantities depend on temperature, we understand that $\sigma_{tot} = \sigma_{tot}(T)$. By combining the results discussed in sections 9.2.2 and 9.2.3, we obtain the qualitative T-dependence of σ_{tot} shown in figure 9.8 for an n-doped semiconductor. This picture is interpreted as follows[10]:
- *in the intrinsic region* the higher the temperature, the larger is the concentration of carriers. A remarkable feature of this region is that the carriers number increases more rapidly than the mobility decreases because of phonon scattering (which, as we know, is the leading scattering mechanism

---

[10] It is of course understood that the n-doping level is kept constant.

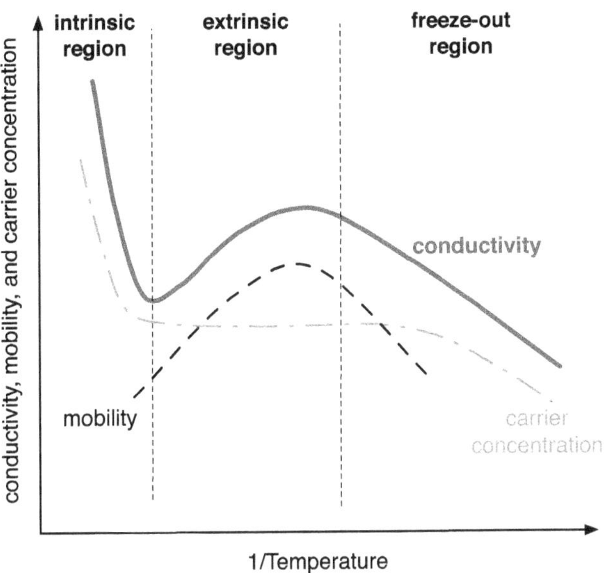

Figure 9.8. Qualitative trend of the conductivity, electron mobility, and electron concentration in conduction band (log scale) in an n-doped semiconductor as function of inverse temperature.

in this temperature range). All in all, conductivity rapidly increases with temperature;

- *in the extrinsic region* the conductivity is mainly affected by the variation of mobility with temperature, instead basically with the concentration of carriers being constant. The conductivity variation with temperature is overall similar to the mobility one;
- *in the freeze-out region* both the concentration of carriers (fewer and fewer electrons are thermally excited) and the mobility (which, in this region, is severely limited by Coulomb scattering) decrease with decreasing temperature. The conductivity follows the same behaviour.

Once again, this case-study highlights general features that are similarly valid in p-doped semiconductors.

9.2.5 Drift current in a strong field regime

A tacit assumption in the discussion so far developed is that both τ_e and τ_h scattering times are independent of the applied electric field. This condition is physically verified only if the drift velocity of carriers is smaller than their thermal velocity $|\mathbf{v}_d| \leqslant |\mathbf{v}_e^{th}|$ (a condition that represents the typical weak field regime situation). If, in contrast, we suppose application of a very high electric field, we enter a different conduction regime: carriers can be accelerated up to largely exceed their thermal velocity. We will then talk about conduction in a *strong field regime*.

We know from basic thermodynamics [10–12] that the velocities of the particles forming a molecular gas are associated with its temperature; by extending this concept to the gas of charge carriers, we guess that the strong field regime corresponds to a situation where the temperature of such a gas is very high, even higher than the temperature of the crystal lattice (determined by the ionic vibrations). *This makes it possible for carriers to transfer energy to the lattice*[11]. In the case we are discussing, the transfer of energy from charge carriers to the lattice is obviously mediated by the electron–phonon scattering events which, unlike what happens in a weak field, must be now treated as inelastic collisions[12]. On the other hand, the very fact that the lattice temperature increases is equivalent to saying that *in a strong field regime the scattering events generate phonons*: the energy needed to generate them is subtracted from the gas of charge carriers by a mechanism that becomes more and more efficient as the accelerating external field increases. Because of this, the drift velocity increases as a function of the applied field, until it eventually reaches a saturation value $\mathbf{v}_{sat}$ which cannot be exceeded because so many phonons are generated to prevent further carrier acceleration.

It is easy to elaborate an order-of-magnitude estimate of the saturation velocity by simply evaluating the energy balance for electrons in a saturation condition[13]: the energy gained per unit time by a single electron because of the acceleration impressed by the external field $\mathbf{E}$ must be equal to the energy lost by emission of a phonon in the same time span. This balance is translated in the equation

$$-e\mathbf{v}_{sat} \cdot \mathbf{E} = \frac{\hbar\omega_{phonon}}{\tau_e}, \tag{9.16}$$

where the term on the right-hand side is given by the ratio between the energy $\hbar\omega_{phonon}$ of the emitted phonon and the average time τ_e separating two consecutive scattering events. Since the saturation velocity is the upmost value of the drift velocity, we can insert into equation (9.16) the definition given in equation (7.5) and thus obtain

$$|\mathbf{v}_{sat}| = \sqrt{\frac{\hbar\omega_{phonon}}{m_e^*}}, \tag{9.17}$$

providing the saturation velocity in terms of the energy of the generated phonon. In figure 9.9 we report the variation of the electron and hole drift velocity as a function of the applied electric field in the paradigmatic cases of silicon and germanium. The saturation velocity is easily extracted from these plots.

9.2.6 Diffusion current

We now consider a situation where no electric field is applied to the semiconductor, but the charge carriers are non uniformly distributed along, say, the z direction.

---

[11] This occurs in analogy to what happens when two ordinary molecular gases at different temperatures come into contact: the warmer one gives energy to the colder one, causing a rise in temperature.

[12] Interestingly enough, this is definitely a non-adiabatic effect.

[13] The exact value of the saturation velocity can only be obtained by a careful calculation of the electron–phonon scattering rate [4].

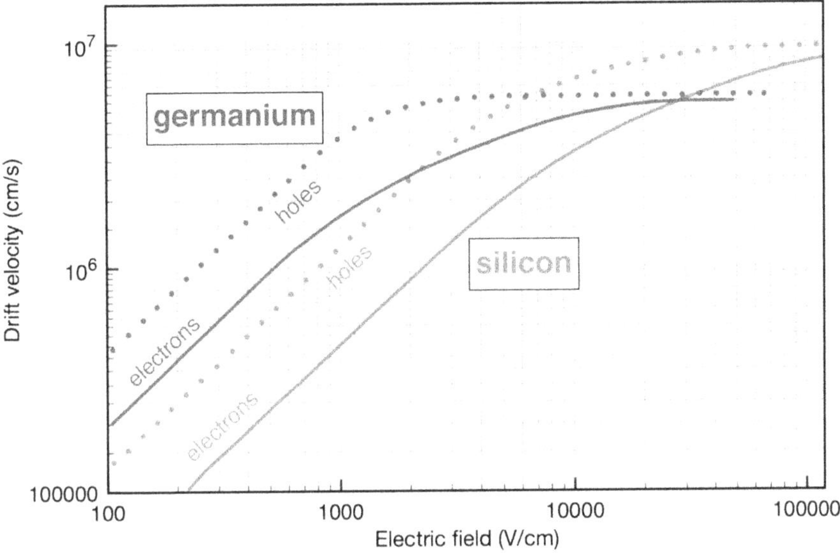

Figure 9.9. Room temperature electron (full lines) and hole (dotted lines) drift velocity in silicon (red) and germanium (blue) as a function of the strong electric field. Data for silicon are taken from [8]. Data for germanium are taken from [13].

To be more specific, we assume that $n_e = n_e(z)$ is the concentration of electrons in CB. By considering an arbitrary section of the sample, we understand that it is crossed by electrons along both z-directions because of their thermal motion. However, the numbers of carriers crossing the section from $-z$ or from $+z$ directions are unequal, since the concentration is not uniform. Accordingly, the net flux of electrons Φ_e (that is, the total number of electrons crossing the section per unit time) is just the balance between the these two currents. Elementary diffusion theory [14] proves that

$$\Phi_e - -D_e \frac{dn_e(z)}{dz}, \tag{9.18}$$

which is referred to as the Fick diffusion law: the net flux is proportional to the concentration gradient. The proportionality constant D_e is known as the *electron diffusivity*; it is easily calculated as

$$D_e = \lambda_e v_e^{th} = \frac{\lambda_e^2}{\tau_e} = (v_e^{th})^2 \tau_e, \tag{9.19}$$

where λ_e is the electron mean fee path, that is, the average distance covered between two consecutive scattering events (occurring at $1/\tau_e$ rate). By inserting in equation (9.19) the result[14] $v_e^{th} = \sqrt{k_B T/m_e^*}$, we obtain the *Einstein relation for the electron diffusivity*

---

[14] We stress that in this case the electrons are assumed to move uni-directionally only along z and, therefore, equipartition imposes that $m_e^*(v_e^{th})^2/2 = k_B T/2$.

$$D_{\mathrm{e}} = \frac{k_{\mathrm{B}}T}{m_{\mathrm{e}}^{*}}\tau_{\mathrm{e}} = \frac{k_{\mathrm{B}}T}{e}\mu_{\mathrm{e}}, \tag{9.20}$$

where we made used of the definition given in equation (9.9). This is quite an important result, predicting that *if electrons have high mobility, then they also have high diffusivity*; furthermore, *diffusivity is found to grow with temperature*. Both features are nicely confirmed by experiments. The diffusion of electrons along their concentration gradient is obviously translated into a *diffusion current* whose density is easily calculated as $(-e)\Phi_{\mathrm{e}}$.

A gradient of hole concentration $n_{\mathrm{h}}(z)$ in VB similarly gives rise to a diffusion current density $(+e)\Phi_{\mathrm{h}}$ of positive charges, which is governed by the very same equations as above. More specifically

$$\Phi_{\mathrm{h}} = -D_{\mathrm{h}}\frac{dn_{\mathrm{h}}(z)}{dz} \quad \text{where} \quad D_{\mathrm{h}} = \frac{k_{\mathrm{B}}T}{e}\mu_{\mathrm{h}}, \tag{9.21}$$

where the hole diffusivity $D_{\mathrm{h}} = (k_{\mathrm{B}}T/e)\mu_{\mathrm{h}}$ has been introduced.

9.2.7 Total current

The most general case that can be given is that of charge transport in a three-dimensional semiconductor with non uniform distribution of carriers and subject to the action of an electric field $\mathbf{E}$ which, for simplicity, we will consider weak. This situation is found in many devices used in information and communication technologies. In these conditions the total current density $\mathbf{J}_{\mathrm{q,tot}}$ is calculated as the sum of the drift and diffusion ones, each provided by both electrons and holes.

By combining the results of section 9.2.1 and section 9.2.6 we can write

$$\mathbf{J}_{\mathrm{q,tot}} = \mathbf{J}_{\mathrm{CB}} + \mathbf{J}_{\mathrm{VB}} \quad \text{with} \quad \begin{cases} \mathbf{J}_{\mathrm{CB}} &= e\mu_{\mathrm{e}}n_{\mathrm{e}}\mathbf{E} + eD_{\mathrm{e}}\nabla n_{\mathrm{e}} \\ \mathbf{J}_{\mathrm{VB}} &= e\mu_{\mathrm{h}}n_{\mathrm{h}}\mathbf{E} - eD_{\mathrm{h}}\nabla n_{\mathrm{h}}, \end{cases} \tag{9.22}$$

where the three-dimensional formulation of the concentration gradients is reported for sake of generality.

9.3 Charge carriers statistics

In order to complete the microscopic theory of charge transport we must face the problem of how to determine in the most general way the carrier concentration in both the conduction and valence band. To this aim, we must separately consider the equilibrium and out-of-equilibrium circumstances. In the first case we will refer to a situation where the concentration of free carriers (whether electrons in CB or holes in VB) is uniquely determined by temperature. The system will be, therefore, in conditions of constant temperature and will not experience any perturbative external action. In contrast, in the second case we will refer to a physical situation where the concentration of free carriers is influenced by external factors (like, for instance, the interaction with an electromagnetic field or the injection of extra carriers from some external source) in addition to thermal effects.

9.3.1 Semiconductors in equilibrium

In a semiconductor in equilibrium at temperature T the density of electrons in conduction band is given by

$$n_e(T) = \int_{E_c}^{+\infty} \bar{g}_{CB}(E)\, n_{FD}(E,\,T)\, dE, \tag{9.23}$$

where, according to the discussion developed in section 9.1.2, we have introduced the notation $\bar{g}_{CB}(E)$ for the CB eDOS. Similarly, the density of holes in valence band is given by

$$n_h(T) = \int_{-\infty}^{E_v} \bar{g}_{VB}(E)\, [1 - n_{FD}(E,\,T)]\, dE, \tag{9.24}$$

where the term in square brackets indicates the probability that a VB state with energy E is not occupied by an electron or, equivalently, it indicates the probability to find there a hole. Obviously, $\bar{g}_{VB}(E)$ is the VB eDOS. In both cases we make use of the parabolic bands approximation, which is just fine for the following developments.

The explicit expressions for the eDOS provided by equations (9.3) and (9.4) must be, respectively, inserted into equations (9.23) and (9.24). In particular, in equation (9.23) all energies belong to the CB so that $E - \mu_c \gg k_B T$;[15] therefore, the Fermi–Dirac distribution can be conveniently approximated as

$$n_{FD}(E,\,T) \sim \exp\left(-\frac{E - \mu_c}{k_B T}\right), \tag{9.25}$$

which leads to

$$\begin{cases} n_e(T) &= N_c \exp\left(-\dfrac{E_c - \mu_c}{k_B T}\right) \\[2mm] N_c &= 2(2\pi m_e^* h^{-2} k_B T)^{3/2}, \end{cases} \tag{9.26}$$

where N_c is known as the *effective density of states in CB*. It is a very important parameter for applications, varying over a wide range of values; for instance, in the case of Si, Ge, and GaAs we have: $N_c^{Si} = 2.7 \times 10^{19}$ cm^{-3}, $N_c^{Ge} = 1.0 \times 10^{19}$ cm^{-3}, and $N_c^{GaAs} = 4.4 \times 10^{17}$ cm^{-3}, respectively.

As for holes, we must consider that in equation (9.24) all energies belong to the VB so that $|E - \mu_c| \gg k_B T$ and, therefore, in this case the Fermi–Dirac distribution can be conveniently approximated as

$$n_{FD}(E,\,T) \sim 1 - \exp\left(-\frac{\mu_c - E}{k_B T}\right), \tag{9.27}$$

---

[15] This is indeed a subtle step: we are implicitly assuming that the Fermi level is at the top of the valence band and that, similarly to metals, the chemical potential and the Fermi energy can be to a good extent identified. These assumptions will soon be readdressed and justified.

which leads to

$$\begin{cases} n_{\mathrm{h}}(T) = N_{\mathrm{v}} \exp\left(-\dfrac{\mu_{\mathrm{c}} - E_{\mathrm{v}}}{k_{\mathrm{B}}T}\right) \\[2mm] N_{\mathrm{v}} = 2(2\pi m_{\mathrm{h}}^{*} h^{-2} k_{\mathrm{B}}T)^{3/2}, \end{cases} \tag{9.28}$$

where N_{v} is known as the *effective density of states in VB*. There results: $N_{\mathrm{v}}^{\mathrm{Si}} = 1.1 \times 10^{19}$ cm^{-3}, $N_{\mathrm{v}}^{\mathrm{Ge}} = 3.9 \times 10^{18}$ cm^{-3}, and $N_{\mathrm{v}}^{\mathrm{GaAs}} = 9.7 \times 10^{18}$ cm^{-3}.

Equations (9.26) and (9.28) provide evidence that *the concentration of free carriers does depend on the position of the chemical potential* μ_{c}, a remarkable result indeed, which forces us to consider in more detail the problem of where exactly the chemical potential is located[16] and, more importantly, how its position depends on temperature and doping condition.

9.3.2 Chemical potential in intrinsic semiconductors

If an intrinsic semiconductor is in equilibrium at temperature T the numbers of electrons in CB and holes in VB are just the same, since for any electron thermally excited a corresponding hole is generated. This means that $n_{\mathrm{e}}(T) = n_{\mathrm{h}}(T)$ or equivalently

$$N_{\mathrm{c}} \exp\left(-\frac{E_{\mathrm{c}} - \mu_{\mathrm{c}}}{k_{\mathrm{B}}T}\right) = N_{\mathrm{v}} \exp\left(-\frac{\mu_{\mathrm{c}} - E_{\mathrm{v}}}{k_{\mathrm{B}}T}\right), \tag{9.29}$$

from which we get

$$\mu_{\mathrm{c}} = \frac{E_{\mathrm{c}} + E_{\mathrm{v}}}{2} - \frac{1}{2}k_{\mathrm{B}}T \ln \frac{N_{\mathrm{c}}}{N_{\mathrm{v}}}, \tag{9.30}$$

defining the position of the chemical potential as a function of temperature. From this result we immediately obtain that $E_{\mathrm{F}} = \lim_{T \to 0} \mu_{\mathrm{c}} = (E_{\mathrm{c}} + E_{\mathrm{v}})/2 = E_{\mathrm{gap}}/2$: *the Fermi level of an intrinsic semiconductor is placed right at the centre of the forbidden gap.* In many contexts the $E_{\mathrm{F}} = E_{\mathrm{gap}}/2$ energy is referred to as the *intrinsic Fermi level.*

The $\mu_{\mathrm{c}} = \mu_{\mathrm{c}}(T)$ dependence predicted by equation (9.30) is in fact very weak since we have that $\ln N_{\mathrm{c}}/N_{\mathrm{v}} \sim \ln m_{\mathrm{e}}^{*}/m_{\mathrm{h}}^{*} \sim 0$: the logarithm of the mass ratio is actually very small since in most semiconductors the electron and hole effective masses are not that much different (see table 9.1).

9.3.3 Chemical potential in doped semiconductors

In doped semiconductors the carrier concentration is determined by the combination of thermal excitation and ionisation phenomena of dopant atoms. In this case it is

---

[16] This information will also teach us where the Fermi level of a semiconductor must be placed. This is not a trivial problem: if we simply adopt the definition of Fermi energy as 'the energy below which any allowed level is fully occupied at $T = 0$ K', we are left with a large ambiguity, since any energy in between the forbidden gap does fulfil this definition.

really needed to make a systematic distinction between thermally generated *intrinsic carriers* and doping related *extrinsic carriers* (see discussion in section 9.1.1). We will separately address the n-doping and p-doping case.

Let us consider first an n-doped semiconductor where N_D identical donors per unit volume have been added, each generating an impurity level with energy E_D just below the CB minimum (see figure 9.3). In equilibrium at temperature T we will find

$$n_e(T) = n_e^{(i)}(T) + n_D(T), \tag{9.31}$$

electrons per unit volume in CB, where $n_e^{(i)}(T)$ is the concentration of intrinsic electrons (that is, electrons promoted from the valence band) and $n_D(T) \leqslant N_D$ is the concentration of extrinsic electrons (that is, electrons promoted from the impurity level). We can distinguish three temperature ranges:

- at *low temperature* (freeze-out region) $n_e^{(i)}(T)$ is negligibly small, while the number of extrinsic electrons is just equal to the number of ionised donors which, in turn, corresponds to the number of empty impurity levels. Combining these arguments with equation (9.28), we can write

$$N_c \exp\left(-\frac{E_c - \mu_c}{k_B T}\right) = N_D[1 - n_{FD}(E_D,\, T)] \sim N_D \exp\left(\frac{E_D - \mu_c}{k_B T}\right), \tag{9.32}$$

  where we have approximated the Fermi–Dirac distribution function as we did in section 9.3.1. This result immediately leads to

$$\mu_c = \frac{E_c + E_D}{2} + \frac{k_B T}{2} \ln \frac{N_D}{N_c}, \tag{9.33}$$

  which indicates that *at $T = 0$ K the Fermi level of an n-doped semiconductor is placed right in between the donor level and the bottom of the conduction band*;
- at *intermediate temperature* (extrinsic region) we still have $n_e^{(i)}(T) \sim 0$, but all donors have been ionised. This is tantamount to writing $n_e(T) = N_D$ so that

$$\mu_c = E_c + k_B T \ln \frac{N_D}{N_c}, \tag{9.34}$$

- at *high temperature* (intrinsic region) the population of intrinsic electrons largely exceeds the extrinsic one, although all donors are ionised: therefore, the semiconductor actually behaves as if it were intrinsic. According to the conclusion reported in section 9.3.2, we can consider μ_c basically independent of temperature and set it at the $E_{gap}/2$ midgap position.

This analysis is fully consistent with experimental evidence, as shown in figure 9.10 in the case of n-doped silicon. The same figure also reports the opposite case of p-doping where similar arguments as before can be developed. More specifically, for a p-doped semiconductor we have

$$n_h(T) = n_h^{(i)}(T) + n_A(T), \tag{9.35}$$

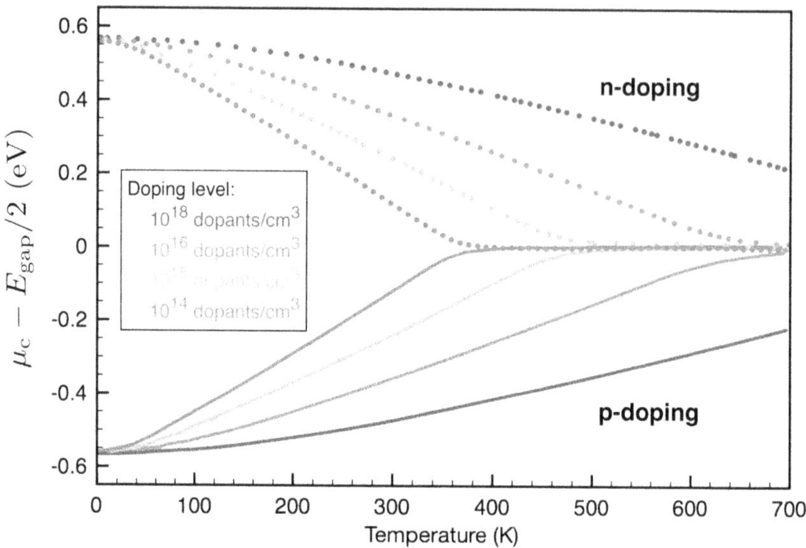

Figure 9.10. Position of the chemical potential μ_c in n-doped (dotted lines) and p-doped (full lines) silicon as function of temperature and doping level. The μ_c position is reported relative to the intrinsic Fermi level $E_{gap}/2 = 0.56\,\text{eV}$.

where $n_h^{(i)}(T)$ and $n_A(T) \leqslant N_A$ are the concentration of intrinsic and extrinsic holes, respectively. It is assumed that N_A acceptors per unit volume have been placed in the material, each generating an impurity level E_A just above the VB maximum (see figure 9.3). By increasing temperature we span the usual three regions:

- at *low temperature* (freeze-out region) the number of intrinsic holes is negligibly small and therefore

$$\mu_c = \frac{E_v + E_A}{2} - \frac{k_B T}{2} \ln \frac{N_A}{N_v}, \tag{9.36}$$

which indicates that *at $T = 0\,K$ the Fermi level of a p-doped semiconductor is placed right in between the acceptor level and the top of the valence band*;

- at *intermediate temperature* (extrinsic region) $n_A(T) = N_A$, while $n_h^{(i)}(T)$ is still vanishingly small. This leads to

$$\mu_c = E_v - k_B T \ln \frac{N_A}{N_v}, \tag{9.37}$$

- at *high temperature* (intrinsic region) the p-doped semiconductor actually behaves as intrinsic.

9.3.4 Law of mass action

By combining the two general expressions (9.26) and (9.28) we obtain the noteworthy result

$$n_{\mathrm{e}}(T)n_{\mathrm{h}}(T) = N_{\mathrm{c}}N_{\mathrm{v}} \exp\left(-\frac{E_{\mathrm{c}} - E_{\mathrm{v}}}{k_{\mathrm{B}}T}\right) = N_{\mathrm{c}}N_{\mathrm{v}} \exp\left(-\frac{E_{\mathrm{gap}}}{k_{\mathrm{B}}T}\right), \qquad (9.38)$$

known as *law of mass action*[17]. It is valid for any semiconductor in whatever intrinsic or extrinsic configuration and states that *the product between the electron and hole concentrations is independent of the chemical potential* or, equivalently, of the doping level; rather, it only depends on the energy gap.

In the specific case of an intrinsic semiconductor, we always have $n_{\mathrm{e}}(T) = n_{\mathrm{h}}(T)$ and, therefore, the law of mass action takes the intriguing form

$$n_{\mathrm{i}}(T) = N_{\mathrm{c}}N_{\mathrm{v}} \exp\left(-\frac{E_{\mathrm{gap}}}{k_{\mathrm{B}}T}\right) \sim T^{3/2} \exp\left(-\frac{E_{\mathrm{gap}}}{2k_{\mathrm{B}}T}\right), \qquad (9.39)$$

which provides the *intrinsic carrier population* $n_{\mathrm{i}}(T)$ in its explicit dependence on temperature. This equation makes it clear why at high temperature the intrinsic carriers largely exceed in number the extrinsic ones.

We finally remark that in a doped semiconductor the number of electrons is not equal to the number of holes: which of the two carrier population is larger depends on the actual doping level. It is therefore useful to distinguish between *majority carriers* and *minority carriers*: in n-doped (p-doped) semiconductors the majority carriers are the electrons (holes), while the minority carriers are the holes (electrons).

9.3.5 Semiconductors out of equilibrium

The external actions affecting the carrier populations, in addition to ordinary thermal excitation effects, are basically the *injection or extraction of carriers* and the *photo-generation or recombination of electron–hole pairs due to photon absorption or emission*.

The injection of carriers is a process that generates an *excess of electrons or holes* (as compared to the concentration that these carriers would have in equilibrium conditions) through a suitable system of electrical contacts coupling the semi-conductor to some external circuit. The same setup is used in the process of carrier extraction. On the other hand, the photo-generation corresponds to the promotion of electrons at higher energy levels (with a corresponding generation of holes in the initial energy level) by absorption of photons with appropriate energy. Several absorption mechanisms are possible [16], among which we recall (i) the *inter-band mechanism*, where an electron is excited directly from the VB to the CB (this process involves photons of energy $\hbar\omega_{\mathrm{photon}} \geqslant E_{\mathrm{gap}}$) and (ii) the transition from a band state to an impurity level or viceversa[18]. Recombinations occur whenever an electron–hole pair is annihilated because of inverse transitions. This can happen through either a *radiative* or a *non-radiative* process. In the first case the recombination

---

[17] This wording is taken from the physical chemistry chapter dealing with chemical reactions [15].

[18] More specifically: an electron initially placed in VB could be promoted to an acceptor or to a donor level, as long as they are unoccupied. Similarly, an electron initially accommodated on an acceptor or donor level could be excited to the CB.

causes the emission of a photon with energy equal to the energy difference between the initial and final levels occupied by the electron. Instead, as far as non-radiative recombinations are concerned, all or part of the energy lost by the electron during the transition to a lower energy level is released in the form of phonon emission. Non-radiative processes, therefore, increase the crystal lattice temperature.

The common feature of all non-equilibrium mechanisms is that *the law of mass action is no longer fulfilled* since either $n_e n_h > n_i^2$ in the injection and photo-generation cases or $n_e n_h < n_i^2$ if extraction or recombination occurs. A new statistics, other than the equilibrium one, is needed: this is an advanced topic of semiconductor theory [1, 2, 4] which we outline only qualitatively by introducing the *carrier generation rate* G and the *carrier recombination rate* R: they, respectively, count the number of carriers generated or destroyed per unit time in the unit volume because of non equilibrium mechanisms[19]. It is a difficult task of solid state theory to provide a microscopic expression for them: we assume they are known. Obviously, the net rate of variation of the carrier populations with respect to equilibrium is given by the difference (G − R), which could be either positive or negative provided that a larger or smaller number of carriers is, respectively, found because of injection, generation, extraction and recombination mechanisms. These phenomena are reflected in a modification of the continuity equations for the electron and hole currents which must now be written as follows

$$\begin{cases} \text{electron current:} \quad \dfrac{\partial n_e}{\partial t} \;=\; G_e - R_e + \dfrac{1}{e}\nabla \mathbf{J}_{CB} \\[2ex] \quad\text{hole current:} \quad \dfrac{\partial n_h}{\partial t} \;=\; G_h - R_h + \dfrac{1}{e}\nabla \mathbf{J}_{VB}, \end{cases} \tag{9.40}$$

where the two density current vectors possibly contain either the drift or diffusion contributions as explained in section 9.2.7.

9.4 Optical absorption

9.4.1 Conceptual framework

The notions developed in the previous chapters allow us to model a semiconductor as a system composed by an electron gas (which determines its electrical conduction properties) penetrated by a phonon gas (which determines its vibrational and thermal properties). If this system is embedded into an electromagnetic (em) radiation field, we must add to such a physical model also a photon gas component. *Investigating the optical properties of a semiconductor is equivalent to studying the mutual interactions among electrons, phonons and photons.* They give rise to two opposite optical phenomena: the semiconductor can (i) absorb photons from the radiation field, a process resulting in the generation of electron–hole pairs which sum to those already existing for ordinary thermal excitation, or (ii) host events of electron–hole recombination, with corresponding loss of energy by emission of

---

[19] We understand that there exist different generation $G_{e,h}$ and recombination $R_{e,h}$ rates for electrons and holes.

photons and/or phonons. In the first case we speak of *absorption*, in the second of *emission of electromagnetic radiation*.

Ultimately, understanding the physics of energy exchange processes among electrons, phonons and photons is a key issue both for the most complete characterisation of the physical properties of a semiconductor and for mastering the principles of operation of optoelectronic devices. In our tutorial introduction we will only address absorption, since it is comparatively more direct to access. Emission phenomena (inter-band emission as well as photo- and electro-luminescence) are treated elsewhere [16].

9.4.2 Phenomenology of optical absorption

When an electromagnetic radiation is directed on a semiconductor, reflection, transmission and absorption are empirically observed. Maxwell equations, through the laws of Snell and Fresnel [16–18], make reason of the first two phenomena. In order to characterise absorption we need, instead, the development of a microscopic theory able to account for any single possible conversion process of photons into excitations of the electron and/or phonon system. To this end, it is useful to consider the prototypical situation shown in figure 9.11 in which an optical absorption experiment is conceptually illustrated: the surface of a semiconductor in equilibrium at room temperature is enlightened by a monochromatic radiation at normal incidence, measuring how its intensity decreases inside the illuminated sample. The physical process that determines the attenuation of the radiation—that is, absorption—is governed by the energy subtraction mechanisms operated by the medium at damage of the radiation.

So, let Φ_{photon} be the flux of photons (that is, the number of photons per unit of time incident on the exposed surface of the semiconductor) each of energy $\hbar\omega$ (where, of course, ω is their angular frequency, equal for all provided that the em radiation is monochromatic). If the incident intensity is $I_0 = \hbar\omega\Phi_{\mathrm{photon}}$, then inside the material it decreases in proportion to the number of absorbed photons. This number, in turn, varies with the depth z with respect to the surface: the deeper the layer, the less is the amount of received photons, as a consequence of all the absorption processes occurring in shallower layers. Therefore, we can indicate by $I(z)$ the intensity of em radiation at a depth z from the surface. Moreover, the

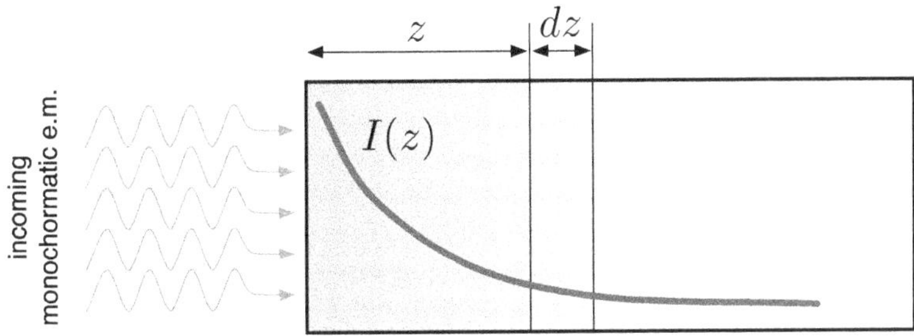

Figure 9.11. Conceptual representation of an optical absorption experiment at normal incidence in a semiconductor. $I(z)$ is the intensity of the absorbed electromagnetic field.

amount of energy absorbed by an inner layer depends on its thickness[20]. By combining these simple observations, we can write *the amount of energy absorbed by a layer of thickness dz at a distance z from the surface* as

$$dI(z) = -\alpha\, I(z)\, dz, \qquad (9.41)$$

where α is known as the *absorption coefficient* of the material[21]. The negative sign that appears in this equation indicates that we are actually calculating a decrease in the number of photons. In essence, α quantifies the effectiveness of the photon absorption process by the crystal. By integration equation (9.41) we immediately obtain the *Beer law*

$$I(z) = I_0 \exp(-\alpha z), \qquad (9.42)$$

which we will use as the theoretical basis to interpret an optical absorption experiment.

Before going into the detail of the microscopic model for α, we must observe that *it depends on the frequency of the incident em radiation*. Although the theoretical demonstration of this result goes beyond the level of treatment that we are developing[22], we can make account of this empirically verified important fact by discussing the main phenomenological results under the following assumptions:

- *the radiation field is weak enough not to alter the pre-existing band structure of the semiconductor*. In other words, we will assume that the only effect of the radiation field consists in promoting transitions between states (which remain unchanged, as compared to the case in which the semiconductor is not irradiated);
- *the vibrational spectrum contains infrared-active optical phonons*, that is, such vibrational modes are capable of generating oscillating electric dipoles that mate with the em field in the infrared spectral region. The coupling results in absorption of infrared photons[23];
- *there exists a population of free carriers that ensure continuous background absorption at all frequencies*. These are the carriers created in VB or in CB by thermal excitation (or even by the same photo-excitation). Since, because of this, both bands become only partially filled, it is possible to promote free carrier transitions between levels within the same band. Therefore, through these transitions it is possible to absorb photons of any energy.

As a last preliminary remark, we stress that the true characteristics of a typical semiconductor band structure make different types of absorption transitions

---

[20] It is intuitive that a thick layer will absorb more photons than a thinner layer.

[21] A simple dimensional analysis shows that it is measured in units $[\mathrm{m}^{-1}]$.

[22] It is in fact necessary to describe the effects of the radiation field on the crystalline electron levels by quantum time-dependent perturbation theory. In this way the transition probability for electrons and phonons to change their quantum state is calculated, obtaining the absorption coefficient [6, 16].

[23] As already extensively discussed in chapter 3, not all semiconductors enjoy this capability: in elemental semiconductors optical vibration modes do not generate oscillating electric dipoles, given the fact that the two basis atoms are equal and, therefore, equally charged (to the first order).

possible, namely: (i) the *inter-band transitions*, corresponding to direct promotion of electrons from the VB to the CB; (ii) *transitions between impurity levels and the closest band*, such as electron excitations from the VB to acceptor levels or from donor levels to the CB; (iii) *intra-band* involving transitions between states lying in the same band. The synopsis of absorption processes is reported in figure 9.12, left.

The typical absorption spectrum of a semiconductor is shown qualitatively in figure 9.13. First of all we observe that $\alpha(\omega)$ is represented in arbitrary units because its quantitative estimation can only be obtained by thoroughly calculating the transition probabilities for any single intra-band and inter-band transition, as well as for any other one occurring between band levels and impurity states or by lattice absorption [6]. In addition, those that in figure 9.13 are schematically displayed as single peaks, in fact appear in a real experimental spectrum as multi-peak structures, each one corresponding to a specific transition of the family of transitions active in the spectral region considered.

9.4.3 Inter-band absorption

We will develop our arguments considering the model case of an intrinsic semiconductor in thermal equilibrium at room temperature and in approximation of non-degenerate parabolic bands. We will focus on the spectral region from near

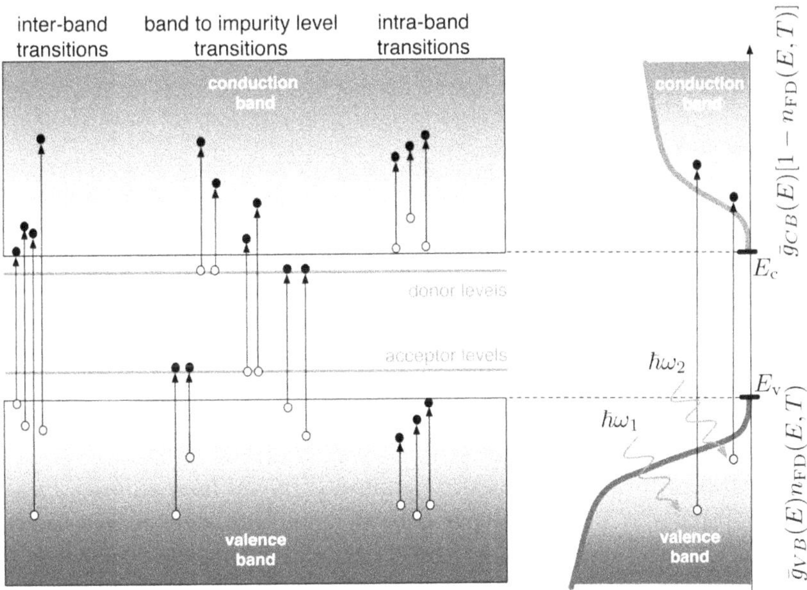

Figure 9.12. Left: Classification of the typical absorption transitions in a model compensated semiconductor at $T > 0$ K, where both donor and acceptor impurities are present (green lines). The final and initial position of the excited electrons are shown by a full and empty dot, respectively. In the latter case the situation is interpreted as a hole generation. Right: the corresponding CB and VB eDOS. A larger number of photons with energy $\hbar\omega_1$ is absorbed since the joint density of states is larger. The colour gradient in the CB and VB bands indicates the number of states which, at the different energies, are available to serve as starting or ending point of the electron transition: the darker the colour, the higher such a number.

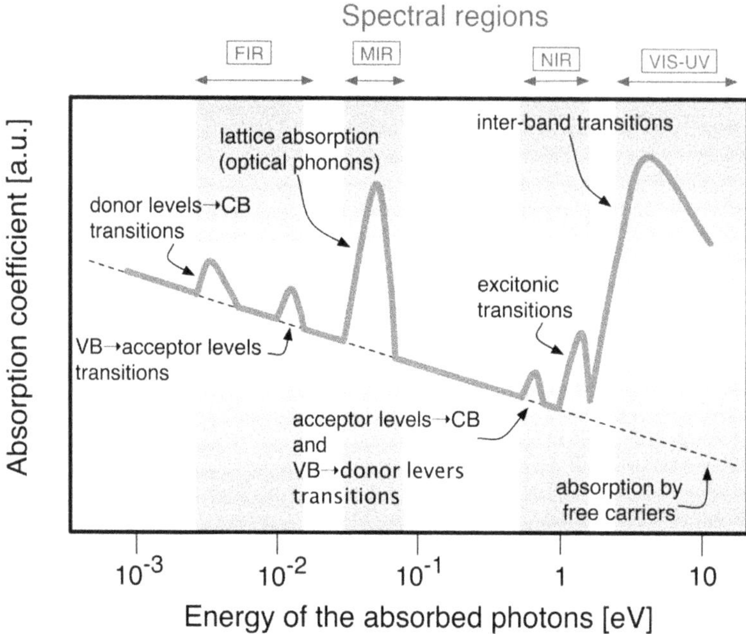

Figure 9.13. Schematic representation of the optical absorption spectrum (red line) of a semiconductor. The intensity and the shape of each peak are purely indicative.

infrared to ultraviolet. Despite its simplicity this model contains all the fundamental physics necessary for a conceptually rigorous treatment of optical absorption, allowing us to extend the main results of the model to all semiconductors. Under these assumptions, the only possible absorption process is associated with the *electron transitions between VB and CB states* since: (i) the intrinsic character of the semiconductor excludes transitions involving impurity states; (ii) the number of free carriers (electrons in CB or holes in VB) is very small, as the assumed temperature value sets a freeze-out condition (see figure 9.7) and, therefore, the continuous background absorption (dashed line in figure 9.13) due to intra-band transitions is negligible; (iii) absorption processes involving phonons are not possible in the selected spectral region, well above the energy scale of lattice vibrations.

As already anticipated, the absorption process of em radiation by a material medium is ultimately due to individual photon–electron interactions[24]. Since they occur in the absence of dissipative phenomena[25], *both the total energy and the total momentum of the electron–photon pair are conserved* during the absorption process. If the radiation is a monochromatic wave of angular frequency ω_{photon} and the photon wavevector is k_{photon}, then we can impose

---

[24] Yet another tacit assumption: we are assuming that only one-photon absorption processes occur. Two(or more)-photon processes are indeed possible, but definitely much less frequent and challenging to describe [16, 19, 20].

[25] In other words, the photon–electron interaction does not involve external actions to the particle pair.

$$E_2 - E_1 = \hbar\omega_{\text{photon}} \quad \text{and} \quad \hbar k_2 - \hbar k_1 = \hbar k_{\text{photon}}, \qquad (9.43)$$

where the initial and final electron states have, respectively, energy E_1 and E_2 and the corresponding wavefunctions have wavevector k_1 and k_2. The momentum conservation constraint lends itself to an important remark: for any radiation frequency in between the infrared and the ultraviolet regions, the photon wavelength λ_{photon} is always $\lambda_{\text{photon}} \gtrsim 250$ nm on the other hand, the typical de Broglie wavelength λ_e of the matter wave describing a crystalline electron is comparable to the lattice constant a_0, so that $\lambda_e \sim 1$ nm. Therefore, *always verified is the condition* $\lambda_{\text{photon}} \gg \lambda_e$. This simple order-of-magnitude estimates clearly shows that *the photon wavevector is always negligibly small as compared to its electron counterpart*[26]. This leads to an important practical consequence: we can to a very good approximation always set

$$\hbar k_2 - \hbar k_1 = \hbar k_{\text{photon}} \sim 0 \quad \rightarrow \quad \Delta k = 0, \qquad (9.44)$$

which dictates that *one-photon inter-band transitions can only be vertical*, that is, they occur with no change of electron momentum.

We eventually observe that the physics of optical absorption transitions depends on the underlying band structure which (i) determines the energy positioning of a pair of initial and final electron states and (ii) associates with any such pair a specific wavevector, fulfilling the conservation law in the form given in equation (9.43). In order to proceed further we must, therefore, consider separately the case of a direct gap semiconductor form the case of an indirect gap one.

Direct transitions
Let us consider a direct gap semiconductor, such as gallium arsenide, under the above simplifying assumptions. The first important characteristic of intra-band processes that take place under irradiation is that *no absorption of photons with energy $\hbar\omega_{\text{photon}} < E_{\text{gap}}$ can actually take place*. Let us consider an electron initially placed anywhere in VB with energy E_1; then, the absorption of a photon of energy $\hbar\omega_{\text{photon}} < E_{\text{gap}}$ would promote it to an energy level $E_2 = E_1 + \hbar\omega_{\text{photon}}$. This level, however, would inevitably end up falling below the bottom of the CB, that is in the forbidden gap. We can describe this situation by saying that *this absorption process is prohibited because the energy E_2 of the final state does not correspond to any allowed crystalline electron state*[27]. Ultimately, the semiconductor is transparent to the electromagnetic radiation component at this frequency. The physical situation is shown schematically in figure 9.14. This result can be summarised by stating that $\alpha(\omega) = 0$ for any $\omega < E_{\text{gap}}/\hbar$.

---

[26] More explicitly: for a photon in the indicated spectral range we have $k_{\text{photon}} = 2\pi/\lambda_{\text{photon}} < 0.025$ nm^{-1}. Instead, for a crystalline electron we have $k_e \sim 2\pi/a_0 \sim 10^1 - 10^2$ nm^{-1}. Therefore, k_{photon} is always some good order-of-magnitude smaller than k_e.

[27] A much less phenomenological approach would require the application of time-dependent perturbation theory, eventually providing a zero probability for such a transition [16, 19, 20].

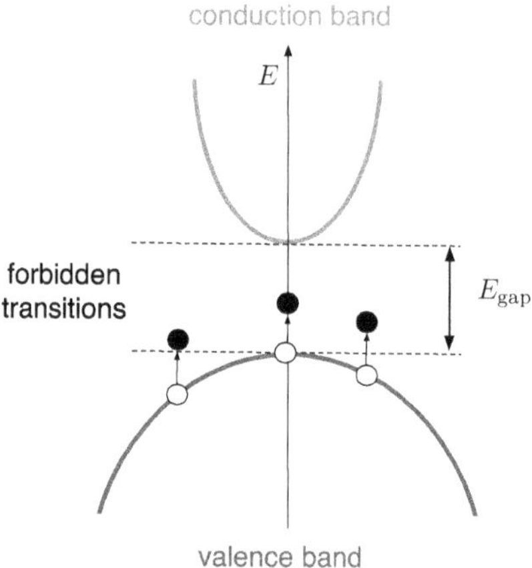

Figure 9.14. These direct optical transitions do not occur since the final state lies in the forbidden gap.

The occurrence of intra-band transitions has a dramatic dependence on the density of the electron states: the absorption, in fact, will be more or less intense due to the greater or lesser number of pairs of initial/final states available to host the corresponding electron transition. The finite temperature eDOS across the forbidden gap is represented in figure 9.12(right), where the eDOS given in equations (9.3) and (9.4) are reported, respectively, weighted by the probability that their levels are empty (CB states must host the promoted electrons) and occupied (VB states must release electrons), respectively. From figure 9.12(right) it is immediate to conclude that the absorption will be low for photons with energy $\hbar\omega_{\mathrm{photon}} \sim E_{\mathrm{gap}}$, given the fact that the number of pairs of full/empty states available, respectively, in proximity of the VB top and CB bottom is low; on the other hand, as ω_{photon} increases the number of pairs accordingly increases and the absorption of photons becomes more and more likely. The situation is described by stating that *near the gap the absorption coefficient depends on the joint density of states in VB and CB*. Here, the actual $\alpha(\omega)$ dependence directly derives from the parabolic bands approximation

$$\alpha(\omega) \sim \left(\hbar\omega - E_{\mathrm{gap}}\right)^{1/2}, \tag{9.45}$$

which indicates the more straight method to determine E_{gap} in direct gap semiconductors: regardless of practical experimental setup[28], just measure $\alpha(\omega)$ over a suitable range of frequencies and plot its square $\alpha^2(\omega)$ as a function of the energy of the absorbed photons. This is done in figure 9.15: the intercept of the linear extrapolation of the experimental data for $\alpha^2(\omega)$ with the energy axis directly provides the semiconductor energy gap.

We conclude by observing that *the momentum conservation constraint makes intra-band transitions between electronic states with different wave vectors prohibited*, even if energetically admissible. In figure 9.16 we illustrate this concept: the

---

[28] Not at all an easy task, indeed.

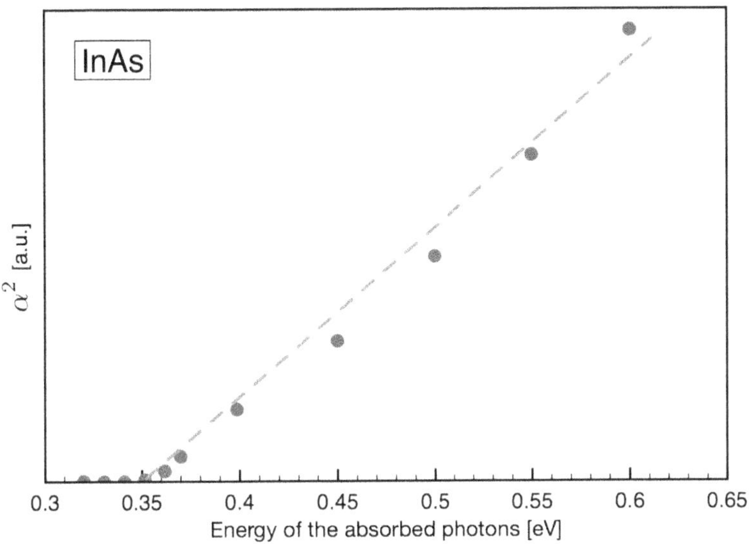

Figure 9.15. Room temperature absorption in InAs. It is reported the square magnitude α^2 of the coefficient for a direct estimation of $E_{\text{gap}} = 0.35$ eV through equation (9.45). Experimental data (blue dots) are taken from [21].

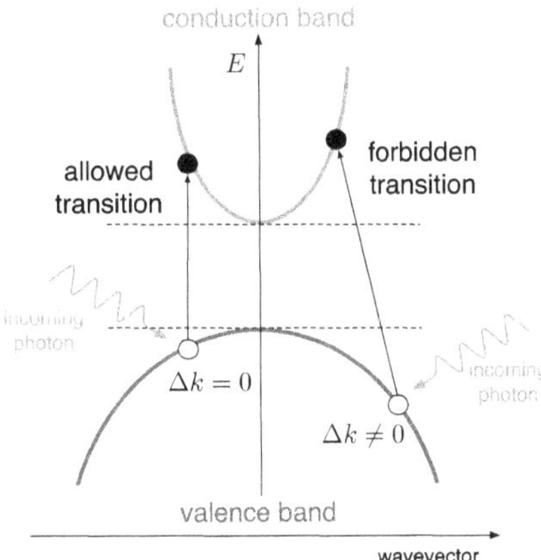

Figure 9.16. Schematic representation of allowed and forbidden optical transitions in a direct gap semiconductor.

transition on the left in the picture is allowed, because it is vertical; conversely, the transition on the right is clearly prohibited because $\Delta k \neq 0$ and, therefore, it would imply a variation of electron wavevector. Ultimately, the two conservation laws reported in equation (9.43) act in practice as *selection rules that discriminate between physically possible absorption processes and prohibited ones.*

Indirect transitions

It is now time to consider the case of an indirect gap semiconductor, such as silicon, always under the same simplifying hypotheses as before. As a starting point of our discussion, we can assume that also in this case the E_{gap} value is an absorption threshold for photons: in fact, the same arguments developed previously are still valid. However, in order to exceed this threshold *it is now necessary to supply to the electron not only the minimum amount of energy sufficient to pass from the VB to the CB, but also the exact amount Δk of wavevector necessary to connect the VB top to the CB bottom* (which in this case are no more vertically aligned). Since the absorbed photon does not carry momentum as imposed by equation (9.44), *the transition can take place only if accompanied by the absorption of a phonon with suitable wave vector $q = \Delta k$*. Of course, the phonon also carries a certain amount of energy $\hbar\omega_{\text{phonon}}$ (where ω_{phonon} is the frequency of the corresponding vibrational mode), which implies that *the absorption threshold is lowered to $E_{\text{gap}} - \hbar\omega_{\text{phonon}}$*. In other words, the electron must simultaneously absorb a photon and a phonon, respectively, from the radiation and thermal bath[29]. The situation is schematically represented in figure 9.17.

Indirect transitions are complex physical processes involving three-particles: an electron, a photon, and a phonon[30]. Their occurrence rate is therefore much lower than in the case of direct transitions [4]: *indirect gap semiconductors are very poorly efficient absorbers of electromagnetic radiation*. We finally observe that also possible is the indirect process involving the absorption of a large energy photon (that brings the electron far above the CB bottom), followed by the emission of a phonon that, at

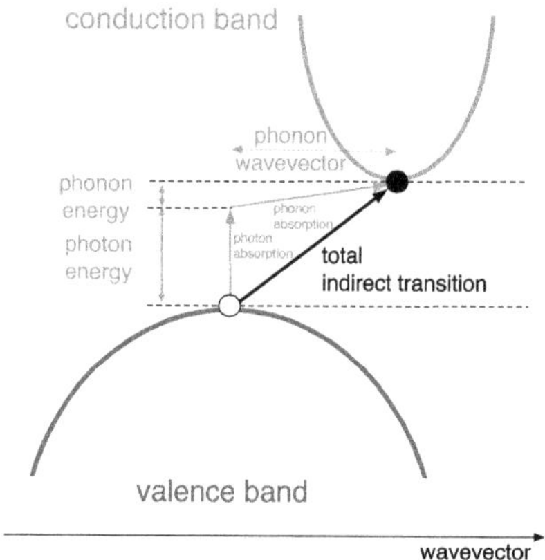

Figure 9.17. Schematic representation of an allowed optical transition in an indirect gap semiconductor.

---

[29] The careful reader should have noticed that this is the second time—after discussing the electron–phonon scattering in transport theory—we deal with phenomena falling beyond the strict adiabatic approximation discussed in section 1.3.4.

[30] Actually, things can be even more complicated than that, as more than two phonons can be actually required. It is said that indirect transitions are processes of a higher order than direct ones.

the same time, allows (i) the excited electron to relax towards the CB bottom level and (ii) the gain of that momentum component necessary to activate the non vertical transition.

The $\alpha(\omega)$ dependence for indirect transitions can no longer be derived in an elementary way as in the previous case. In fact, only an accurate perturbative calculation [4] shows that this dependence is

$$\alpha(\omega) \sim \left[\hbar\omega - \left(E_{\text{gap}} + \hbar\omega_{\text{phonon}} \right) \right]^2, \tag{9.46}$$

where for sake of simplicity we have assumed a one-phonon event. This result is nicely consistent with experimental evidence, as reported in figure 9.18 in the case of germanium. Thanks to equation (9.46) the energy gap value is determined by a simple extrapolation, once that absorption shoulder generated by the phonon-assisted mechanism at energies slightly lower than E_{gap} is properly considered.

9.4.4 Excitons

Each inter-band transition generates an electron–hole pair. Since the two particles are generated at the same point within the semiconductor and have opposite electric charge, they can give life to *a bound electron–hole system, referred to as exciton,* which (in some cases) is free to move with the host crystal. As shown graphically in figure 9.19, we must distinguish between Wannier excitons and Frenkel excitons: the former are formed in semiconductors and, in fact, are delocalised; the latter, instead, are formed in insulating materials and are located near specific lattice sites, which

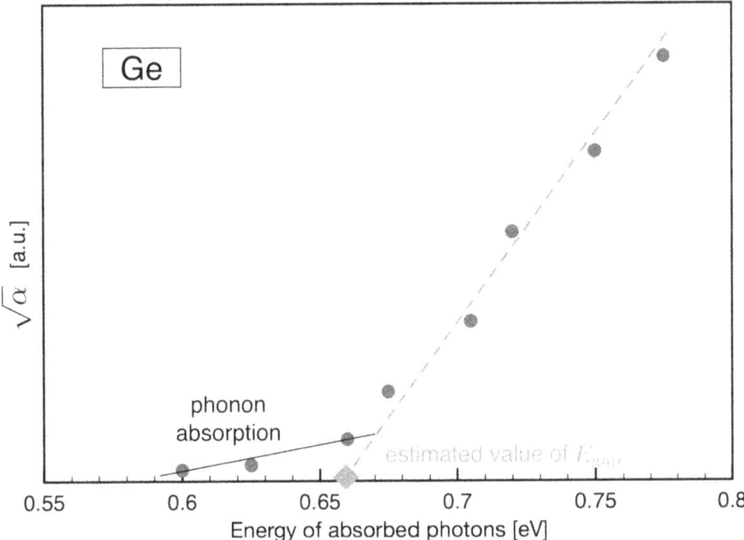

Figure 9.18. Room temperature absorption in Ge. It is reported the square root $\sqrt{\alpha}$ of the coefficient for a direct estimation of $E_{\text{gap}} = 0.66$ eV through equation (9.46) by extrapolation (magenta dashed line), once that tell-tale sign of phonon absorption (black full line) has been duly taken into account. Experimental data (blue dots) are taken from [21].

Figure 9.19. Pictorial view of Wannier and Frenkel excitons in a model semiconductor whose atoms are represented by the grey spheres.

act as exciton traps. We will only address the case of Wannier excitons, without further specification.

In order to describe the basic physics of an exciton we can use a rather simple model by extending the results known for the hydrogen atom [5] to the case of the electron–hole bound pair. In doing so, we must have only two shrewdnesses. First of all, we must use an *effective exciton mass value* μ_x properly defined as

$$\frac{1}{\mu_x} = \frac{1}{m_e} + \frac{1}{m_h},\tag{9.47}$$

and then we must consider that the bound pair moves within a crystal and not in vacuum: consequently, the intensity of the Coulomb electron–hole interaction will be decreased, compared to that of the electron–proton pair of the hydrogen atom, by a factor ϵ_r^{-1} provided by the relative permittivity of the semiconductor. By introducing these corrections to the hydrogenic levels, we immediately get for the exciton energy E_x this result

$$E_x = -\frac{\mu_x}{m_e \epsilon_r^2}\mathcal{R}_H\,\frac{1}{n^2}\quad\text{with}\quad n = 1, 2, 3, \cdots,\tag{9.48}$$

where $\mathcal{R}_H$ is the Rydberg constant for hydrogen. Breaking news: *an exciton represents a quantum system with a discrete energy spectrum.* Typical values of the binding energy of an exciton in its ground-state (corresponding to the $n = 1$ state in equation (9.48)) fall in the range 10^{-3}–10^{-2} eV. Therefore, at room temperature[31] it is very difficult to observe excitons in a semiconductor: we must cool the system at very low temperatures, so that the thermal excitations become inefficient to dissolve the electron–hole pair.

---

[31] The energy of the thermal bath is of the order of $k_B T \sim 0.026$ eV at room temperature.

The exciton size is defined by its electron–hole orbital radius

$$r_{\mathrm{x}} = \frac{m_{\mathrm{e}}\epsilon_{\mathrm{r}}}{\mu_{\mathrm{x}}}\, a_{\mathrm{H}}\, n^2 \qquad \text{with} \quad n = 1, 2, 3, \ldots, \tag{9.49}$$

where $a_{\mathrm{H}} = 0.529$ Å is the Bohr radius. Typical values in semiconductor materials are in the range 1–10 nm.

Measurements provide evidence of some general trends in most elemental and compound semiconductors (both type III–V and type II–VI): (i) it is easier to observe excitons in direct gap semiconductors; (ii) the larger the forbidden energy gap, the larger the exciton binding energy; (iii) as the forbidden energy gap increases, the electron–hole orbital radius decreases. The first evidence is due to the fact that, in order to form a bound pair, the electron and the hole must have the same velocity[32]. In direct gap semiconductors this condition is fully verified by the zone-centre transitions, i.e. those occurring at $k = 0$ (see for example the case of GaAs shown in figure 8.8): vertical inter-band transitions occur between the VB top and the CB bottom at Γ; this is tantamount to saying that the transitions interest states for both of which the corresponding group velocities are zero. The last two characteristics are instead due to the empirically observed fact that upon increasing the value of the forbidden gap (i) the relative dielectric constant ϵ_{r} increases and (ii) the exciton effective mass μ_{x} decreases.

Excitons affect the optical spectrum of a semiconductor in the region of energies close to the absorption threshold, as shown in figure 9.20. We will discuss this phenomenon in the particularly simple case of a direct gap semiconductor in which the vertical transition takes place at the zone-centre: this is the case of GaAs. The amount of energy stored in an exciton is given by the difference between the energy required to form the electron–hole pair and the energy needed to permanently bind

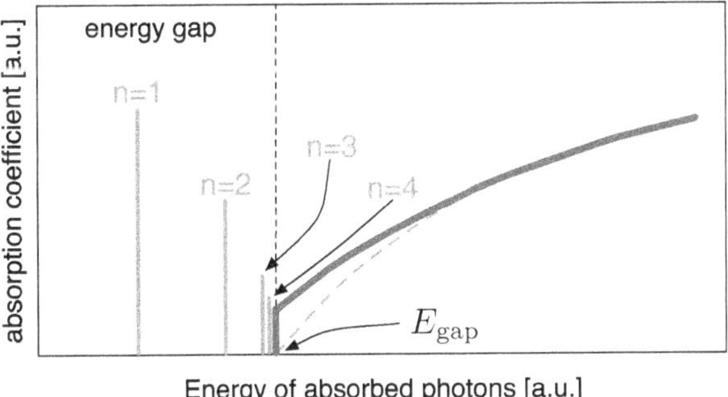

Figure 9.20. Absorption spectrum (conceptual rendering) of a direct gap intrinsic semiconductor nearby the absorption threshold. Blue full line: inter-band absorption; red lines: exciton absorption. Dashed magenta line: the residual absorption subtracted by exciton effects. The label n is defined in equation (9.50).

---

[32] If the two velocities were different, the two particles would inevitably end up moving away, thus weakening their binding and eventually dissolving the pair.

the two particles. In our study-case, the first energy is equal to E_{gap}, while the second is given by equation (9.48). A simple relation for *the discrete exciton spectrum $E_{\text{x},n}$* is accordingly calculated

$$E_{\text{x},n} = E_{\text{gap}} - \frac{\mu_{\text{x}}}{m_{\text{e}}\epsilon_{\text{r}}^2}\, \mathcal{R}_{\text{H}}\, \frac{1}{n^2} \quad \text{with} \quad n = 1, 2, 3, \dots , \tag{9.50}$$

showing that *an exciton can in fact absorb photons at discrete energies, each smaller than the energy gap.* Similarly to atomic hydrogen spectroscopy, these processes generate *a discrete absorption spectrum* that adds to the ordinary inter-band spectrum and, in fact, lowers the intrinsic absorption threshold of the semiconductor.

References

[1] Balkanski M and Wallis R F 1989 *Semiconductor Physics and Applications* (Oxford: Oxford University Press)

[2] Cardona M and Yu P Y 2010 *Fundamentals of Semiconductors* (Heidelberg: Springer)

[3] Grundmann M 2010 *The Physics of Semiconductors* (Heidelberg: Springer)

[4] Seeger K 1989 *Semiconductor Physics* (Heidelberg: Springer)

[5] Colombo L 2019 *Atomic and Molecular Physics: A Primer* (Bristol: IOP Publishing)

[6] Wolfe C M, Holonyak N Jr and Stillman G E 1989 *Physical Properties of Semiconductors* (Englewood Cliffs, NJ: Prentice-Hall)

[7] Wolfe C M, Stillman G E and Lindley W T J 1970 *Appl. Phys.* **41** 3088

[8] Jacoboni C, Canali C, Ottaviani G and Alberigi Quaranta A 1977 *Solid-State Electron.* **20** 77–89

[9] Sze S M 1981 *Physics of Semiconductor Devices* (New York: Wiley)

[10] Swendsen R H 2012 *An Introduction to Statistical Mechanics and Thermodynamics* (Oxford: Oxford University Press)

[11] Callen H 1985 *Thermodynamics and An Introduction to Thermostatistics* (New York: Wiley)

[12] Dittman R and Zemansky M 1997 *Heat and Thermodynamics* 7th edn (New York: McGraw-Hill)

[13] Jacobini C, Nava F, Canali C and Ottaviani G 1981 *Phys. Rev.* B **24** 1014

[14] Reif F 1987 *Fundamentals of Statistical and Thermal Physics* (New York: McGraw-Hill)

[15] Pauling L 1970 *General Chemistry* (New York: Dover)

[16] Fox M 2001 *Optical Properties of Solids* (Oxford: Oxford University Press)

[17] Feynman R P, Leighton R B and Sands M 1963 *The Feynman Lectures on Physics* (Reading, MA: Addison-Wesley)

[18] Kenyon I R 2008 *The Light Fantastic—A Modern Introduction to Classical and Quantum Optics* (Oxford: Oxford Science Publications)

[19] Grosso G and Pastori Parravicini G 2014 *Solid State Physics* 2nd edn (Oxford: Academic)

[20] Bassani F and Pastori-Parravicini G 1975 *Electronic States and Optical Transitions in Solids* (Oxford: Pergamon)

[21] Palik E D 1985 *Handbook of the Optical Constants of Solids* (San Diego, CA: Academic)

IOP Publishing

Solid State Physics
A primer
Luciano Colombo

Chapter 10

Density functional theory

Syllabus—*Density functional theory is outlined on the basis of the fundamental Hohenberg–Kohn theorem and the practical Kohn–Sham procedure. The theorem first provides the mathematical background for replacing the total electron wavefunction with the electron density as the key object for the quantum theory of a system of interacting electrons subject to an external potential. Next, the Kohn–Sham procedure effectively reduces the overwhelmingly complicated many-body electron problem to a much simpler one corresponding to a fictitious system of non-interacting electrons, characterised by the same density as the real one. The combination of theorem and procedure leads to defining the exchange-correlation energy, whose calculation is here developed under the leading local density approximation. The density functional theory is nowadays referred to as the standard model for a fully parameter-free quantum theory of the solid state.*

10.1 Setting the problem and cleaning up the formalism

In chapter 1 we elaborated the most fundamental quantum mechanical formulation of solid state theory, leading to the formidable total Schrödinger problem given in equation (1.13), which is here reported for convenience

$$\left[-\frac{\hbar^2}{2}\sum_{\alpha}\frac{1}{M_{\alpha}}\nabla_{\alpha}^2 - \frac{\hbar^2}{2m_{\mathrm{e}}}\sum_i\nabla_i^2 + \hat{V}_{\mathrm{ne}}(\mathbf{r},\,\mathbf{R}) + \hat{V}_{\mathrm{nn}}(\mathbf{R}) + \hat{V}_{\mathrm{ee}}(\mathbf{r}) \right]\Phi(\mathbf{r},\,\mathbf{R}) = E_T\Phi(\mathbf{r},\,\mathbf{R}), \quad (10.1)$$

where the meaning of each symbol has been previously defined[1]. Thanks to the adiabatic approximation (see section 1.3.4), the total wavefunction is cast in the form of the product between the total ionic and electronic wavefunctions $\Phi(\mathbf{r},\,\mathbf{R}) = \Psi_{\mathrm{n}}(\mathbf{R})\Psi_{\mathrm{e}}^{(\mathbf{R})}(\mathbf{r})$ and this allows us to write two separate constitutive equations for the electron and ion problems, respectively; in particular, the

---

[1] We recall that in this equation $\mathbf{r}$ and $\mathbf{R}$ are the full set of positions of all electrons and all ions, respectively.

doi:10.1088/978-0-7503-2265-2ch10

eigenvalue problem for the electrons is set by the following equation (previously labelled as equation (1.15))

$$\left[-\frac{\hbar^2}{2m_e} \sum_i \nabla_i^2 + \hat{V}_{ne}(\mathbf{r}, \mathbf{R}) + \hat{V}_{ee}(\mathbf{r}) \right] \Psi_e^{(\mathbf{R})}(\mathbf{r}) = E_e^{(\mathbf{R})}\Psi_e^{(\mathbf{R})}(\mathbf{r}), \qquad (10.2)$$

where $\Psi_e^{(\mathbf{R})}(\mathbf{r})$ and $E_e^{(\mathbf{R})}$ are the total electron wavefunction and energy, respectively, when ions are clamped in the $\mathbf{R}$ configuration. This equation contains a first important conceptual achievement: *in solid state physics, the electron eigenvalue problem must be solved for each possible ion displacement pattern*: this, of course, reflects a very demanding task.

The estimation of the workload required to solve equation (10.2) is even more daunting if we consider that no analytical solution is indeed conceivable and, therefore, we must seek a numerical one. Basically, this requires that we partition the real space into a grid of discrete points, so that the derivatives appearing in equation (10.2) can be calculated by finite-difference methods. We better understand the key issue of the numerical approach by considering the practical case of a crystal containing N_c atoms in its unit cell of volume V_c. We know that such a volume is $V_c \sim \mathcal{O}(\text{Å}^3)$ and, therefore, it is straightforward to admit that an accurate discretisation for the numerical calculation should consist in a grid of points at least spaced by ~ 0.1 Å along any Cartesian direction. This implies that we need at least $N_p \sim \mathcal{O}(10^3)$ points. Counting N_{val} valence electrons for each basis atom, the numerical calculation would require $N_p^{N_c N_{val}}$ complex numbers. In the case of, say, silicon $N_c = 2$, $N_{val} = 4$ so that $\mathcal{O}(10^{24})$ numbers should be calculated. Since the problem given in equation (10.2) is typically translated into matrix operations[2], it is clearly impossible to perform them with arrays of $\mathcal{O}(10^{24})$ rank. The single-particle approximation was elaborated precisely to bypass the numerical bottleneck just described. Here we are instead going to follow a rather different approach, which will consist in *replacing the total electron wavefunction by the electron density as the key quantity of the solid state theory*.

In order to keep the formalism as clean as possible we are going to omit some (annoying) details: at this stage of the discussion, the following simplifications of the system of symbols should not generate any ambiguity. First of all, we will leave out the indication of the functional dependence of $\Psi_e^{(\mathbf{R})}(\mathbf{r})$ on the set of all electronic coordinates $\mathbf{r}$, as well as its parametric dependence on the ionic configuration $\mathbf{R}$. Also, since the adiabatic approximation has been operated and we are focussing just on the electronic problem, there is no need to label either the wavefunction or the energy. In short: $\Psi_e^{(\mathbf{R})}(\mathbf{r}) \to \Psi$ and $E_e^{(\mathbf{R})} \to E$.

Another major simplification comes by switching from the SI units adopted so far to the *atomic units system* where distances are measured in units of the Bohr radius

---

[2] The electron energies and wavefunctions are numerically calculated respectively as the eigenvalues and eigenvectors of the matrix representation of the Hamiltonian operator.

a_0 [1], masses in units of the electron mass m_e, and energies in units $e^2/(4\pi\epsilon_0 a_0)$ which is referred to as Hartree. By this choice we get a twofold advantage: (i) atomic units are very conveniently proportioned, so that most of the solid state properties turn out to have almost unitary values (for instance, first nearest neighbours distances in crystals are of the order of 1 a.u. of length) and (ii) the electron Schrödinger equation assumes a very simple form, only depending on pure numbers[3].

By combining all these changes in the notation, we can rewrite equation (10.2) in the more compact form

$$\left[-\frac{1}{2}\sum_i \nabla_i^2 + \sum_i \hat{V}_{\text{ext}}(\mathbf{r}_i) + \sum_{i>j} \frac{1}{|\mathbf{r}_i - \mathbf{r}_j|} \right] \Psi = E\Psi, \tag{10.3}$$

where it has been introduced the operator

$$\hat{V}_{\text{ext}}(\mathbf{r}_i) = -\sum_\alpha \frac{Q_\alpha}{|\mathbf{r}_i - \mathbf{R}_\alpha|}, \tag{10.4}$$

describing the *external potential* $V_{\text{ext}}(\mathbf{r}_i)$ acting on each ith electron because of ionic Coulomb attraction. This wording stresses the fact that this kind of electrostatic interaction does involve the ions, which are objects *external to the electron system.* Equation (10.3) is very elegant since it clearly indicates that the only parameters needed to solve it are the Q_α charges, namely: we only need to know which chemical species are present in the crystal of interest in order to proceed with the solution of the electron eigenvalue problem. For this reason, this way of proceeding is known as a *first-principles or ab initio method*: the reader should appreciate the conceptual difference with respect to the tight-binding method developed in appendix G, where two- and three-centre energy integrals were considered disposable constants to fit.

The notation here presented will be used throughout this chapter. We made this choice so as to match our formalism to the common practice [2–6]: as a matter of fact, the density functional theory literature is prevalently written in atomic units and makes use of the distinction between electron–ion interactions (external potential term) and electron–electron interactions (Hartree term).

10.2 The Hohenberg–Kohn theorem

We start from a seemingly trivial statement: equation (10.3) provides evidence that any change in E is associated with a change in the electron wavefunction Ψ. Formally, this is tantamount to saying that *the energy is a functional of* Ψ this mathematical concept is outlined in appendix I.

The key achievement of the resulting formalism—which is referred to as the *density functional theory* (DFT)—is a theorem firstly proved by Hohenberg and Kohn in 1964 which states that *the ground-state energy E_{GS} of an electron–ion system*[4] *is uniquely defined by a functional of the corresponding electron density ρ_{GS}*

---

[3] This means that it can be solved once for all, regardless the adopted unit system.
[4] Throughout this chapter we assume spinless electrons.

$$E_{GS} = F[\rho_{GS}], \tag{10.5}$$

where, interestingly enough, ρ_{GS} is a function of just three variables: those needed to specify the space position where such a density must be evaluated. We remark that, for any quantum state, the electron density is defined as

$$\rho(\mathbf{r}) = N \int |\Psi|^2 \, d\mathbf{r}_2 d\mathbf{r}_3 d\mathbf{r}_4 \cdots d\mathbf{r}_N, \tag{10.6}$$

where N is the total number of electrons so that

$$\int \rho(\mathbf{r})d\mathbf{r} = N, \tag{10.7}$$

which corresponds to a standard normalisation condition.

The proof of the Hohenberg–Kohn theorem proceeds in two steps. We first observe that by changing the ionic configuration $\mathbf{R} \to \mathbf{R}'$ we correspondingly change the external potential $V_{\text{ext}} \to V'_{\text{ext}}$ or, in other words, we affect the Schrödinger problem cast in equation (10.3). Therefore, we conclude that in *any quantum state the external potential uniquely determines the total electron wavefunction* and so the electron density. We now introduce an arbitrary assumption by supposing that the very same ground-state electron density ρ_{GS} can be found for two unalike ionic configurations, respectively, generating the external potentials V_{ext} and V'_{ext} with $V_{\text{ext}} \neq V'_{\text{ext}}$. By setting

$$\hat{H} = \left[-\frac{1}{2} \sum_i \nabla_i^2 + \sum_i \hat{V}_{\text{ext}}(\mathbf{r}_i) + \sum_{i>j} \frac{1}{|\mathbf{r}_i - \mathbf{r}_j|} \right]$$
$$= \hat{T} + \sum_i \hat{V}_{\text{ext}}(\mathbf{r}_i) + \hat{V}_{\text{Hartree}} \tag{10.8}$$

$$\hat{H}' = \left[-\frac{1}{2} \sum_i \nabla_i^2 + \sum_i \hat{V}'_{\text{ext}}(\mathbf{r}_i) + \sum_{i>j} \frac{1}{|\mathbf{r}_i - \mathbf{r}_j|} \right]$$
$$= \hat{T} + \sum_i \hat{V}'_{\text{ext}}(\mathbf{r}_i) + \hat{V}_{\text{Hartree}}, \tag{10.9}$$

we can write two eigenvalue problems

$$\hat{H}\Psi = E\Psi \quad \text{and} \quad \hat{H}'\Psi' = E'\Psi', \tag{10.10}$$

where $\Psi \neq \Psi'$ and $E \neq E'$. For further convenience we have introduced the total electron kinetic energy operator $\hat{T}$ and the Hartree operator $\hat{V}_{\text{Hartree}}$ associated to the electron–electron Coulomb interaction energy. By using the eigenfunction Ψ of the first Schrödinger problem appearing in equation (10.10), we can write[5]

---

[5] This proof is based on the standard variational principle of quantum mechanics.

$$\int \Psi^* \left[-\frac{1}{2} \sum_i \nabla_i^2 + \sum_{i>j} \frac{1}{|\mathbf{r}_i - \mathbf{r}_j|} \right] \Psi \, d\mathbf{r}_1 \cdots d\mathbf{r}_N$$

$$+ \int \Psi^* \left[\sum_i \hat{V}_{\text{ext}}(\mathbf{r}_i) \right] \Psi \, d\mathbf{r}_1 \cdots d\mathbf{r}_N = E$$

$$\int \Psi^* \left[-\frac{1}{2} \sum_i \nabla_i^2 + \sum_{i>j} \frac{1}{|\mathbf{r}_i - \mathbf{r}_j|} \right] \Psi \, d\mathbf{r}_1 \cdots d\mathbf{r}_N$$

$$+ \int \Psi^* \left[\sum_i \hat{V}'_{\text{ext}}(\mathbf{r}_i) \right] \Psi \, d\mathbf{r}_1 \cdots d\mathbf{r}_N > E',$$

$$(10.11)$$

where the second inequality holds since Ψ is not the ground-state eigenfunction of $\hat{H}'$. These two relations can be reformulated in terms of the ground-state electron density by taking into account the definition given in equation (10.6)

$$\int \Psi^* \left[-\frac{1}{2} \sum_i \nabla_i^2 + \sum_{i>j} \frac{1}{|\mathbf{r}_i - \mathbf{r}_j|} \right] \Psi \, d\mathbf{r}_1 \cdots d\mathbf{r}_N + \int \rho_{\text{GS}}(\mathbf{r}) V_{\text{ext}}(\mathbf{r}) \, d\mathbf{r} = E$$

$$\int \Psi^* \left[-\frac{1}{2} \sum_i \nabla_i^2 + \sum_{i>j} \frac{1}{|\mathbf{r}_i - \mathbf{r}_j|} \right] \Psi \, d\mathbf{r}_1 \cdots d\mathbf{r}_N + \int \rho_{\text{GS}}(\mathbf{r}) V'_{\text{ext}}(\mathbf{r}) \, d\mathbf{r} > E'.$$

$$(10.12)$$

By combining these results we get

$$E - E' > \int \rho_{GS}(\mathbf{r})[V_{\text{ext}}(\mathbf{r}) - V'_{\text{ext}}(\mathbf{r})] \, d\mathbf{r}. \qquad (10.13)$$

The same procedure could be repeated by first calculating the ground state energy E' with the eigenfunction Ψ' and then by calculating the expectation value of $\hat{H}$ in the same state; this leads to

$$E' - E > \int \rho_{GS}(\mathbf{r})[V'_{\text{ext}}(\mathbf{r}) - V_{\text{ext}}(\mathbf{r})] \, d\mathbf{r}. \qquad (10.14)$$

By summing equations (10.13) and (10.14) we obtain the mathematical contradiction $0 > 0$: indeed an intriguing result! Since all calculations are correct, the only possible logical conclusion is that *our initial assumption that two unalike ionic configurations correspond to the very same ground-state electron density is false.*

In summary, the Hohenberg–Kohn theorem states that *the ground-state energy of an electron system is uniquely determined by the corresponding ground-state electron density* which only depends on three space coordinates. Through the use of a functional of the electron density, this beautiful formal result dramatically reduces the workload of the theory[6] by reducing the N-electron problem—defined by $3N$

space coordinates—to a different problem ruled by just three space coordinates. It is worth noting another important issue: since the knowledge of the ground state energy enables the calculation of many other properties, the formulation of the theorem can be meaningfully extended up to understand that the ground state electron density uniquely determines the properties of such a state[7].

In conclusion, we notice that the theory we developed so far does not provide any practical information about the functional $F[\rho_{GS}]$: we still lack any information about its mathematical form and the practical way to calculate it. This task will be developed in the next section.

10.3 The Kohn–Sham equations

In order to work out the electron density functional, we can combine equation (10.5) with equation (10.12)(top)

$$F[\rho] = \int \rho(\mathbf{r}) V_{\text{ext}}(\mathbf{r})\, d\mathbf{r} + \int \Psi^* \left(\hat{T} + \hat{V}_{\text{Hartree}} \right) \Psi\, d\mathbf{r}_1 \cdots d\mathbf{r}_N, \qquad (10.15)$$

and notice that the dependence upon the electron density is explicit in the first term describing the external potential, while it is only implicit[8] in the kinetic energy and Hartree terms, thus making the mathematical problem really very convoluted.

The problem is solved by *the Kohn–Sham procedure* [2, 3] which relies on suitable independent-electron wavefunctions obtained by introducing *a fictitious system of non-interacting electrons characterised by the same density as the real one*. We are developing this procedure in the following, omitting most of the mathematical subtleties which are addressed in more advanced books [4, 5].

The key idea is threefold: (i) first, we introduce new single-particle *Kohn–Sham orbitals* $\{\varphi_{\text{KS},i}(\mathbf{r})\}_{i=1,2,\ldots,N}$ to be determined in the following and write

$$\rho(\mathbf{r}) = \sum_i \left| \varphi_{\text{KS},i}(\mathbf{r}) \right|^2, \qquad (10.16)$$

for the electron density of the corresponding fictitious system of independent electrons; (ii) next, we calculate its kinetic and Hartree contributions to the total energy, as well as the contribution due to the external potential; (iii) finally, we add a new term to the total energy functional $F[\rho]$ accounting for the difference between the fictitious independent-electron system described by the Kohn–Sham orbitals and the real one described by a many-body multi-electron wavefunction. This new term, hereafter indicated by $E_{\text{xc}}[\rho]$, is cast in the form of a (still unknown) functional of the

---

[6] More precisely: the workload associated with the numerical implementation of the theory.

[7] We stress that this result is only valid for the ground state, since its energy is the lowest possible one: a feature which has been explicitly exploited in our formal proof of the Hohenberg–Kohn theorem. For a generic quantum state other than the ground one, a similar conclusion cannot be drawn.

[8] More precisely, it is the wavefunction itself to be a functional of the density: $\Psi[\rho]$.

electron density and it is referred to as the *exchange-correlation energy*. This procedure leads to

$$F[\rho] = \int \rho(\mathbf{r}) V_{\text{ext}}(\mathbf{r}) \, d\mathbf{r}$$
$$- \frac{1}{2} \sum_i \int \varphi^*_{\text{KS},i}(\mathbf{r}) \nabla^2 \varphi_{\text{KS},i}(\mathbf{r}) \, d\mathbf{r} \tag{10.17}$$
$$+ \frac{1}{2} \int \int \frac{\rho(\mathbf{r})\rho(\mathbf{r}')}{|\mathbf{r} - \mathbf{r}'|} \, d\mathbf{r} d\mathbf{r}'$$
$$+ E_{\text{xc}}[\rho],$$

where the first three contributions on the right-hand side define *the total energy of an idealised system of independent electrons*[9]. The last term instead carries all the physical information needed to pass from the idealised to the real system: its is customary to set $E_{\text{xc}}[\rho] = E_{\text{x}}[\rho] + E_{\text{c}}[\rho]$ and treat separately the exchange and the correlation contributions, as outlined in the following.

According to the Hohenberg–Kohn theorem, *the ground-state electron density $\rho_{\text{GS}}(\mathbf{r})$ is that which minimises the energy $E = F[\rho]$*: this statement is formalised by a variational principle, that is by setting to zero its functional derivative (see appendix I)

$$\left. \frac{\delta F[\rho]}{\delta \rho} \right|_{\rho=\rho_{\text{GS}}} = 0. \tag{10.18}$$

By using the chain rule for the derivative of a composite function, we can write

$$\left. \frac{\delta F[\rho]}{\delta \varphi^*_{\text{KS},i}} \right|_{\rho=\rho_{\text{GS}}} = \left. \frac{\delta F[\rho]}{\delta \rho} \right|_{\rho=\rho_{\text{GS}}} \frac{\delta \rho}{\delta \varphi^*_{\text{KS},i}} = \left. \frac{\delta F[\rho]}{\delta \rho} \right|_{\rho=\rho_{\text{GS}}} \varphi_{\text{KS},i} = 0, \tag{10.19}$$

which represents an interesting conclusion: *the functional derivative of the total energy with respect to the KS orbitals is zero in the ground-state*. In order to proceed we must request such orbitals to be orthonormal

$$\int \varphi^*_{\text{KS},i}(\mathbf{r})\varphi_{\text{KS},j}(\mathbf{r}) \, d\mathbf{r} = \delta_{ij}, \tag{10.20}$$

so that the KS electron density provided in equation (10.16) is properly normalised. As usual, this constraint to equation (10.19) is enforced (i) by introducing Lagrange multipliers ξ_{ij}, (ii) by defining the constrained energy functional F'

$$F' = F - \sum_{ij} \xi_{ik}\left[\int \varphi^*_{\text{KS},i}(\mathbf{r})\varphi_{\text{KS},j}(\mathbf{r}) \, d\mathbf{r} - \delta_{ij}\right], \tag{10.21}$$

---

[9] Once again, we remark that they correspond to the external potential (electron–ion Coulomb interaction term), the total kinetic energy, and the Hartree (electron–electron Coulomb interaction term) contributions, respectively.

and (iii) by eventually imposing its minimum condition

$$\left. \frac{\delta F'[\rho]}{\delta \varphi_{\mathrm{KS},i}^{*}} \right|_{\rho=\rho_{\mathrm{GS}}} = 0. \tag{10.22}$$

This leads to

$$
\begin{aligned}
\sum_{j} \xi_{ij}\, \varphi_{\mathrm{KS},j} &= \left. \frac{\delta F[\rho]}{\delta \varphi_{\mathrm{KS},i}^{*}} \right|_{\rho=\rho_{\mathrm{GS}}} \\
&= -\frac{1}{2}\nabla^2 \varphi_{\mathrm{KS},i} \\
&\quad + \frac{\delta}{\delta\rho}\left[\int \rho(\mathbf{r})V_{\mathrm{ext}}(\mathbf{r})\, d\mathbf{r}\right]\varphi_{\mathrm{KS},i}(\mathbf{r}) \\
&\quad + \frac{1}{2}\frac{\delta}{\delta\rho}\left[\int\int \frac{\rho(\mathbf{r})\rho(\mathbf{r}')}{|\mathbf{r}-\mathbf{r}'|}\, d\mathbf{r}d\mathbf{r}'\right]\varphi_{\mathrm{KS},i}(\mathbf{r}) \\
&\quad + \frac{\delta E_{\mathrm{xc}}[\rho]}{\delta\rho}\varphi_{\mathrm{KS},i}(\mathbf{r}),
\end{aligned}
\tag{10.23}
$$

where we have made use of equation (10.19). By calculating the functional derivatives (see appendix I)

$$
\begin{aligned}
\frac{\delta}{\delta\rho}\left[\int \rho(\mathbf{r})V_{\mathrm{ext}}(\mathbf{r})\, d\mathbf{r}\right] &= V_{\mathrm{ext}}(\mathbf{r}) \\
\frac{\delta}{\delta\rho}\left[\int\int \frac{\rho(\mathbf{r})\rho(\mathbf{r}')}{|\mathbf{r}-\mathbf{r}'|}\, d\mathbf{r}d\mathbf{r}'\right] &= 2\int \frac{\rho(\mathbf{r}')}{|\mathbf{r}-\mathbf{r}'|}\, d\mathbf{r}',
\end{aligned}
\tag{10.24}
$$

we obtain

$$\sum_{j} \xi_{ij}\, \varphi_{\mathrm{KS},j} = \left[-\frac{1}{2}\nabla^2 + V_{\mathrm{ext}}(\mathbf{r}) + \int \frac{\rho(\mathbf{r}')}{|\mathbf{r}-\mathbf{r}'|}\, d\mathbf{r}' + \frac{\delta E_{\mathrm{xc}}}{\delta\rho}\right]\varphi_{\mathrm{KS},i}(\mathbf{r}), \tag{10.25}$$

which translates into a set of *single-particle Kohn–Sham equations*

$$\left[-\frac{1}{2}\nabla^2 + V_{\mathrm{ext}}(\mathbf{r}) + V_{\mathrm{Hartree}}(\mathbf{r}) + V_{\mathrm{xc}}(\mathbf{r})\right]\varphi_{\mathrm{KS},i}(\mathbf{r}) = E_i \varphi_{\mathrm{KS},i}(\mathbf{r}), \tag{10.26}$$

where the new term $V_{\mathrm{xc}}(\mathbf{r})$ is defined as

$$V_{\mathrm{xc}}(\mathbf{r}) = \left. \frac{\delta E_{\mathrm{xc}}[\rho]}{\delta\rho} \right|_{\rho=\rho_{\mathrm{GS}}}, \tag{10.27}$$

and it is called *exchange-correlation potential*.

The achievement represented by the equations (10.26) is more conceptual than practical, since *we do not really know the exchange-correlation potential yet* and this information is needed in order to provide the exact ground-state density and energy of the real many-body electron system. Without this correction we still operate at the level of the independent electron approximation (although at a higher erudition with respect to the tight-binding theory): something we definitely want to improve. For this reason, we need to develop some knowledge about the way to construct $E_{xc}[\rho]$: this is the aim of the next section. Here we can nevertheless anticipate a major conceptual breakthrough: provided that the exchange-correlation problem has been solved in some way or another, *the set of equations* (10.26) *plays a key role in the first-principles theory of the solid state* for a twofold reason, namely: (i) they allow us to construct the electron density as shown in equation (10.16); and (ii) they make it possible to calculate many ground-state physical properties of crystalline systems, through the prediction of its total energy $E_{GS} = F[\rho_{GS}]$. This is often recognised by addressing these equations as the *Kohn–Sham theory of the solid state*.

10.4 The exchange-correlation functional

The construction of the most accurate exchange-correlation functional for a many-body electron system is an active and lively topic of contemporary solid state theory. Presenting a full account of the various ongoing efforts pointing in this direction is surely beyond the scope of the present introductory chapter to DFT and we forward the avid reader to more advanced textbooks [4–6]. Here we limit ourselves to treating the problem at the lowest possible sophistication, known as the *local density approximation* (LDA), which has historically represented a first widely-adopted practical implementation of DFT and it still represents a meaningful benchmark.

The starting point is to approximate the real electron system as a *homogeneous electron gas* (HEG), namely a fictitious system where the ion positive charges are uniformly distributed in space—so as to form a positive background—and the electron density is similarly uniform. This model allows us to focus just on electrons and their mutual interactions, without taking into explicit consideration the role of ions[10]. The exchange energy E_x^{HEG} of this model system is exactly obtained by a standard quantum mechanical calculation [2]

$$E_x^{HEG} = -\frac{3}{4}\left(\frac{3}{\pi}\right)^{1/3} V \, [\rho^{HEG}]^{4/3}, \tag{10.28}$$

where $\rho^{HEG} = N/V$ is the density of an N-electron homogeneous gas with volume V. On the other hand, there is no exact result for the correlation energy E_c^{HEG}. A numerical procedure has therefore become the common practice: the correlation energy is obtained by the computed value of the HEG total energy (which is the output of the numerical solution of its complete Schrödinger equation) from which all the known contributions (kinetic, Hartree, and exchange energy) are subtracted. A standard parametrisation formula is used for the result [3–5]

---

[10] It is important to remark that the HEG is not the free electron gas discussed in chapter 7, since Coulomb repulsions between electrons are now considered.

$$E_{\mathrm{c}}^{\mathrm{HEG}} = V \; \rho^{\mathrm{HEG}} \times \begin{cases} 0.002 \; r_s \ln r_s - 0.116 \; r_s + 0.031\,1 \ln r_s - 0.048 & \text{if } r_s < 1 \\ -0.142\,3 \; (0.333\,4 \; r_s + 1.052\,9 \; \sqrt{r_s} + 1)^{-1} & \text{if } r_s \geqslant 1, \end{cases} \tag{10.29}$$

where $r_s = (3/4\pi\rho^{\mathrm{HEG}})^{1/3}$.

The electron density in a real crystal is of course not the same as in a homogeneous electron gas (simply because it cannot be as uniform) and, consequently, it would be a too drastic approximation just using the same HEG exchange and correlation energies throughout the system. Rather, we could divide the crystal into infinitesimal volume elements $d\mathbf{r}$ and attribute each of them a different *local density value* $\rho(\mathbf{r})$ to be used in the HEG expressions for the exchange and correlation energy functionals. The total exchange and correlation energies are then obtained by integrating over the crystal volume. This is easy to show in the case of the exchange term (which has the exact result reported in equation (10.28)) that this way of proceeding leads to

$$E_x^{\mathrm{LDA}}[\rho(\mathbf{r})] = \frac{1}{V} \int_V E_x^{\mathrm{HEG}}[\rho(\mathbf{r})] \; d\mathbf{r} = -\frac{3}{4}\left(\frac{3}{\pi}\right)^{1/3} \int_V [\rho(\mathbf{r})]^{4/3} \; d\mathbf{r}, \tag{10.30}$$

which provides the *exchange energy functional in the local density approximation*. The corresponding exchange potential $V_x(\mathbf{r})$ is obtained by a similar functional derivation as in equation (10.27) (see appendix I)

$$V_x^{\mathrm{LDA}}(\mathbf{r}) = \frac{\delta E_x^{\mathrm{LDA}}}{\delta \rho(\mathbf{r})} = -\left(\frac{3}{\pi}\right)^{1/3} [\rho(\mathbf{r})]^{1/3}, \tag{10.31}$$

which represents a really important practical result: *in order to calculate the exchange potential in position* $\mathbf{r}$, *we only need to know the electron density in that point*.

A similar conclusion is drawn for the correlation potential, although the algebra is more complicated since $E_{\mathrm{c}}^{\mathrm{HEG}}$ is not available in analytical form, as reported in equation (10.29). It is anyway understood that $E_{\mathrm{xc}}^{\mathrm{LDA}} = E_x^{\mathrm{LDA}} + E_{\mathrm{c}}^{\mathrm{LDA}}$.

Nowadays a number of improvements going beyond the local density approximation are indeed available including, but not limited to, the more fundamental description taking into account the electron spin or the more accurate treatment of non-homogeneous electron densities (like, for instance, the gradient correction approach) [2–5].

10.5 The practical implementation and applications

Everything is now ready to solve the equations (10.26) since we have computable expressions for any energy term there contained. The problem is that both the Hartree and the exchange-correlation terms do depend on the electron density or, equivalently, on the very solutions of the Kohn–Sham equations, as clearly suggested by equation (10.16). The solution to this impasse is offered by the iterative method presented in section 1.4.1: some initial electron density is guessed (for instance, ρ could be defined as the sum of the electron densities of isolated atoms) and the corresponding Kohn–Sham equations solved; next, a new density is generated by using the just calculated Kohn–Sham orbitals and confronted with the previously guessed one: if they differ above an agreed accuracy threshold, the new density is used to write another set of Kohn–Sham equations to be again solved; this loop is repeated until the electron densities of the nth and $(n-1)$th iterations agree to within the chosen accuracy.

Once that full convergence is eventually proclaimed, the resulting orbitals, ground-state total energy and electron density can be used to calculate, in a totally parameter-free fashion, a large variety of solid state properties, including (but not limited to): equilibrium crystalline structures; cohesive energy and elastic moduli; vibrational spectra and phonon properties (like, for instance, the lattice thermal conductivity); the band structure and electron properties (like, for instance, the electrical conductivity)[11]. For instance, the phonon dispersion relations reported in figures 3.7–3.9 have been obtained by diagonalising a dynamical matrix whose entries are the second derivatives of the crystal total energy E_T [7] which, as discussed in section 1.3.4, has been calculated as $E_T = F[\rho_{GS}(\mathbf{r})] + U(\mathbf{R})$. The same DFT vibrational frequencies can be used to predict the heat capacity $C_V^{\text{quantum}}(T)$ according to equation (4.17) or the thermal conductivity $\kappa_l(T)$ according to equation (4.40).

In conclusion, DFT is nowadays looked upon as the 'standard model' for condensed matter physics: as a matter of fact, it is the most widespread method for calculating the quantum-mechanical properties of molecules, solids, nano-structures, surfaces and even of liquids. In addition, all superior theoretical schemes realistically applicable on a large scale (like, for instance, methods based on hybrid functionals, or the time-dependent DFT) are based on the density functional theory [2–6].

References

[1] Colombo L 2019 *Atomic and Molecular Physics: A Primer* (Bristol: IOP Publishing)

[2] Giustino F 2014 *Materials Modelling using Density Functional Theory* (Oxford: Oxford University Press)

[3] Sholl D S and Steckel J A 2009 *Density Functional Theory—A Practical Introduction* (New York: Wiley)

[4] Engel E and Dreizler R M 2011 *Density Functional Theory* (Berlin: Springer)

[5] Martin R M 2012 *Electronic Structure—Basic Theory and Practical Methods* (Cambridge: Cambridge University Press)

[6] Parr R G and Yang W 1989 *Density Functional Theory of Atoms and Molecules* (Oxford: Oxford Science Publications)

[7] Baroni S, de Gironcoli S, Dal Corso A and Giannozzi P 2001 Phonons and related crystal properties from density-functional perturbation theory *Rev. Mod. Phys.* **73** 515

---

[11] On the other hand, since DFT is a ground-state theory it is unable to accurately predict the band gap of insulators or any optical property, like the absorption coefficient. These properties can be calculated by using time-dependent DFT or many-body techniques.

Part IV

Concluding remarks

IOP Publishing

Solid State Physics
A primer
Luciano Colombo

Chapter 11

What is missing in this 'Primer'

A 'Primer' cannot be, by definition, a complete and in-depth introduction to any topic. This is especially true for condensed matter physics, which is likely the most rich, diverse, interdisciplinary, and fast-developing area of physical sciences[1]. This volume is no exception and, therefore, it must be acknowledged that many important issues have not been dealt with. We feel dutifully committed to mentioning them.

A concise list of missing arguments, as well as of topics which have not been thoroughly treated, is the following:
1. crystallography treated by group theory,
2. Hartree–Fock theory for the many-body electron problem,
3. dielectric screening,
4. non-adiabatic phenomena,
5. a more detailed account for optical properties in metals and insulators,
6. magnetic field effects on the electron gas,
7. magnetic properties and magnetic ordering,
8. superconductivity,
9. surface physics,
10. physics of low-dimensional solid state systems.

Most of the above topics are covered by the textbooks quoted in the bibliography appended to each chapter: for further information, we refer the reader to these manuals. We observe, however, that their level is still introductory, although often more thorough than found in this Primer. The cutting-edge research in modern condensed matter physics is quite far away: at that level, it is preferable to deal directly with the scientific literature published in specialised journals of solid state and/or materials physics.

---

[1] An interest account for this is reported in the paper by Sinatra R, Deville P, Szell M, Wang D and Brabási A-L 2015 *Nature Physics* **11** 971, where a thorough analysis based on Web of Science data reveals the growth and multidisciplinarity of any physics subdiscipline over the last 100 years, or so.

Part V

Appendices

IOP Publishing

Solid State Physics
A primer
Luciano Colombo

Appendix A

The periodic table of elements

By applying in conjunction the Pauli principle and the Hund rules [1], we can determine the ground-state electronic configurations of all atoms and, accordingly, draw the modern release of the Mendeleev periodic table of elements. It is reported in figure A.1, where the color shadowing represents the progressive filling of the atomic s-, p-, d-, and f-shells.

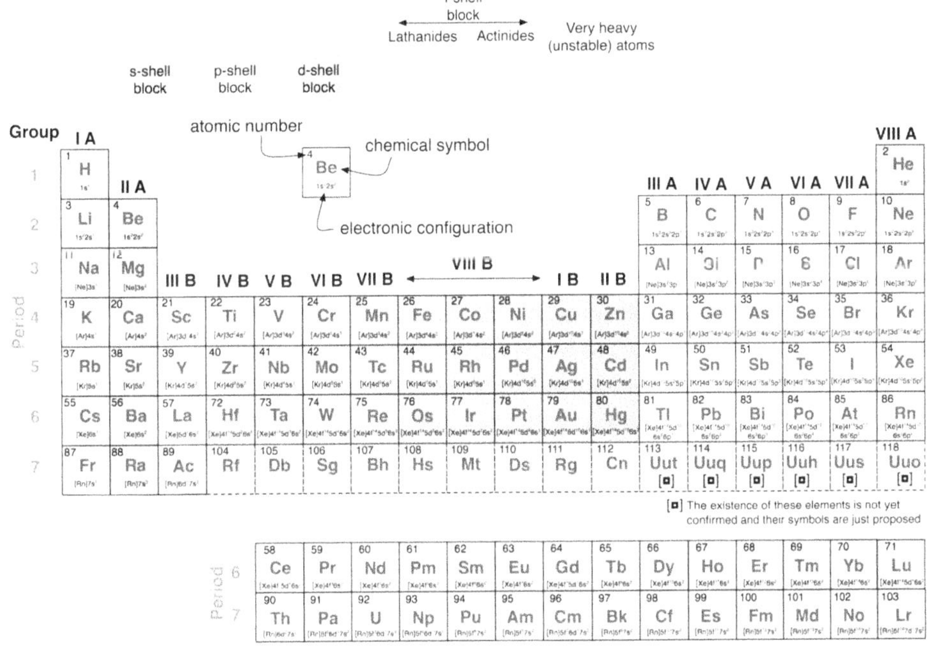

Figure A.1. The periodic table of the elements.

Reference

[1] Colombo L 2019 *Atomic and Molecular Physics: A Primer* (Bristol: IOP Publishing)

IOP Publishing

Solid State Physics
A primer
Luciano Colombo

Appendix B

Alloys, polycrystals, and quasi-crystals

B.1 Alloys

Let us consider a diamond lattice and imagine randomly occupying its sites with Si and Ge atoms. By setting x the concentration of Si atoms, with the constraint that $x \leqslant 1$, it turns out that the corresponding concentration of lattice sites occupied by Ge atoms must be equal to $(1 - x)$. The resulting crystal structure is neither diamond-like (because we have two different chemical species) nor zincblende-like (because the two species are not orderly distributed). Therefore, we will refer to this new atomic architecture as Si_xGe_{1-x} *alloy*. A similar situation is found if we start with a zincblende lattice and imagine occupying (i) the anionic sublattice with As atoms and (ii) the cationic sublattice by a random distribution of Al atoms in concentration x and Ga atoms in concentration $(1 - x)$. Once again, the constraint $x \leqslant 1$ must hold. In this case the structure is in part ordered (the underlying lattice is fcc with a two-atom basis) and in part random (the chemical distribution is random, even if it respects the stoichiometry of a typical III–V semiconductor). We will refer to this new atomic architecture as $Al_xGa_{1-x}As$ *alloy*. The two alloys are pictorially represented in figure B.1.

Metal alloys also exist, sometimes referred to as bulk metallic glasses. They are classified in three most common structures: (i) substitutional alloy (a bcc or fcc lattice is randomly decorated by two or more metal species); (ii) interstitial alloy (where a number of interstitial defects are found within a normal metal lattice, corresponding to chemical species other than the host one); and (iii) substitutional–interstitial alloy (where the two previous situations are both found). Interstitial defects can be either metallic or non-metallic, like e.g. in the mostly relevant case of steel: an alloy made by iron, silicon, and carbon. In general, the resulting alloy has properties that differ, even largely, from those of the pure metals, such as increased strength or hardness. This is the realm of physical metallurgy [1, 2].

The lattice constant of an alloy a_0^{alloy} is predicted by the Vegard empirical law which in the simple case of a binary mixture reads as

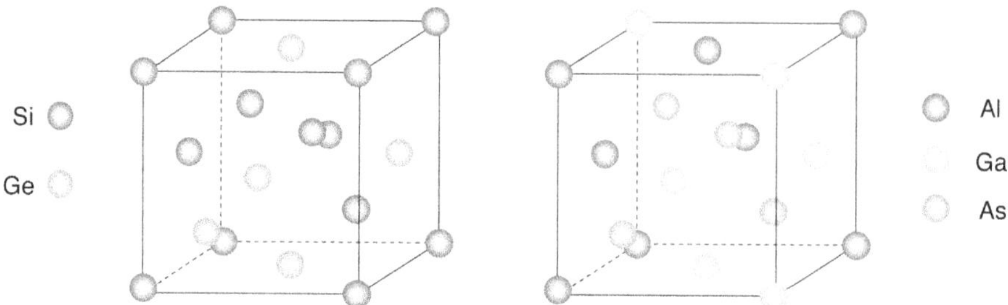

Figure B.1. A Si_xGe_{1-x} alloy (left) and a $Al_xGe_{1-x}As$ alloy (right). See figure 2.12 for the corresponding diamond and zincblend lattice, respectively.

$$a_0^{\text{alloy}} = x\, a_0^{\text{a}} + (1 - x)\, a_0^{\text{b}} + \xi\, x(1 - x), \tag{B.1}$$

where a_0^{a} and a_0^{b} are the lattice constants of the two elemental solids made with the same atoms found in the alloy[1]. The last term $\xi\, x(1 - x)$ (where ξ is a pure phenomenological constant to be determined, case by case, in an empirical way) is often omitted, so approximating the variation to a linear law: in many cases this is a more than good approximation.

B.2 Polycrystals

Polycrystals are solid objects made by an assembly of randomly oriented finite crystals (usually referred to as crystallites or grains). Crystallites have typical dimensions ranging from the nm to the μm length scale which correspond to nanocrystalline or microcrystalline systems, respectively. Interface regions between facing grains usually have high structural disorder: here, grain boundaries are defined by the misalignment among crystallites, generating a non-crystalline bonding interface environment or, possibly, even a thin amorphous layer.

B.3 Quasi-crystals

Quasi-crystals are atomic architectures with symmetry properties in between a crystal and a liquid. They are experimentally obtained by rapidly cooling binary molten mixtures (typically, inter-metallic alloys). The intermediate crystalline-liquid character of their crystal structure is exploited by the twofold feature that (i) it lacks translational invariance, but (ii) it contains a regular pattern of similar structural units.

The most effective way to visualise a quasi-crystal is by following the Penrose tiling construction shown in figure B.2. Two rhombic building blocks are chosen, respectively with a pair of internal angles of (108°, 72°) and (144°, 36°). These rhombuses are combined as shown in the figure, so as to generate a structure that, although lacking translational invariance, presents a regular repetition of equally oriented decagons. This regular patter defines families of parallel lines, always

---

[1] With reference to figure 2.22 they are the Si and Ge lattice constants for the alloy Si_xGe_{1-x} or the AlAs and GaAs lattice constants for the alloy $Al_xGa_{1-x}As$.

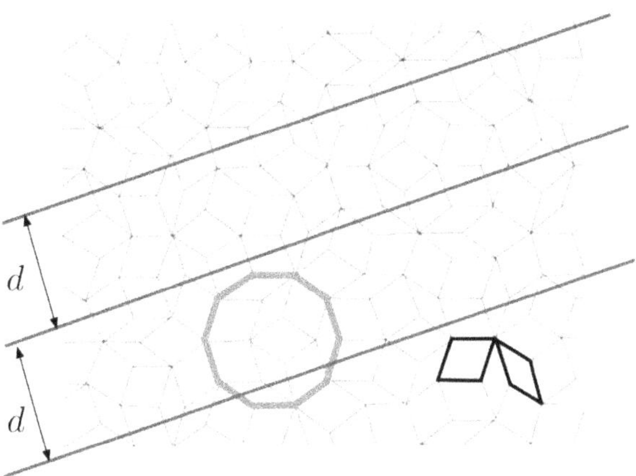

Figure B.2. A two-dimensional quasi-crystal obtained by combining the two structural units shown as shaded rhombic building blocks. The red decagon represents the structural unit which is found with regular orientation, so that periodic parallel lines (shown in blue colour) are found. Note that similar families of parallel planes are found in the quasi-crystal, forming a 72° angle to each other.

forming a 72° angle to each other. The resulting lattice displays a five-fold rotational symmetry, which is indeed forbidden in normal two-dimensional lattices, as discussed in section 2.3.2.

The three-dimensional counterpart of the quasi-crystal shown in figure B.2 is obtained by placing atoms with icosahedral symmetry[2]. This structure is obtained provided that a tetrahedral arrangement of atoms is used as basic building block in the attempt to realise a close-packed structure: the icosahedron is obtained by letting 20 tetrahedra to share a common vertex. Slight distortions of such icosahedra must be allowed to obtain the three-dimensional quasi-crystal, thus preventing long-range translational invariance. X-ray diffraction provides experimental evidence that some inter-metallic alloys display a three-dimensional five-fold symmetry.

References

[1] Abbashian R, Abbashian L and Reed-Hill R E 2009 *Physical Metallurgy Principles* 4th edn (Stanford, CT: Cengage Learning)
[2] Smallman R E and Ngan A H W 2007 *Physical Metallurgy and Advanced Materials* 7th edn (Oxford: Butterworth-Heinemann)

---

[2] We recall that icosahedrons are solids with six C_5 rotation axes (Schoenflies nomenclature).

IOP Publishing

Solid State Physics
A primer
Luciano Colombo

Appendix C

Essential thermodynamics

C.1 Basic definitions

Let us consider a condensed matter system and agree to describe it according to classical physics. By assigning the individual positions and momenta of its elementary constituents[1], we define a *microstate* of the system. As easily understood, the number of microscopic degrees of freedom needed to set a specific microstate is really very high. On the other hand, any macroscopic configuration of the same system considered as a whole can be precisely defined by a comparatively much smaller number of *thermodynamic parameters*, like for instance its volume V or the number of moles n_k for each kth chemical species found in the system (more thermodynamic parameters are rigorously defined below). We will address such a configuration as a *macrostate*. The key idea underlying thermodynamics is to replace the description of a system based on microstates with a representation of its physics based on macrostates. In essence, thermodynamics is a sophisticated averaging procedure of microscopic degrees of freedom [1–3].

In general, a macrostate can be realised in a myriad of different micro-modes[2]. If we name Ω such a number and set [3]

$$S = k_B \ln \Omega, \tag{C.1}$$

we introduce an operative definition of the system *entropy S* which is, therefore, a direct estimation of all the possible microstates compatible with the assigned set of thermodynamic parameters. The *equilibrium state* is defined as the *state with maximum entropy*; in other words, it represents the macrostate with the maximum number of micro-realisations. When a system is at equilibrium, no time variation of its thermodynamic parameters is observed. However, we remark that it undergoes

---

[1] We could also define a set of generalised coordinates and conjugated momenta: our reasoning is unaffected by this choice.

[2] Just think about a mole of a monoatomic gas in a container with rigid and fixed walls: each possible macrostate can be obtained by several choices of the particle positions within the container.

doi:10.1088/978-0-7503-2265-2ch14

microscopic evolution: basically, it can explore all those different microstates which correspond to the same set of thermodynamic parameters defining the state of equilibrium.

Under the constrain which limits our theory to equilibrium states only, it is nevertheless possible to consider *changes of thermodynamic state* caused by some mechanical, chemical, or thermal action occurring so slowly that the system can reach the equilibrium at each intermediate step. In this case we speak of *quasi-static transformations* which, by construction, are a sequence of equilibrium states. In general, a transformation is 'slow enough' to become quasi-static if it occurs over a time span much larger than the typical relaxation time of the system.

C.2 Internal energy

We name *internal energy* $\mathcal{U}$ of a thermodynamic system its total energy content; it corresponds to the work needed to form such an aggregate. The internal energy is a *state function* since its value is uniquely defined by the thermodynamic parameters defining the macrostate; we write

$$\mathcal{U} = \mathcal{U}(S, V, \{n_k\}), \tag{C.2}$$

where $\{n_k\}$ represents in compact form the set of numbers defining how many moles for each $k = 1, 2, 3, \ldots$ different chemical species form the system. Upon differentiation

$$d\mathcal{U} = \left.\frac{\partial \mathcal{U}}{\partial S}\right|_{V,\{n_k\}} dS + \left.\frac{\partial \mathcal{U}}{\partial V}\right|_{S,\{n_k\}} dV + \sum_k \left.\frac{\partial \mathcal{U}}{\partial n_k}\right|_{V,S,\{n_j\}_{j\neq k}} dn_k, \tag{C.3}$$

we get the definition of the following thermodynamic parameters [1, 2]

$$\text{temperature:} \quad T = \left.\frac{\partial \mathcal{U}}{\partial S}\right|_{V,\{n_k\}}$$

$$\text{pressure:} \quad P = -\left.\frac{\partial \mathcal{U}}{\partial V}\right|_{S,\{n_k\}} \tag{C.4}$$

$$\text{chemical potential:} \quad \mu_k = \left.\frac{\partial \mathcal{U}}{\partial n_k}\right|_{V,S,\{n_j\}_{j\neq k}},$$

which are known as *intensive parameters* or *state variables*; on the other hand, internal energy, entropy, volume, and mole numbers are *extensive parameters*. Through these definitions we can rewrite equation (C.3) in a more compact and elegant way

$$d\mathcal{U} = T\, dS - P\, dV + \sum_k \mu_k\, dn_k, \tag{C.5}$$

which basically represents an energy balance since the quantities $-PdV$, TdS, and $\mu_k dn_k$, respectively, represent the mechanical, non-mechanical, and chemical work exchanged during a quasi-static infinitesimal transformation. The exchanged non-mechanical work is referred to as the *exchanged heat* and it is formally written as $dQ = TdS$.

At equilibrium the thermodynamic parameters are such that $\mathcal{U}$ is minimum. Furthermore, they obey restrictions that can be easily obtained by considering the simple case of a monoatomic system made by two sub-systems which we label by '1' and '2'. We consider three different situations where, respectively, the two sub-systems can (i) exchange just heat, or (ii) exchange heat and mechanical work, or (iii) exchange heat and mass. Then, it is easy to prove that in the three cases at equilibrium we have (i) $T_1 = T_2$, or (ii) $T_1 = T_2$ and $P_1 = P_2$, or (iii) $T_1 = T_2$ and $\mu_1 = \mu_2$, where of course $T_{1,2}$, $P_{1,2}$, and $\mu_{1,2}$ are the temperature, pressure, and chemical potential of the two sub-systems. These results prove that the formal definitions provided in equations (C.4) agree with their phenomenological counterparts.

Equation (C.2) is a *fundamental thermodynamic equation* since it contains any information about the system[3]. Intensive parameters are related by *equations of state*: a simple example is given by $PV = nRT$, namely the equation of state for a monoatomic ideal gas. While useful for many applications, we must nevertheless understand that each equation of state only contains a limited amount of information. As a matter of fact, the full set of equations of state valid for a system is able to give the same knowledge provided by its fundamental equation $\mathcal{U} = \mathcal{U}(S,V,\{n_k\})$.

C.3 Thermodynamic potentials

Extensive and intensive parameters have been so far used as independent and derived quantities, respectively. In many circumstances, however, it is more appropriate to use intensive parameters as independent variables (since it is much easier to measure them) and to replace the internal energy by some new *thermodynamic potential*.

Formally, this is done by calculating the *Legendre transformation* $\mathrm{L}_X[\mathcal{U}]$ of the internal energy with respect to a suitable intensive parameter X. Accordingly, we define [1–3]

$$\text{Helmholtz free energy:} \quad \mathrm{L}_T[\mathcal{U}] = \mathcal{U} - TS = \mathcal{F}(T,V,\{n_k\}) \tag{C.6}$$

$$\text{enthalpy:} \quad \mathrm{L}_P[\mathcal{U}] = \mathcal{U} + PV = \mathcal{H}(S,P,\{n_k\}) \tag{C.7}$$

$$\text{Gibbs free energy:} \quad \mathrm{L}_{T,P}[\mathcal{U}] = \mathcal{U} - TS + PV = \mathcal{G}(T,P,\{n_k\}), \tag{C.8}$$

representing, respectively, the work exchanged quasi-statically during a transformation occurring at constant temperature (Helmholtz free energy), at constant pressure (enthalpy), and at constant temperature and pressure (Gibbs free energy).

The three thermodynamic potentials allow us to define the equilibrium condition when the system is constrained. For instance, if the system is coupled to a heat reservoir (that is, its temperature is kept constant), then the equilibrium state is that

---

[3] It is also possible to choose entropy $S = S(\mathcal{U},V,\{n_k\})$ as the state function of the system. This allows for a different formal development of thermodynamics.

with minimum Helmholtz free energy. On the other hand, if the system is kept at constant pressure (that is, it undergoes the mechanical action of a pressure reservoir) then the equilibrium is defined by the minimum of the enthalpy. Finally, the minimum of the Gibbs free energy defines the equilibrium state of a system coupled to the double action of a heat and a pressure reservoir.

C.4 Some thermodynamic materials properties

Through the derivatives of the internal energy reported in equations (C.4), we have defined the relevant thermodynamic parameters of temperature, pressure, and chemical potential. Other physical properties of solids can be similarly defined. In the following, *we will always assume that the numbers of moles are kept constant*: for brevity, we will omit formally indicating this constraint.

It is known experimentally that, in general, a solid increases its volume upon heating. The *thermal expansion coefficient* β defines the fractional volume increase per unitary increment of temperature at constant pressure[4]

$$\beta = \frac{1}{V}\frac{\partial V}{\partial T}\bigg|_P = -\frac{1}{V}\frac{\partial V}{\partial P}\bigg|_T \frac{\partial P}{\partial T}\bigg|_V = \frac{1}{B}\frac{\partial P}{\partial T}\bigg|_V, \tag{C.9}$$

where in the right-hand side of the above equation the *bulk modulus* has been introduced

$$B = -V\frac{\partial P}{\partial V}\bigg|_T, \tag{C.10}$$

with $B > 0$ since $(\partial P/\partial V)_T$ is negative for condensed matter systems; its inverse function

$$\alpha_T = B^{-1} = -\frac{1}{V}\frac{\partial V}{\partial P}\bigg|_T, \tag{C.11}$$

is referred to as the *isothermal compressibility*: it provides the fractional volume decrease per unitary increment of pressure at constant temperature.

More robust experimental evidence is that, by quasi-statically supplying heat to a thermodynamic system, we generally observe an increase of its temperature. We define the amount of heat we need to supply in order to produce a unitary increase of the temperature, while keeping the volume constant, as

$$C_V(T) = \frac{\partial \mathcal{U}}{\partial T}\bigg|_V = T\frac{\partial S}{\partial T}\bigg|_V = \frac{\partial Q}{\partial T}\bigg|_V, \tag{C.12}$$

---

[4] In deriving this equation we make use of a nontrivial mathematical identity. Let us consider a function $f = f(x,y,z)$ under the constraint $f(x,y,z) = $ constant. Then, the variables (x,y,z) are not independent since the constraint defines an implicit relationship among them. Accordingly, their partial derivatives satisfy a number of relations, among which $(\partial x/\partial y)_z = -(\partial z/\partial y)_x/(\partial z/\partial x)_y = -(\partial x/\partial z)_y\,(\partial z/\partial y)_x$ that we used to obtain equation (C.9). We also use the identities $(\partial x/\partial y)_z = 1/(\partial y/\partial x)_z$ and $(\partial x/\partial y)_z = (\partial x/\partial w)_z/(\partial y/\partial w)_z$.

and we refer to this quantity as the *constant-volume heat capacity* of the system. $C_V(T)$ is an extensive property, while its corresponding intensive property is referred to as the *constant-volume specific heat* and it is calculated dividing $C_V(T)$ by the system mass[5]. Similarly we define the *constant-pressure heat capacity* as

$$C_P(T) = \left.\frac{\partial \mathcal{U}}{\partial T}\right|_P = T\left.\frac{\partial S}{\partial T}\right|_P = \left.\frac{\partial Q}{\partial T}\right|_P. \tag{C.13}$$

For any real crystal we have [1, 2]

$$\begin{aligned}
C_P(T) &= C_V(T) + TV\,\frac{\beta^2}{\alpha_T} \\
&= C_V(T) - T\,\frac{(\partial P/\partial V\,|_V)^2}{\partial P/\partial V\,|_T} \\
&= C_V(T) - T\left(\left.\frac{\partial V}{\partial T}\right|_P\right)^2 \left.\frac{\partial P}{\partial V}\right|_T
\end{aligned} \tag{C.14}$$

that is $C_P(T) > C_V(T)$: we need to supply a larger amount of heat in constant-pressure conditions since part of the dispensed thermal energy is spent in mechanical work.

References

[1] Callen H 1985 *Thermodynamics and An Introduction to Thermostatistics* (New York: Wiley)
[2] Dittman R and Zemansky M 1997 *Heat and Thermodynamics* 7th edn (New York: McGraw-Hill)
[3] Goodstein D L 1985 *States of Matter* (New York: Dover)

---

[5] Sometimes the specific heat is defined per unit of volume, instead of per unit of mass.

IOP Publishing

Solid State Physics
A primer
Luciano Colombo

Appendix D

Force constant models for lattice dynamics

In this appendix we briefly describe some *force constant models* which have been extensively used in lattice dynamics.

In the phenomenological approach, the force constant models are basically used to calculate the entries of the dynamical matrix, whose eigenvalues eventually provide the vibrational frequencies. The most relevant common feature is that they exploit a number of *empirical interactions*, each depending on *unknown parameters* that must be *fitted on some experimental information*.

The actual form of such empirical interactions is guessed according to the character of the interatomic bonding found within the crystal of interest. Quite a few different analytical forms for such interactions have been proposed over the years, including (but not limited to): pairwise or many-body potentials, as well as central or bond-angle dependent potentials [1–3]. However, in one way or another, under the harmonic approximation they are reduced to *a system of effective springs connecting either massive point-like ions or massless objects mimicking the charge distribution of the valence electrons*[1]. The common feature of empirical force fields is that they are *short-ranged*: interactions only extend to just the nearest coordination shells. For this reason the force constant models are commonly labelled as first-next-neighbours or second-next-neighbours (and so on ...) interactions models. Normally, the more extended the interactions, the more accurate is the calculation of the vibrational spectrum. The price to pay is the increasing number of fitting parameters, a feature that makes the resulting dynamical model definitely not a fundamental one (and, therefore, poorly transferable to systems unalike the fitted one).

In the case of ionic crystals, *long-range Coulomb interactions* are added to short-range ones. Ions are considered as point-like objects (and, therefore, no atomic polarisation effects are usually included at this level) with an *effective dynamical charge* to be empirically determined [2, 4]. The evaluation of the corresponding

---

[1] The object representing the electron charge distribution is assumed to be massless in order to enforce the adiabatic approximation.

electrostatic forces requires laborious lattice sums: indeed a very tricky task, which is typically solved by the Ewald construction [5].

Before outlining the main physical features of some popular force constant models, it is important to remark that the modern theory of lattice dynamics no longer relies on such empirical models. Rather, a more fundamental and superior first principles approach is followed: a fully quantum mechanical calculation of the crystal total energy is performed, from which force constants are eventually obtained. This approach requires the more advanced theory outlined in chapter 10, as well as a nontrivial numerical implementation based on high-performance computing algorithms [6]. For this reason empirical force constant models are nowadays still useful as pedagogical tools: they allow a very intuitive approach to the calculation of the vibrational properties of a crystal, easily translated into a low-complexity computational scheme.

D.1 The rigid ion model

By far the most simple approach is based on the *rigid ion model*, which over the years has been extensively applied to metals as well as to ionic crystals. In the first case just central pair interactions between ions are used, up to a pre-assigned order of neighbours (in fact the range of interactions works as an additional free parameter of the model), as shown in figure D.1(top).

In view of its simplicity, the rigid ion model has been occasionally adopted to calculate vibrational frequencies in covalently bonded solids. In this case both Coulomb and short-range forces are needed, the latter both central and non-central (to stabilise, for instance, the tetrahedral coordination in crystals with a diamond-like structure).

The next level of increasing sophistication consists in guessing a lattice potential energy depending both on ionic displacements and bond angle distortions. The two sets of force constants, respectively, describe bond-stretching and bond-bending vibrations. This solution is referred to as the *valence force field*.

D.2 The shell model

A significant step forward in accuracy is represented by the *shell model*, existing in different versions for ionic and covalent crystals. Here an ion is described as a non-polarisable core (formed by the nucleus and the core electrons) around which valence electrons are modelled as a massless shell. Each ion core is coupled to its shell by isotropic springs, as shown in figure D.1(bottom, left). Cores and shells have different electric charge, so that their sum corresponds to the net charge of the ion.

The new key feature of this model is that ion polarisation is allowed by means of the relative core–shell displacements. Similarly to the rigid ion case, core–core short-range interactions are also considered. In addition, shell–shell and core–shell interactions between nearest neighbouring ions are included, each described by a suitable spring. The accuracy is significantly improved in the *breathing shell model* version, where radial deformability of the electron shell is allowed as well (this is not shown in figure D.1).

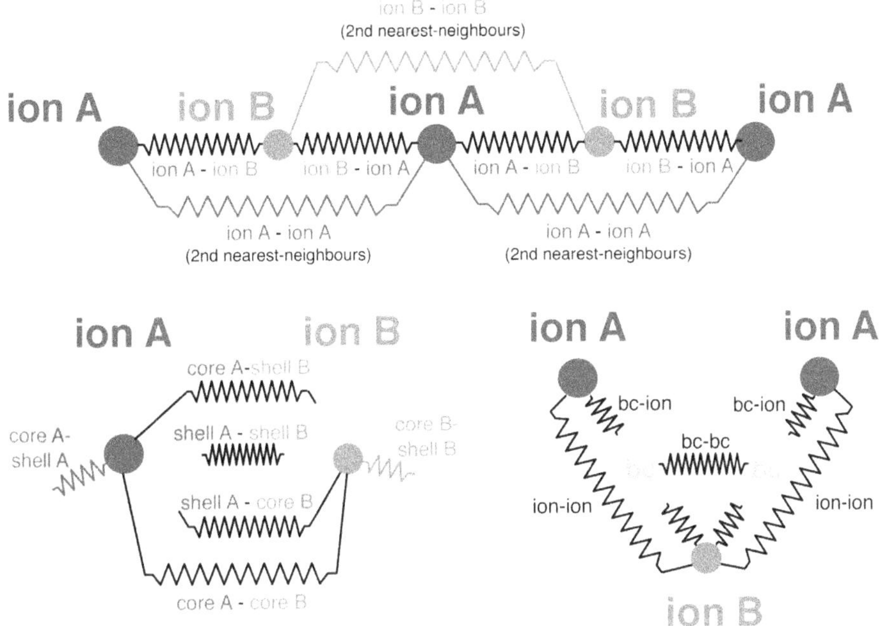

Figure D.1. Schematic representation of the system of short-range effective harmonic springs for a two-ion crystal (ion A and ion B), as predicted by the rigid ion model (top; it is shown the case of interactions up to the second-next-neighbours), in the shell model (bottom, left) and in the bond charge model (bottom right; it is shown the case of an elemental covalent solid). Long-range Coulomb interactions can be added, if requested.

Long-range Coulomb interactions are included in this scheme by considering each core–shell pair as a point-like object and, therefore, by performing lattice sums similarly to the case of the rigid ion model.

D.3 The bond charge model

The shell model can hardly be applied to covalent crystals, where *the valence charge is shared between nearest neighbouring ions* rather than split among them (see figure 2.22). In these crystals the maximum of the valence charge is located somewhere along the bond. More specifically, this maximum is found at the midpoint of the bond in elemental covalent crystals (like diamond, silicon or germanium), while it is slightly shifted towards the anion in such a way as to divide the bond in two segments whose lengths are found in the 3:5 or 2:6 ratio in III–V or II–VI materials, respectively. In figure D.1(bottom, right) the case of an elemental crystal is shown.

The valence charge is treated as a massless *bond charge* (bc) which is coupled to neighbouring ions and other bc's by effective springs. This mimics the incomplete screening of ionic charges and yields effective non-central short-range interactions between anions and cations. This is crucially important in order to stabilise the tetrahedral coordination.

Long-range Coulomb interactions are included also in this scheme by considering each ion and bc as a point-like object and, therefore, by performing lattice sums similarly to the case of the rigid ion model.

References

[1] Brüesch P 1982 *Phonons: Theory and Experiment* vol I (Berlin: Springer)
[2] Srivastava G P 1990 *The Physics of Phonons* (Bristol: Adam Higler)
[3] Finnis M 2003 *Interatomic Forces* (Oxford: Oxford University Press)
[4] Böttger H 1983 *Principles of the Theory of Lattice Dynamics* (Berlin: Akademie)
[5] Kittel C 1996 *Introduction to Solid State Physics* 7th edn (Hoboken, NJ: Wiley)
[6] Baroni S, de Gironcoli S, Dal Corso A and Giannozzi P 2001 Phonons and related crystal properties from density-functional perturbation theory *Rev. Mod. Phys.* **73** 515

Solid State Physics
A primer
Luciano Colombo

Appendix E

Quantum statistics

E.1 Identical particles

Two particles are identical provided they have the very same characteristics (i.e. the same intrinsic physical properties like, for instance, the mass, the charge, the spin, and so on). Identical particles can be distinguished, if treated according to classical mechanics, by taking into consideration their positions. On the other hand, according to quantum mechanics we cannot assign a well defined position to any particle and this prevents us distinguishing among them: in short, in quantum mechanics *identical particles* are also *indistinguishable*.

Let us then consider the Hamiltonian operator $\hat{H}$ describing a set of N *identical and indistinguishable particles*: it is invariant upon interchanging any two particles. However, the same operation could differently affect its eigenfunction $\Phi(\mathbf{r}_1, \mathbf{r}_2, \ldots, \mathbf{r}_N)$ describing the state of the system. This is full of consequences worthy of consideration.

Let $\hat{P}_{\mu \leftrightarrow \nu}$ the *permutation operator* which acts on the eigenfunctions of $\hat{H}$ by swapping the particles labelled by the indices $\mu \leftrightarrow \nu$. Because of the above invariance, we have

$$[\hat{H}, \hat{P}_{\mu \leftrightarrow \nu}] = 0, \tag{E.1}$$

and, therefore, $\Phi(\mathbf{r}_1, \mathbf{r}_2, \ldots, \mathbf{r}_N)$ *is a wavefunction of either* $\hat{H}$ *and* $\hat{P}_{\mu \leftrightarrow \nu}$. It is easy to prove that

$$\hat{P}_{\mu \leftrightarrow \nu} \Phi(\mathbf{r}_1, \mathbf{r}_2, \cdots, \mathbf{r}_N) = \pm \Phi(\mathbf{r}_1, \mathbf{r}_2, \cdots, \mathbf{r}_N), \tag{E.2}$$

or, equivalently: any wavefunction describing a set of identical and indistinguishable particles is either symmetric or antisymmetric under particle interchange:

$$\hat{P}_{\mu \leftrightarrow \nu} \Phi = +\Phi \qquad \text{symmetric wavefunction,} \tag{E.3}$$

$$\hat{P}_{\mu \leftrightarrow \nu} \Phi = -\Phi \qquad \text{antisymmetric wavefunction.} \tag{E.4}$$

doi:10.1088/978-0-7503-2265-2ch16   E-1

The symmetry property of a wavefunction represents an intrinsic property of the specific set of particles considered since it cannot be altered by any external action. Identical and indistinguishable particles are said to be *fermions* or *bosons* if their collective quantum states are described by antisymmetric or symmetric wavefunctions, respectively [1–4].

The antisymmetric character of the fermion wavefunction has a very important consequence which we will derive under the assumption that a system of N fermions can be treated within a single-particle approximation; in other words, we will assume that the complete problem

$$\hat{H}\ \Phi(\mathbf{r}_1, \mathbf{r}_2, \ldots, \mathbf{r}_N) = E_T\ \Phi(\mathbf{r}_1, \mathbf{r}_2, \ldots, \mathbf{r}_N), \tag{E.5}$$

can be reduced to N different single-particle problems

$$\hat{h}_i\ \phi_{n_i}(\mathbf{r}_i) = E_i \phi_{n_i}(\mathbf{r}_i), \tag{E.6}$$

where $i = 1, 2, \ldots, N$ labels the fermions and n_i is a suitable set of quantum numbers describing the single-particle state with energy E_i. Of course we have $E_T = \sum_i E_i$.

In order to exploit its antisymmetric character we can write the total wavefunction in the form of a Slater determinant

$$\Phi(\mathbf{r}_1, \mathbf{r}_2, \ldots, \mathbf{r}_N) = \frac{1}{\sqrt{N!}} \begin{vmatrix} \varphi_{n_1}(\mathbf{r}_1) & \varphi_{n_1}(\mathbf{r}_2) & \cdots & \varphi_{n_1}(\mathbf{r}_N) \\ \varphi_{n_2}(\mathbf{r}_1) & \varphi_{n_2}(\mathbf{r}_2) & \cdots & \varphi_{n_2}(\mathbf{r}_N) \\ \cdots & \cdots & \cdots & \cdots \\ \varphi_{n_N}(\mathbf{r}_1) & \varphi_{n_N}(\mathbf{r}_2) & \cdots & \varphi_{n_N}(\mathbf{r}_N) \end{vmatrix}. \tag{E.7}$$

This form of $\Phi(\mathbf{r}_1, \mathbf{r}_2, \ldots, \mathbf{r}_N)$ naturally embodies the anti-symmetric character of the state function, since by swapping two columns the Slater determinant will change sign. In addition, we remark that a determinant vanishes if two rows are equal. This implies that a system of identical and indistinguishable fermions cannot occupy a many-body state where two single-particle states are equal. This statement is commonly referred to as *Pauli principle*; it states that it is impossible to have two fermions with the same set of quantum numbers.

The rationalisation of a large number of different experimental investigations leads to the conclusion that electrons, protons, and neutrons are fermions, while photons and phonons are bosons [5]. The fermion or boson character of an elementary particle is ultimately determined by its spin: fermions have half odd integer spin $\hbar/2$, $3\hbar/2$, … while bosons have integer spin 0, $\hbar$, ….

In quantum statistics [6, 7] the ultimate goal is to find the equilibrium state of a system of fermions or bosons. In this appendix we adopt the single-particle picture and assume a discrete energy spectrum, where single-particle states can possibly be degenerate. The idea is to calculate at first the number of different and distinguishable ways to distribute the particles among the available energy levels; next, the most probable partition is determined as that with the maximum number of different and distinguishable realisations; eventually, the equilibrium state is identified with such a most probable distribution.

We will separately consider the case of fermions and bosons. Since we are mainly interested in the physics of electrons and phonons, we will impose that the total number of fermions is conserved, while the number of bosons can vary.

E.2 Fermi–Dirac statistics

Let us consider a system of fermions at temperature T. We name by ϵ_i the energy of the ith single-particle state with degeneracy d_i. The number of different and distinguishable ways of placing $n_i \leqslant d_i$ fermions on this level is

$$\frac{d_i(d_i - 1)(d_i - 2) \cdots (d_i - n_i + 1)}{n_i!} = \frac{d_i!}{n_i!(d_i - n_i)!}. \tag{E.8}$$

Accordingly, the product

$$P_{\text{fermion}} = \prod_i \frac{d_i!}{n_i!(d_i - n_i)!} \tag{E.9}$$

provides the total number of different and distinguishable ways of distributing n_1, n_2, n_3, ... fermions on the levels with energy E_1, E_2, E_3, ... and degeneracy d_1, d_2, d_3, ..., respectively. The equilibrium distribution for the fermion system is found by maximising P_{fermion} under the two constraints

$$N = \sum_i n_i \qquad E_T = \sum_i n_i E_i, \tag{E.10}$$

corresponding to the conservation of the number of particles and system total energy, respectively. The standard way to proceed is maximising the function $\ln P_{\text{fermion}}$ instead of P_{fermion}, since we can take profit from the Stirling formula $\ln x! = x \ln x - x$ which is valid for $x \gg 1$. The two conservation conditions are compensated by two Lagrange multipliers. This leads to

$$n_i(T) = \frac{d_i}{\exp\left[(E_i - \mu_c)/k_B T\right] + 1} = n_{\text{FD}}(E_i, T), \tag{E.11}$$

which is known as the Fermi–Dirac distribution law [6, 7]; μ_c is referred to as the *chemical potential* of the fermion system and it is itself a function of the temperature $\mu_c = \mu_c(T)$. It is easy to prove that at zero temperature we find $n_i = d_i$ for all states with energy up to $E = \mu_c$, while for all states with energy $E > \mu_c$ we have $n_i = 0$. In other words, *at zero temperature all levels with energy below the chemical potential are fully occupied, while those with energy higher than the chemical potential are totally empty.* The zero-temperature value of the chemical potential $\mu_c(T = 0) = E_F$ is known as the Fermi energy.

E.3 Bose–Einstein statistics

Let us consider a system of bosons at temperature T. We name E_i the energy of the ith single-particle state with degeneracy d_i. Since in this case there is no limit in the number of particles that can be accommodated on a single level, the number of different and distinguishable ways to arrange n_i bosons on this level is

$$\frac{(n_i + d_i - 1)!}{n_i!(d_i - 1)!}. \tag{E.12}$$

Accordingly, the product

$$P_{\text{boson}} = \prod_i \frac{(n_i + d_i - 1)!}{n_i!(d_i - 1)!}, \tag{E.13}$$

provides the total number of different and distinguishable ways of distributing n_1, n_2, n_3, ... bosons on the levels with energy E_1, E_2, E_3, ... and degeneracy d_1, d_2, d_3, ..., respectively. The equilibrium distribution for the boson system is found by maximising P_{fermion} under the only constraint

$$E_T = \sum_i n_i E_i, \tag{E.14}$$

since, by direct reference to phonons and photons, we assumed that *the number of particles is not conserved* in this case[1]. We can proceed similarly to the case of fermions[2] obtaining

$$n_i(T) = \frac{d_i}{\exp\left[E_i/k_\mathrm{B}T\right] - 1} = n_{\mathrm{BE}}(E_i, T), \tag{E.15}$$

which is known as the Bose–Einstein distribution law [6, 7].

References

[1] Sakurai J J and Napolitano J 2011 *Modern Quantum Mechanics* 2nd edn (Reading, MA: Addison-Wesley)

[2] Miller D A B 2008 *Quantum Mechanics for Scientists and Engineers* (New York: Cambridge University Press)

[3] Griffiths D J and Schroeter D F 2018 *Introduction to Quantum Mechanics* 3rd edn (Cambridge: Cambridge University Press)

[4] Bransden B H and Joachain C J 2000 *Quantum Mechanics* (Upper Saddle River, NJ: Prentice-Hall)

---

[1] In our discussion we will treat only two kinds of boson particles, namely phonons and photons. In both cases the elementary processes we will be interested in generate as well as annihilate particles: for instance, by increasing/decreasing the crystal temperature we create or annihilate phonons; the same situations are found when considering anharmonicity, that is phonon–phonon interactions; finally, the absorption/emission of electromagnetic radiation by a crystalline solid is described in the quasi-classical picture in terms of photon annihilation/generation, respectively.

[2] In this case, however, just one single Lagrange multiplier is needed, providing the Boltzmann factor $1/k_\mathrm{B}T$.

[5] Eisberg R and Resnick R 1985 *Quantum Physics of Atoms, Molecules, Solids, Nuclei, and Particles* 2nd edn (Hoboken, NJ: Wiley)

[6] Glazer M and Wark J 2001 *Statistical Mechanics—A Survival Guide* (Oxford: Oxford University Press)

[7] Swendsen R H 2012 *An Introduction to Statistical Mechanics and Thermodynamics* (Oxford: Oxford University Press)

IOP Publishing

Solid State Physics
A primer
Luciano Colombo

Appendix F

Fermi–Dirac integrals and Sommerfeld expansion

At finite temperature the free electron gas is described by the Fermi–Dirac distribution law[1]

$$n_{FD}(E, T) = \frac{1}{1 + \exp[(E - \mu_c)/k_B T]},$$ (F.1)

providing the probability that the quantum state at energy E is occupied. In this law μ_c is the chemical potential which is itself a function of temperature; in particular, at $T = 0$ K we have $\mu_c = E_F$, where E_F is known as Fermi energy; for most metals $E_F \sim 10^4$ K (see table 7.2) and, therefore, in practice we always have $T \ll T_F = E_F/k_B$.

When developing the quantum physics of the free electron gas, we often meet integrals of the form

$$I(E) = \int_0^{+\infty} f(E)\, n_{FD}(E, T)\, dE,$$ (F.2)

where $f(E)$ is a generic function of the energy E: for instance, in section 7.3.2 two different forms of $f(E)$ are used to calculate the total number of electrons and the energy of the gas, respectively. These integrals belong to the family of *Fermi–Dirac integrals* and do not benefit from a general analytical expression. Their calculation proceeds through an expansion[2] [2, 3] which relies on the assumption that $\mu_c \sim E_F \gg k_B T$ for any temperature of physical interest. This assumption will be justified later by the results we will obtain.

---

[1] In this appendix we assume spin-resolved energy levels and, therefore, no indication of the degeneracy degree is reported (see appendix E for more details).
[2] The asymptotic form of the Fermi–Dirac integrals was originally published in [1].

doi:10.1088/978-0-7503-2265-2ch17

The calculation starts by splitting the integral given in equation (F.2) in two parts

$$I(E) = \int_0^{\mu_c} f(E)\, n_{FD}(E,\,T)\, dE + \int_{\mu_c}^{+\infty} f(E)\, n_{FD}(E,\,T)\, dE. \tag{F.3}$$

Next, the Fermi–Dirac distribution law is conveniently rewritten as

$$n_{FD}(E,\,T) = \frac{1}{1 + \exp[(E - \mu_c)/k_B T]} = 1 - \left\{1 + \exp[-(E - \mu_c)/k_B T]\right\}^{-1}, \tag{F.4}$$

and the $I(E)$ is accordingly recast in the form of a sum of three terms

$$\begin{aligned}
I(E) = {}& \int_0^{\mu_c} f(E)\, dE \\
& - \int_0^{\mu_c} f(E) \left\{1 + \exp[-(E - \mu_c)/k_B T]\right\}^{-1} dE \\
& + \int_{\mu_c}^{+\infty} f(E) \left\{1 + \exp[(E - \mu_c)/k_B T]\right\}^{-1} dE,
\end{aligned} \tag{F.5}$$

which can be separately evaluated by introducing the integration variables $\xi = -(E - \mu_c)/k_B T$ in the second integral and $\zeta = +(E - \mu_c)/k_B T$ in the third integral so that

$$\begin{aligned}
I(E) = {}& \int_0^{\mu_c} f(E)\, dE \\
& - k_B T \int_0^{\mu_c/k_B T} f(\mu_c - k_B T\xi)\, [1 + \exp(\xi)]^{-1}\, d\xi \\
& + k_B T \int_0^{+\infty} f(\mu_c + k_B T\zeta)\, [1 + \exp(\zeta)]^{-1}\, d\zeta.
\end{aligned} \tag{F.6}$$

We now consider the upper limit of the second integral: by assuming that $\mu_c \sim E_F$, we obtain that $\mu_c/k_B T \cdot E_F/k_B T \gg 1$ at any temperature T of physical interest. Therefore, to a very good approximation we can extend the upper limit of such an integral up to infinity. This procedure leads to rewriting equation (F.6) in the following new form

$$\begin{aligned}
I(E) = {}& \int_0^{\mu_c} f(E)\, dE \\
& + k_B T \int_0^{+\infty} \left[f(\mu_c + k_B T\xi) - f(\mu_c - k_B T\xi)\right][1 + \exp(\xi)]^{-1}\, d\xi.
\end{aligned} \tag{F.7}$$

The function $f(\mu_c + k_B T\xi)$ is further expanded in powers[3] of ξ

---

[3] This is a subtle point: basically, we are assuming that this function is analytical and well-behaved nearby $E = \mu_c$. While this is indeed the case for a free electron gas, small corrections are likely expected in more realistic situations where, for instance, the electronic density of states could be somewhere singular or show rapid variations in this energy range.

$$f(\mu_c + k_B T \xi) = \sum_{n=0}^{+\infty} \frac{1}{n!} f^{(n)}(\mu_c) (k_B T)^n \xi^n, \tag{F.8}$$

where $f^{(n)}(\mu_c)$ indicates the nth order derivative of the f function calculated at μ_c. This leads to[4]

$$I(E) = \int_0^{\mu_c} f(E)\, dE$$

$$+ 2 \sum_{m=0}^{+\infty} \frac{1}{(2m+1)!} f^{(2m+1)}(\mu_c) (k_B T)^{2(m+1)} \int_0^{+\infty} \xi^{2m+1} [1 + \exp(\xi)]^{-1}\, d\xi. \tag{F.9}$$

The integrals appearing in this equation can be calculated exactly [4], eventually providing the general expression

$$I(E) = \int_0^{\mu_c} f(E)\, dE + \frac{\pi^2}{6} f^{(1)}(\mu_c) (k_B T)^2 + \frac{7\pi^4}{360} f^{(3)}(\mu_c) (k_B T)^4 + \cdots, \tag{F.10}$$

for the Fermi–Dirac integrals in the Sommerfeld expansion.

In order to accept this general result, we must verify the assumption $\mu_c \sim E_F$: this check is carried out in section 7.3.2.

References

[1] Sommerfeld A 1927 *Z. Physik* **47** 1
[2] Ashcroft N W and Mermin N D 1976 *Solid State Physics* (London: Holt-Saunders)
[3] Grosso G and Pastori Parravicini G 2014 *Solid State Physics* 2nd edn (Oxford: Academic)
[4] Gradshteyn I S and Ryzhik I M 1980 table of Integrals *Series, and Products* (New York: Academic)

---

[4] When comparing equation (F.9) to equation (F.7), it must be understood that $m = (n-1)/2$ since, by replacing equation (F.8) in equation (F.7) we easily recognise that the second integral vanishes for any even value of n.

IOP Publishing

Solid State Physics
A primer
Luciano Colombo

Appendix G

The tight-binding theory

G.1 From atomic orbitals to Bloch sums

In the spirit of the single-particle picture described in section 1.4.1, the many-body electron problem is reduced to the problem of one electron moving in the average field V_{cfp} due to the other valence electrons and to the ions (clamped at their equilibrium position). So as not to burden the formalism too much, we will omit any explicit reference to the actual ionic configuration $\mathbf{R}$ in the electron wavefunction and energy symbols. The reduced quantum one-electron Hamiltonian operator reads as

$$\hat{H} = -\frac{\hbar^2}{2m_{\text{e}}}\nabla^2 + \hat{V}_{\text{cfp}}(\mathbf{r}), \tag{G.1}$$

whose eigensolution $\Phi_{n\mathbf{k}}(\mathbf{r})$ is provided by the Schrödinger equation

$$\hat{H}\Phi_{n\mathbf{k}}(\mathbf{r}) = E_n(\mathbf{k})\Phi_{n\mathbf{k}}(\mathbf{r}), \tag{G.2}$$

where $\mathbf{k}$ is the electron wavevector, n is the band index, and $E_n(\mathbf{k})$ is the one-electron energy. Since the Hamiltonian $\hat{H}$ is invariant upon lattice translations, the one-electron wavefunction must obey the Bloch theorem (see section 6.3)

$$\Phi_{n\mathbf{k}}(\mathbf{r} + \mathbf{R}_\text{l}) = e^{i\mathbf{k}\cdot\mathbf{R}_\text{l}}\Phi_{n\mathbf{k}}(\mathbf{r}), \tag{G.3}$$

where $\mathbf{R}_\text{l}$ is a lattice vector. The tight-binding method consists in writing the crystalline wavefunction as a linear combination of atomic orbitals (LCAO) $\{\varphi_{\alpha\text{lb}}(\mathbf{r}) = \varphi_\alpha(\mathbf{r} - \mathbf{R}_\text{l} - \mathbf{R}_\text{b})\}$, where consistently with the notation used elsewhere in this Primer $\mathbf{R}_\text{b}$ will indicate the ion positions within the basis, while the label α will stand for the full set of quantum numbers defining the atomic orbital. The LCAO expansion must of course fulfil the Bloch theorem and, therefore, we preliminarily introduce the Bloch sums

doi:10.1088/978-0-7503-2265-2ch18

$$\varphi_{\alpha \mathrm{bk}}^{\mathrm{Bloch}}(\mathbf{r}) = \frac{1}{\sqrt{N}} \sum_{\mathrm{l}} e^{i\mathbf{k}\cdot\mathbf{R}_{\mathrm{l}}} \varphi_\alpha(\mathbf{r} - \mathbf{R}_{\mathrm{l}} - \mathbf{R}_{\mathrm{b}}), \qquad (\mathrm{G.4})$$

where N is the number of unit cells contained in the region which the periodic Born–von Karman boundary condition is applied to, so that

$$\begin{aligned}
\Phi_{n\mathbf{k}}(\mathbf{r}) &= \frac{1}{\sqrt{N_{\mathrm{b}}}} \sum_{\alpha\mathrm{b}} \tilde{B}_{n\alpha\mathrm{b}} \varphi_{\alpha\mathrm{bk}}^{\mathrm{Bloch}}(\mathbf{r}) \\
&= \frac{1}{\sqrt{NN_{\mathrm{b}}}} \sum_{\alpha\mathrm{lb}} e^{i\mathbf{k}\cdot\mathbf{R}_{\mathrm{l}}} \tilde{B}_{n\alpha\mathrm{b}} \varphi_\alpha(\mathbf{r} - \mathbf{R}_{\mathrm{l}} - \mathbf{R}_{\mathrm{b}}),
\end{aligned} \qquad (\mathrm{G.5})$$

where N_{b} is the number of atoms in the lattice basis. The $\tilde{B}$ coefficients are set to

$$B_{n\alpha\mathrm{lb}}(\mathbf{k}) = e^{i\mathbf{k}\cdot\mathbf{R}_{\mathrm{l}}} \tilde{B}_{n\alpha\mathrm{b}}, \qquad (\mathrm{G.6})$$

for further convenience.

Unfortunately, it is impossible to straightforwardly implement the tight-binding method so formulated because of the very large number of integrals that should be computed, corresponding either to scalar products between two Bloch sums or to matrix elements of the Hamiltonian. This is basically due to the fact that the $\{\varphi_{\alpha\mathrm{lb}}(\mathbf{r})\}$ basis set is non orthogonal, since atomic wavefunctions centred on different lattice positions in principle are not orthogonal.

We can remove the non-orthogonality problem by introducing a new set of Löwdin orthogonalized orbitals. To this aim, let us preliminarily define the *overlap integral* between two atomic orbitals as

$$S_{\alpha'\mathrm{l'b'},\alpha\mathrm{lb}} = \langle \varphi_{\alpha'\mathrm{l'b'}}(\mathbf{r}) | \varphi_{\alpha\mathrm{lb}}(\mathbf{r}) \rangle - \delta_{\alpha'\alpha} \delta_{(\mathrm{l'b'})(\mathrm{lb})}, \qquad (\mathrm{G.7})$$

and recast the Schrödinger problem in matrix notation

$$\mathbb{H}\, \mathbb{B}_n(\mathbf{k}) = E_n(\mathbf{k})\, (\mathbb{S} + \mathbb{1})\, \mathbb{B}_n(\mathbf{k}), \qquad (\mathrm{G.8})$$

where $\mathbb{S}$ is the overlap matrix whose entries are defined in equation (G.7), while $\mathbb{H}$ is the Hamiltonian matrix whose elements are defined as

$$H_{\alpha'\mathrm{l'b'},\alpha\mathrm{lb}} = \langle \varphi_{\alpha'\mathrm{l'b'}}(\mathbf{r}) | \hat{H} | \varphi_{\alpha\mathrm{lb}}(\mathbf{r}) \rangle, \qquad (\mathrm{G.9})$$

and the Dirac notation has been used. Obviously, the entries of the column vectors $\mathbb{B}_n(\mathbf{k})$ are given in equation (G.6), while $\mathbb{1}$ is the unit matrix.

By now introducing a new set of $\mathbb{C}_n(\mathbf{k}) = \{C_{n\alpha\mathrm{lb}}(\mathbf{k})\}$ coefficients

$$\mathbb{B}_n(\mathbf{k}) = (\mathbb{S} + \mathbb{1})^{-1/2} \mathbb{C}_n(\mathbf{k}), \qquad (\mathrm{G.10})$$

equation (G.8) is easily recast in the following form

$$\mathbb{H}_{\mathrm{L}} \mathbb{C}_n(\mathbf{k}) = E_n(\mathbf{k}) \mathbb{C}_n(\mathbf{k}), \qquad (\mathrm{G.11})$$

where the label L reminds that a Löwdin matrix transformation

$$\mathbb{H}_{\mathrm{L}} = (\mathbb{S} + \mathbb{1})^{-1/2}\, \mathbb{H}\, (\mathbb{S} + \mathbb{1})^{-1/2}, \qquad (\mathrm{G.12})$$

has been executed. The crystalline wavefunction is now accordingly written as

$$\Phi_{n\mathbf{k}}(\mathbf{r}) = \frac{1}{\sqrt{NN_b}} \sum_{\alpha l b} B_{n\alpha l b}(\mathbf{k}) \varphi_\alpha(\mathbf{r} - \mathbf{R}_l - \mathbf{R}_b)$$

$$= \frac{1}{\sqrt{NN_b}} \sum_{\alpha l b} \sum_{\alpha' l' b'} (\mathbb{S} + \mathbb{1})^{-1/2}_{\alpha l b, \alpha' l' b'} \, C_{n\alpha' l' b'}(\mathbf{k}) \, \varphi_\alpha(\mathbf{r} - \mathbf{R}_l - \mathbf{R}_b) \qquad (G.13)$$

$$= \frac{1}{\sqrt{NN_b}} \sum_{\alpha' l' b'} C_{n\alpha' l' b'}(\mathbf{k}) \psi_{\alpha'}(\mathbf{r} - \mathbf{R}_{l'} - \mathbf{R}_{b'}),$$

where the new *Löwdin orbitals* have been introduced

$$\psi_{\alpha' l' b'}(\mathbf{r}) = \psi_{\alpha'}(\mathbf{r} - \mathbf{R}_{l'} - \mathbf{R}_{b'}) = \sum_{\alpha l b}(\mathbb{S} + \mathbb{1})^{-1/2}_{\alpha l b, \alpha' l' b'} \, \varphi_\alpha(\mathbf{r} - \mathbf{R}_l - \mathbf{R}_b). \qquad (G.14)$$

The Löwdin orbitals are mutually orthogonal (the proof is trivial) and have the same symmetry properties (that is the same *s*- or *p*- or *d*-character) of the original non-orthogonal atomic wavefunctions which they are derived from.

G.2 The two-centre approximation

The formal solution of the eigenvalue problem leads to the following secular problem

$$\sum_{\alpha l b} \left[H_{\alpha' l' b', \alpha l b} - E_n(\mathbf{k}) \delta_{\alpha' \alpha} \delta_{(l' b')(l b)} \right] C_{n\alpha l b}(\mathbf{k}) = 0, \qquad (G.15)$$

where the matrix elements of the Hamiltonian are calculated by using the Löwdin orbitals as

$$H_{\alpha' l' b', \alpha l b} = \langle \psi_{\alpha' l' b'}(\mathbf{r}) | \hat{H} | \psi_{\alpha l b}(\mathbf{r}) \rangle. \qquad (G.16)$$

In principle, provided that the Löwdin orbitals are known as a result of an atomic physics calculation, these terms can be computed exactly, thus providing a parameter-free theory for the calculation of the band structure of a solid. However, this approach appears of little sense[1] since it is rather cumbersome. Some further handling of the theory is therefore worthy at this stage, aimed at transforming the present formal theory into a practical tool. Two strategies are in fact possible: either we can choose suitable *approximations* which allow us to focus just on the leading terms and to disregard other computationally-intensive (but less important) contributions; or we can introduce *empirical features* into the present theory so to avoid expensive calculations.

By assuming that the crystal field potential $V_{\text{cfp}}(\mathbf{r})$ is the sum of spherical potentials $V_{lb}(\mathbf{r})$ located at each $(\mathbf{R}_l + \mathbf{R}_b)$ lattice site, we can rewrite equation (G.1) as

---

[1] The evaluation of the merits and limitations of the LCAO approach represents a classical topic of quantum chemistry [1–3].

$$\hat{H} = -\frac{\hbar^2}{2m_e}\nabla^2 + \hat{V}_{lb}(\mathbf{r}) + \sum_{l'b'\neq lb} \hat{V}_{l'b'}(\mathbf{r}) + \sum_{l''b''\neq l'b',lb} \hat{V}_{l''b''}(\mathbf{r}), \qquad (G.17)$$

and accordingly separate three different contributions

$$\begin{aligned} H_{\alpha'l'b',\alpha lb} &= \langle \psi_{\alpha'l'b'}(\mathbf{r})| -\frac{\hbar^2}{2m_e}\nabla^2 + \hat{V}_{lb}(\mathbf{r})|\psi_{\alpha lb}(\mathbf{r})\rangle \\ &+ \langle \psi_{\alpha'l'b'}(\mathbf{r})| \sum_{l'b'\neq lb} \hat{V}_{l'b'}(\mathbf{r})|\psi_{\alpha lb}(\mathbf{r})\rangle \\ &+ \langle \psi_{\alpha'l'b'}(\mathbf{r})| \sum_{l''b''\neq l'b',lb} \hat{V}_{l''b''}(\mathbf{r})|\psi_{\alpha lb}(\mathbf{r})\rangle, \end{aligned} \qquad (G.18)$$

which are separately handled. The first *intra-atomic* term is easily computed thanks to the orthogonality of the Löwdin orbitals

$$\langle \psi_{\alpha'l'b'}(\mathbf{r})| -\frac{\hbar^2}{2m_e}\nabla^2 + \hat{V}_{lb}(\mathbf{r})|\psi_{\alpha lb}(\mathbf{r})\rangle = E_\alpha \delta_{\alpha'\alpha}\delta_{(l'b')(lb)}, \qquad (G.19)$$

while the remaining two contributions are called, respectively, *two-centre* and *three-centre* energy integrals for obvious reasons.

It is quite reasonable to assume that three-center integrals are comparatively smaller than two-centre ones, because of the localized character of the Löwdin orbitals. A *two-center approximation* is accordingly introduced, assuming that three-centre energy integrals can be simply disregarded. This drastic guess was originally introduced by Slater and Koster in 1954 [4] to formally treat the Hamiltonian matrix elements as if we had a sort of 'diatomic molecule' system. By introducing the distance vector

$$\mathbf{t} = \mathbf{R}_{l'b'} - \mathbf{R}_{lb}, \qquad (G.20)$$

a sort of 'molecular axis' between any pair of atoms is naturally set with two spherical coordinates systems centred on the two atomic positions connected by such a vector and the z-axes parallel to $\mathbf{t}$, as indicated in figure G.1. In this double frame of reference, the angular parts of the orbitals $\psi_{\alpha l'b'}(\mathbf{r})$ and $\psi_{\alpha lb}(\mathbf{r})$ have the form of a

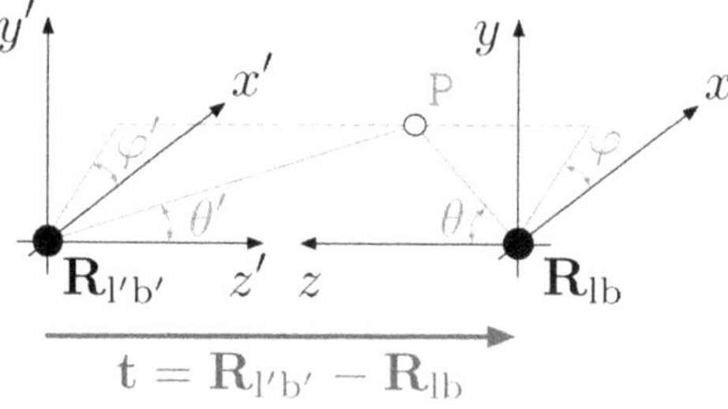

Figure G.1. Definition of the two spherical coordinates systems for the equivalent 'diatomic molecule' formed by atoms at $\mathbf{R}_{l'b'}$ (left) and $\mathbf{R}_{lb}$ (right), respectively. Any point P is defined by different polar angles θ and θ', but by a single azimuth angle since $\varphi = \varphi'$ by construction.

spherical harmonic function, both depending on the same azimuth angle φ. Therefore, it turns out that the angular functions depending on such an azimuth angle combine so that the matrix element of the Hamiltonian becomes proportional to $\int_0^{2\pi} e^{i(m-m')\varphi} d\varphi$, where m and m' are the magnetic quantum numbers of the two orbitals: this implies that it is zero unless $m = m'$ and, by adopting the same notation as in atomic physics, the σ, π, d, ... symbols are used to label the non-vanishing matrix elements corresponding to $m = m' = 0$, 1, 2, ..., respectively. The radial and θ-dependent angular parts of the same integrals are commonly indicated as ss, or sp, or pp, $\cdots$ according to the actual symmetry of the two involved orbitals. The resulting complete labelling for the two-centre hopping integrals is $(ss\sigma)$, or $(sp\sigma)$, or $(pp\sigma)$, and so on. The practical application of this procedure is schematically illustrated in figure G.2 in the case of s- and p-like orbitals.

In order to pass from the Slater–Koster 'diatomic molecule' scheme to a full crystalline picture, any pair of spherical harmonic functions (defined with respect the $\mathbf{t}$ axes) should be rotated so as to match the same Cartesian frame of reference used to span the crystal lattice. This is done by means of the direction cosines $(\hat{\mathrm{l}}, \hat{\mathrm{m}}, \hat{\mathrm{n}})$ defining each $\mathbf{t}$ vector with respect to such a frame. The procedure has been worked out once and for all for any possible pair of orbitals [4], and its result is summarised in table G.1. It should be noted that all hopping integrals do depend on the interatomic distance between the two selected atomic centres: therefore, one should calculate different values for such quantities between first-, second-, third-, $\cdots$ nearest neighbours.

G.3 Calculating the hopping energy integrals

Despite the two-centre approximation, the computational effort required to compute the Hamiltonian matrix elements still remains a formidable task: in order to set up a practical tool, we need to further reduce the complexity of the tight-binding method. A key step in this direction is to consider the two-centre integrals as disposable constants to fit on the basis of some simplifying approximations.

First of all, only close-neighbours interactions are typically taken into account. This is tantamount to stating that, among all two-centre integrals, those

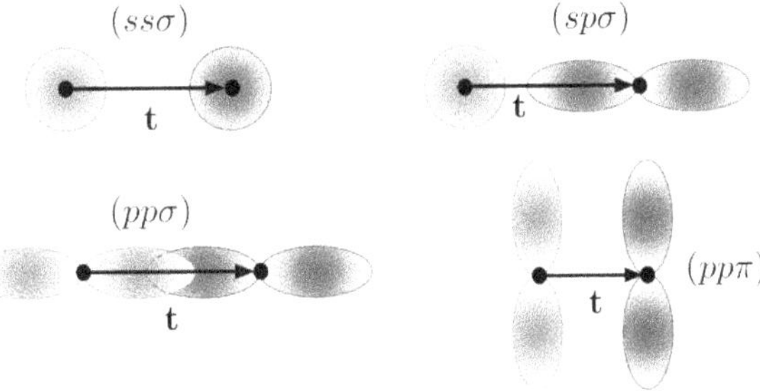

Figure G.2. Overlaps between s- and p-like orbitals within the Slater–Koster 'diatomic molecule' scheme. The bold arrow indicates the distance vector $\mathbf{t} = \mathbf{R}_{\mathrm{l'b'}} - \mathbf{R}_{\mathrm{lb}}$ between two atoms. The standard labelling for the two-centre hoppings is shown.

Table G.1. Two-centre energy integrals. The $(\hat{l}, \hat{m}, \hat{n})$ direction cosines define each **t** vector with respect to unique crystalline frame of reference. The Löwdin orbitals are symbolized according to the standard notation valid for usual atomic orbitals.

Orbital pair	Orbital symmetry	Energy integral
s–s	s, s	$(ss\sigma)$
s–p	s, p_x	$\hat{l}(sp\sigma)$
p–p	p_x, p_x	$\hat{l}^2(pp\sigma) + (1 - \hat{l}^2)(pp\pi)$
	p_x, p_y	$\hat{l}\hat{m}(pp\sigma) - \hat{l}\hat{m}(pp\pi)$
	p_x, p_z	$\hat{l}\hat{n}(pp\sigma) - \hat{l}\hat{n}(pp\pi)$
s–d	s, d_{xy}	$\sqrt{3}\,\hat{l}\hat{m}(sd\sigma)$
	$s, d_{x^2-y^2}$	$\frac{1}{2}\sqrt{3}\,(\hat{l}^2 - \hat{m}^2)(sd\sigma)$
	$s, d_{3z^2-r^2}$	$[\hat{n}^2 - \frac{1}{2}(\hat{l}^2 + \hat{m}^2)](sd\sigma)$
p–d	p_x, d_{xy}	$\sqrt{3}\,\hat{l}^2\hat{m}(pd\sigma) + m(1 - 2\hat{l}^2)(pd\pi)$
	p_x, d_{yz}	$\sqrt{3}\,\hat{l}\hat{m}\hat{n}(pd\sigma) - 2\hat{l}\hat{m}\hat{n}(pd\pi)$
	p_x, d_{zx}	$\sqrt{3}\,\hat{l}^2\hat{n}(pd\sigma) + \hat{n}(1 - 2\hat{l}^2)(pd\pi)$
	$p_x, d_{x^2-y^2}$	$\frac{1}{2}\sqrt{3}\,\hat{l}(\hat{l}^2 - \hat{m}^2)(pd\sigma) + \hat{l}(1 - \hat{l}^2 + \hat{m}^2)(pd\pi)$
	$p_y, d_{x^2-y^2}$	$\frac{1}{2}\sqrt{3}\,\hat{m}(\hat{l}^2 - \hat{m}^2)(pd\sigma) + \hat{m}(1 + \hat{l}^2 - \hat{m}^2)(pd\pi)$
	$p_z, d_{x^2-y^2}$	$\frac{1}{2}\sqrt{3}\,\hat{n}(\hat{l}^2 - \hat{m}^2)(pd\sigma) + \hat{n}(\hat{l}^2 - \hat{m}^2)(pd\pi)$
	$p_x, d_{3z^2-r^2}$	$\hat{l}[\hat{n}^2 - \frac{1}{2}(\hat{l}^2 + \hat{m}^2)](pd\sigma) - \sqrt{3}\,\hat{l}\hat{n}^2(pd\pi)$
	$p_y, d_{3z^2-r^2}$	$\hat{m}[\hat{n}^2 - \frac{1}{2}(\hat{l}^2 + \hat{m}^2)](pd\sigma) - \sqrt{3}\,\hat{m}\hat{n}^2(pd\pi)$
	$p_z, d_{3z^2-r^2}$	$\hat{n}[\hat{n}^2 - \frac{1}{2}(\hat{l}^2 + \hat{m}^2)](pd\sigma) - \sqrt{3}\,\hat{n}(\hat{l}^2 + \hat{m}^2)(pd\pi)$
d–d	d_{xy}, d_{xy}	$3\hat{l}^2\hat{m}^2(dd\sigma) + (\hat{l}^2 + \hat{m}^2 - 4\hat{l}^2\hat{m}^2)(dd\pi) + (\hat{n}^2 + \hat{l}^2\hat{m}^2)(dd\delta)$
	d_{xy}, d_{yz}	$3\hat{l}\hat{m}^2\hat{n}(dd\sigma) + \hat{l}\hat{n}(1 - 4\hat{m}^2)(dd\pi) + \hat{l}\hat{n}(\hat{m}^2 - 1)(dd\delta)$
	d_{xy}, d_{zx}	$3\hat{l}^2\hat{m}\hat{n}(dd\sigma) + \hat{m}\hat{n}(1 - 4\hat{l}^2)(dd\pi) + \hat{m}\hat{n}(\hat{l}^2 - 1)(dd\delta)$
	$d_{xy}, d_{x^2-y^2}$	$\frac{3}{2}\hat{l}\hat{m}(\hat{l}^2 - \hat{m}^2)(dd\sigma) + 2\hat{l}\hat{m}(\hat{m}^2 - \hat{l}^2)(dd\pi) + \frac{1}{2}\hat{l}\hat{m}(\hat{l}^2 - \hat{m}^2)(dd\delta)$
	$d_{yz}, d_{x^2-y^2}$	$\frac{3}{2}\hat{m}\hat{n}(\hat{l}^2 - \hat{m}^2)(dd\sigma) - \hat{m}\hat{n}[1 + 2(\hat{l}^2 - \hat{m}^2)](dd\pi)$ $+\hat{m}\hat{n}[1 + \frac{1}{2}(\hat{l}^2 - \hat{m}^2)](dd\delta)$
	$d_{zx}, d_{x^2-y^2}$	$\frac{3}{2}\hat{n}\hat{l}(\hat{l}^2 - \hat{m}^2)(dd\sigma) + \hat{n}\hat{l}[1 - 2(\hat{l}^2 - \hat{m}^2)](dd\pi)$ $-\hat{n}\hat{l}[1 - \frac{1}{2}(\hat{l}^2 - \hat{m}^2)](dd\delta)$
	$d_{xy}, d_{3z^2-r^2}$	$\sqrt{3}\,\hat{l}\hat{m}[\hat{n}^2 - \frac{1}{2}(\hat{l}^2 + \hat{m}^2)](dd\sigma) - 2\sqrt{3}\,\hat{l}\hat{m}\hat{n}^2(dd\pi)$ $+\frac{1}{2}\sqrt{3}\,\hat{l}\hat{m}(1 + \hat{n}^2)](dd\delta)$
	$d_{yz}, d_{3z^2-r^2}$	$\sqrt{3}\,\hat{m}\hat{n}[\hat{n}^2 - \frac{1}{2}(\hat{l}^2 + \hat{m}^2)](dd\sigma) + \sqrt{3}\,\hat{m}\hat{n}(\hat{l}^2 + \hat{m}^2 - \hat{n}^2)(dd\pi)$ $-\frac{1}{2}\sqrt{3}\,\hat{m}\hat{n}(\hat{l}^2 + \hat{m}^2)(dd\delta)$
	$d_{zx}, d_{3z^2-r^2}$	$\sqrt{3}\,\hat{l}\hat{n}[\hat{n}^2 - \frac{1}{2}(\hat{l}^2 + \hat{m}^2)](dd\sigma) + \sqrt{3}\,\hat{l}\hat{n}(\hat{l}^2 + \hat{m}^2 - \hat{n}^2)(dd\pi)$ $-\frac{1}{2}\sqrt{3}\,\hat{l}\hat{n}(\hat{l}^2 + \hat{m}^2)(dd\delta)$
	$d_{x^2-y^2}, d_{x^2-y^2}$	$\frac{3}{4}(\hat{l}^2 - \hat{m}^2)^2(dd\sigma) + [\hat{l}^2 + \hat{m}^2 - (\hat{l}^2 - \hat{m}^2)^2](dd\pi)$ $+[\hat{n}^2 + \frac{1}{4}(\hat{l}^2 - \hat{m}^2)^2)](dd\delta))$

$$d_{x^2-y^2},\ d_{3z^2-r^2} \qquad \tfrac{1}{2}\sqrt{3}\,(\hat{l}^2 - \hat{m}^2)[\hat{n}^2 - \tfrac{1}{2}(\hat{l}^2 + \hat{m}^2)](dd\sigma) + \sqrt{3}\,\hat{n}^2(\hat{m}^2 - \hat{l}^2)(dd\pi)$$
$$+\tfrac{1}{4}\sqrt{3}\,(1 + \hat{n}^2)(\hat{l}^2 - \hat{m}^2)(dd\delta)$$

$$d_{3z^2-r^2},\ d_{3z^2-r^2} \qquad [\hat{n}^2 - \tfrac{1}{2}(\hat{l}^2 + \hat{m}^2)]^2(dd\sigma) + 3\hat{n}^2(\hat{l}^2 + \hat{m}^2)(dd\pi)$$
$$+\tfrac{3}{4}(\hat{l}^2 + \hat{m}^2)^2(dd\delta)$$

---

corresponding to a distance between atom $(\mathbf{R}_{l'} - \mathbf{R}_{b'})$ and atom $(\mathbf{R}_l - \mathbf{R}_b)$ larger than a given cutoff are set to zero. This approximation is obviously justified by the localised character of the basis set functions (a true characteristic for both atomic orbitals and Löwdin orbitals) and greatly reduces the number of parameters to fit: in many applications, just first-next-neighbours interactions are considered. Next, a minimal basis set is used for the LCAO expansion. This means that we only consider those orbitals whose energy is close to the energy of the electronic states of the crystalline system we are interested in. In many applications, this procedure basically amounts to considering only those orbitals with energy in the range containing the topmost valence states and lowest conduction states. Such an approximation reduces both the number of parameters to fit and the order of the secular problem to be numerically solved.

These approximations obviously reflect in some important limitation of the tight-binding theory, namely (i) we can explore just a part of the energy spectrum of the crystal and (ii) we have no knowledge about the one-electron wavefunction. In fact, by introducing empirical parameters to replace the Hamiltonian matrix elements, we lose any chance to actually perform the LCAO expansion: accordingly, we cannot compute any physical property directly depending on $\Phi_{n\mathbf{k}}(\mathbf{r})$. This is, for example, the case of the electron density which obviously is quite an important quantity in solid state physics. Above all, however, the method turns out to be *semi-empirical*. While the reassuring prefix 'semi' stands for the fact that the resulting model is firmly rooted in a rigorous quantum treatment of crystalline electronic states, the 'empirical' character implies that its accuracy, reliability and transferability need to be proved from case to case. In any case, the tight-binding theory still has a great pedagogical value and it is really useful for quick-and-easy band-structure calculations.

The fitted two-centre integrals are determined on a suitable database of physical information obtained either from experiments or by more sophisticated first-principles calculations. Slater–Koster parameters are nowadays available for a large variety of elemental as well as compound materials, as for instance summarised in [5, 6].

G.4 Tight binding at work

As a showcase application of the tight-binding theory, let us consider the specific case of silicon (a crystal with diamond structure and a two-atom basis) and let us proceed under the two-centre and first-next-neighbours approximations. If we just

Table G.2. Two-centre integrals according to different sp^3 parametrizations for silicon (units of eV).

	Ref. [1]	Ref. [2]	Ref. [3]	Ref. [4]
e_s	−5.250	−6.535	−5.200	−4.478
e_p	+1.200	+1.760	+1.200	+2.552
$(ss\sigma)$	−2.038	−1.820	−1.940	−1.913
$(sp\sigma)$	+1.745	+1.960	+1.750	+2.594
$(pp\sigma)$	+2.750	+3.060	+3.050	+4.464
$(pp\pi)$	−1.075	−0.870	−1.080	−1.275

Ref. [1]: Kwon I, Biswas R, Wang C Z, Ho K M, and Soukoulis C M 1994 *Phys. Rev.* B **49** 7242.
Ref. [2]: Goodwin L, Skinner A J, and Pettifor D G 1989 *Europhys. Lett.* **9** 701.
Ref. [3]: Chadi D J and Cohen M L 1975 *Phys. Status Solidi* B **68** 405.
Ref. [4]: This is the set of 'universal tight-binding parameters' discussed in [5].

focus on the topmost valence states, we can adopt a minimal sp^3 basis-set for each atom of the basis. The corresponding energy integrals have been fitted by many authors and some parametrizations are reported in table G.2 where we have introduced the notation e_s and e_p for the on-site integrals corresponding to ss and pp orbital pairs, respectively.

The two-centre energy integrals of the crystalline diamond structure are easily computed using tables G.1 and G.2 and by taking into account that the needed cosine directors are defined for the four atomic pairs formed by the first atom in position $(0, 0, 0)a_0$ and the second atom in positions $(1/4, 1/4, 1/4)a_0$, $(1/4, -1/4, -1/4)a_0$, $(-1/4, 1/4, -1/4)a_0$, and $(-1/4, -1/4, 1/4)a_0$, respectively. The lattice constant of silicon has been named a_0. The Bloch phase factors $\exp(i\mathbf{k} \cdot \mathbf{t}_i)$ with $i = 1, \ldots, 4$ are calculated with position vectors corresponding to the four first-next-neighbours of the pivot atom placed at the origin. In summary, the Hamiltonian matrix to be diagonalised for each $\mathbf{k}$ value within the 1BZ in order to get the valence band dispersions $E_{\mathrm{VB}}(\mathbf{k})$ is

$$\hat{H}(\mathbf{k}) = \begin{bmatrix}
e_s & 0 & 0 & 0 & E_{ss}g_0 & E_{sp}g_1 & E_{sp}g_2 & E_{sp}g_3 \\
0 & e_p & 0 & 0 & -E_{sp}g_1 & E_{p_x,p_x}g_0 & E_{p_x,p_y}g_3 & E_{p_x,p_y}g_2 \\
0 & 0 & e_p & 0 & -E_{sp}g_2 & E_{p_x,p_y}g_3 & E_{p_x,p_x}g_0 & E_{p_x,p_y}g_1 \\
0 & 0 & 0 & e_p & -E_{sp}g_3 & E_{p_x,p_y}g_2 & E_{p_x,p_y}g_1 & E_{p_x,p_x}g_0 \\
E_{ss}g_0^* & -E_{sp}g_1^* & -E_{sp}g_2^* & -E_{sp}g_3^* & e_s & 0 & 0 & 0 \\
E_{sp}g_1^* & E_{p_x,p_x}g_0^* & E_{p_x,p_y}g_3^* & E_{p_x,p_y}g_2^* & 0 & e_p & 0 & 0 \\
E_{sp}g_2^* & E_{p_x,p_y}g_3^* & E_{p_x,p_x}g_0^* & E_{p_x,p_y}g_1^* & 0 & 0 & e_p & 0 \\
E_{sp}g_3^* & E_{p_x,p_y}g_2^* & E_{p_x,p_y}g_1^* & E_{p_x,p_x}g_0^* & 0 & 0 & 0 & e_p
\end{bmatrix} \quad (G.21)$$

where the e's and E's terms are, respectively, the on-site and two-centre energy integrals reported in the right column of table G.1. The $\mathbf{k}$-dependence enters through the geometrical g-factors given by

$$g_0(\mathbf{k}) = \exp(i\mathbf{k} \cdot \mathbf{t}_1) + \exp(i\mathbf{k} \cdot \mathbf{t}_2) + \exp(i\mathbf{k} \cdot \mathbf{t}_3) + \exp(i\mathbf{k} \cdot \mathbf{t}_4)$$
$$g_1(\mathbf{k}) = \exp(i\mathbf{k} \cdot \mathbf{t}_1) + \exp(i\mathbf{k} \cdot \mathbf{t}_2) - \exp(i\mathbf{k} \cdot \mathbf{t}_3) - \exp(i\mathbf{k} \cdot \mathbf{t}_4)$$
$$g_2(\mathbf{k}) = \exp(i\mathbf{k} \cdot \mathbf{t}_1) - \exp(i\mathbf{k} \cdot \mathbf{t}_2) + \exp(i\mathbf{k} \cdot \mathbf{t}_3) - \exp(i\mathbf{k} \cdot \mathbf{t}_4)$$
$$g_3(\mathbf{k}) = \exp(i\mathbf{k} \cdot \mathbf{t}_1) - \exp(i\mathbf{k} \cdot \mathbf{t}_2) - \exp(i\mathbf{k} \cdot \mathbf{t}_3) + \exp(i\mathbf{k} \cdot \mathbf{t}_4)$$

$$(G.22)$$

In figure G.3 the valence bands for crystalline silicon are reported, as obtained from the four different sets of empirical tight-binding parameters reported in table G.2, where the on-site terms have been shifted so as to align at zero energy the top valence state at Γ. This is a standard choice which is allowed since the total width of the valence band is rather governed by the difference $|e_s - e_p|$ than by their absolute values.

While the basic features are equally provided by the four sets of parameters (like for instance, their general topology and degeneracy), some details vary from case to case. Nevertheless, it is important to stress that the semi-empirical tight-binding method has not the ambition to provide an extreme degree of accuracy. Actually, it is not realistic to pretend so, in view of the important approximations we adopted. Nevertheless, it provides in any case a more than decent description of valence states.

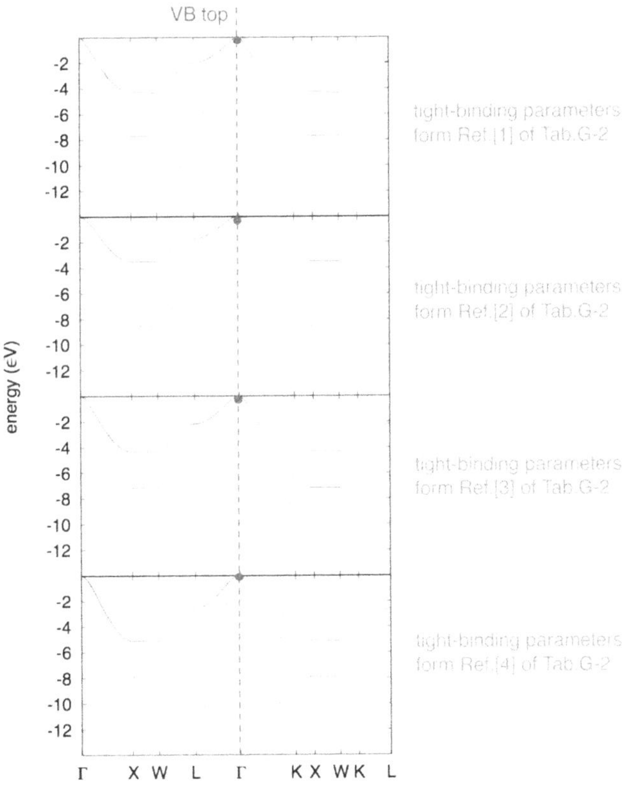

Figure G.3. The valence band dispersions of crystalline silicon calculated with the tight-binding parameters given in table G.2. The blue dots show the position of the top of the valence band (VB): the on-site terms have been shifted so as to align it at zero energy.

Finally, it is important to remark that one should be very heedful in comparing the values appearing in table G.2: while there is a surely meaningful trend in the variation of the different two-centre integrals upon orbital symmetry (which is common to all sets of parameters), it is hard to attach any clear physical connotation to such numbers.

References

[1] Szabo A and Ostlund N S 1996 *Modern Quantum Chemistry* (New York: Dover)
[2] Jensen F 2006 *Introduction to Computational Chemistry* 2nd edn (Hoboken, NJ: Wiley)
[3] Simons J 2003 *An Introduction to Theoretical Chemistry* (Cambridge: Cambridge University Press)
[4] Slater J C and Koster G F 1954 Simplified LCAO method for the periodic potential problem *Phys. Rev.* **94** 1498
[5] Harrison W A 1980 *Electronic Structure and the Properties of Solids* (New York: Dover)
[6] Papacostantopoulos D A 2015 *Handbook of the Band Structure of Elemental Solids* 2nd edn (New York: Springer)

IOP Publishing

Solid State Physics
A primer
Luciano Colombo

Appendix H

Stress and temperature effects on the energy gap

The intrinsic energy gap of an insulating material is affected by the variation of the interatomic distances. This is especially clear in the tight-binding picture: the elements of the one-electron Hamiltonian matrix and of the overlap matrices (see equations (8.26) and (8.27), respectively) crucially depend on the relative distance between all those neighbouring atoms that are included in the model. In turn, interatomic distances are changed by deforming the lattice by means of some applied stress or by increasing its temperature. Their net effect on E_{gap} must be differently treated.

Let us first consider the effect of a mechanical action which, for simplicity, we will assume to be a simple hydrostatic stress. If V_0 is the equilibrium volume and ΔV is its stress-induced variation, then the quantity $\epsilon = \Delta V / V_0$ quantitatively represents the hydrostatic deformation, as extensively discussed in chapter 5. Upon either compression or extension, the intrinsic energy gap value linearly changes with the deformation ϵ since it is experimentally found [1] that

$$E_{\mathrm{CB}}^{\mathrm{bottom}}(\epsilon) = E_{\mathrm{CB}}^{\mathrm{bottom}}(0) + a_{\mathrm{CB}}\,\epsilon$$
$$E_{\mathrm{VB}}^{\mathrm{top}}(\epsilon) = E_{\mathrm{VB}}^{\mathrm{top}}(0) + a_{\mathrm{VB}}\,\epsilon, \tag{H.1}$$

and therefore

$$E_{\mathrm{gap}}(\epsilon) = E_{\mathrm{gap}}(0) + (a_{\mathrm{CB}} - a_{\mathrm{VB}})\,\epsilon, \tag{H.2}$$

where $E_{\mathrm{CB}}^{\mathrm{bottom}}(0)$ and $E_{\mathrm{VB}}^{\mathrm{top}}(0)$ are the zero stress (that is, equilibrium) positions of the minimum and maximum of the conduction and valence band, respectively, and $E_{\mathrm{gap}}(0)$ is the corresponding intrinsic gap.

The two coefficients $a_{\mathrm{CB}} < 0$ and $a_{\mathrm{VB}} > 0$ are referred to as the *deformation potentials* for the two bands. In the simplified discussion here outlined the deformation potentials are scalar quantities, while in fact they are tensors: a more in-depth investigation can be found elsewhere [2–4]. A simple dimensional analysis of the above equations shows that the deformation potentials have the dimensions of an energy, while it is in general found that $|a_{\mathrm{CB}}| \neq |a_{\mathrm{VB}}|$. By adopting the parabolic

doi:10.1088/978-0-7503-2265-2ch19

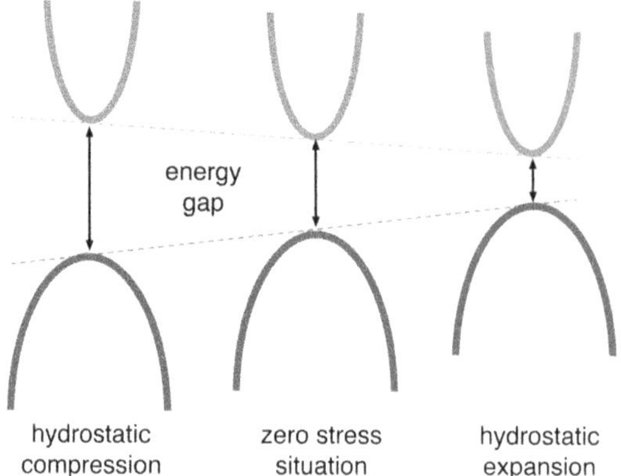

Figure H.1. Hydrostatic stress effects on the energy gap of a direct gap material. It is assumed $a_{\mathrm{CB}} < 0$ and $a_{\mathrm{VB}} > 0$ with $|a_{\mathrm{CB}}| \neq |a_{\mathrm{VB}}|$. Compression and expansion, respectively, correspond to a deformation $\epsilon < 0$ and $\epsilon > 0$, with $\epsilon = \Delta V / V_0$ (where V_0 is the undeformed crystal volume).

band approximation for a model direct gap material, this body of information is conceptualised in figure H.1.

Finally, as for the temperature-dependence of the energy gap, we understand it is ruled over by thermal expansion phenomena (see section 4.2.1). For most semiconductor materials the following phenomenological law [1] is found valid

$$E_{\mathrm{gap}}(T) = E_{\mathrm{gap}}^{(0)} - \frac{\xi_1 T^2}{\xi_2 + T},\qquad\text{(H.3)}$$

where $E_{\mathrm{gap}}^{(0)}$ is the intrinsic gap measured at $T = 0$ K, while ξ_1 and ξ_2 are materials specific empirical constants: typically, $\xi_1 = \mathcal{O}(10^{-4})$ eV K^{-1} and $\xi_2 = \mathcal{O}(10^2)$ K.

References

[1] Grundmann M 2010 *The Physics of Semiconductors* (Heidelberg: Springer)
[2] Cardona M and Yu P Y 2010 *Fundamentals of Semiconductors* (Heidelberg: Springer)
[3] Seeger K 1989 *Semiconductor Physics* (Heidelberg: Springer)
[4] Balkanski M and Wallis R F 1989 *Semiconductor Physics and Applications* (Oxford: Oxford University Press)

IOP Publishing

Solid State Physics
A primer
Luciano Colombo

Appendix I

Functionals and functional derivatives

In mathematics a function $f = f(x)$ is a rule where an input number x returns another number $f(x)$. On the other hand, a *functional* is a rule where an input function $f(x)$ returns a number $F[f(x)]$.

The *functional derivative* [1] of $F[f(x)]$ with respect to $f(x)$ is denoted by the symbol $\delta F/\delta f$ and it is defined through the equation

$$\int \frac{\delta F[f(x)]}{f(x)}\eta(x)dx = \frac{d}{d\epsilon}F[f(x) + \epsilon\eta(x)]_{\epsilon=0} \tag{I.1}$$

where $\eta(x)$ is an arbitrary function and ϵ a real number. Let us then consider some functional derivative of physical interest; for direct use in chapter 10 we adopt the atomic units. In any of the following worked examples let $\rho(\mathbf{r})$ be the function describing the electron density at position $\mathbf{r}$.

The functional associated with the Coulomb interaction between an electron charge distribution $\rho(\mathbf{r})$ and a point-like nucleus sitting in position $\mathbf{R}_\alpha$ is written as

$$V_{\text{Coulomb}}[\rho] = \int \frac{Q_\alpha\rho(\mathbf{r})}{|\mathbf{r} - \mathbf{R}_\alpha|}\, d\mathbf{r}, \tag{I.2}$$

where Q_α is the nuclear charge. By applying the definition given in equation (I.1) we calculate

$$\int \frac{\delta V_{\text{Coulomb}}}{\delta\rho(\mathbf{r})}\eta(\mathbf{r})d\mathbf{r} = \frac{d}{d\epsilon}\left[\int \frac{Q_\alpha[\rho(\mathbf{r}) + \epsilon\eta(\mathbf{r})]}{|\mathbf{r} - \mathbf{R}_\alpha|}d\mathbf{r}\right]_{\epsilon=0}$$
$$= \int \frac{Q_\alpha\eta(\mathbf{r})}{|\mathbf{r} - \mathbf{R}_\alpha|}d\mathbf{r}, \tag{I.3}$$

which leads to

$$\frac{\delta V_{\text{Coulomb}}}{\delta\rho(\mathbf{r})} = \frac{Q_\alpha}{|\mathbf{r} - \mathbf{R}_\alpha|}, \tag{I.4}$$

doi:10.1088/978-0-7503-2265-2ch20

which is easily interpreted: the functional derivative of the Coulomb functional is the ordinary electrostatic potential generated by the nucleus.

The functional associated with the Hartree energy of a system of interacting electrons is written as

$$V_{\text{Hartree}}[\rho] = \frac{1}{2} \int \int \frac{\rho(\mathbf{r})\rho(\mathbf{r}')}{|\mathbf{r} - \mathbf{r}'|} \, d\mathbf{r} d\mathbf{r}'. \tag{I.5}$$

By applying the definition given in equation (I.1) we calculate

$$\int \frac{\delta V_{\text{Hartree}}}{\delta\rho(\mathbf{r})} \eta(\mathbf{r}) d\mathbf{r} = \frac{1}{2} \frac{d}{d\epsilon} \left[\int \int \frac{[\rho(\mathbf{r}) + \epsilon\eta(\mathbf{r})][\rho(\mathbf{r}') + \epsilon\eta(\mathbf{r}')]}{|\mathbf{r} - \mathbf{r}'|} d\mathbf{r} d\mathbf{r}' \right]_{\epsilon=0}$$
$$= \int \int \frac{\rho(\mathbf{r}')\eta(\mathbf{r})}{|\mathbf{r} - \mathbf{r}'|} d\mathbf{r} d\mathbf{r}', \tag{I.6}$$

which leads to

$$\frac{\delta V_{\text{Hartree}}}{\delta\rho(\mathbf{r})} = \int \frac{\rho(\mathbf{r}')}{|\mathbf{r} - \mathbf{r}'|} d\mathbf{r}', \tag{I.7}$$

which, once again, shows that the functional derivative of the Hartree functional is the electrostatic potential generated by the charge distribution.

Finally, let us consider the general power-law functional

$$F[\rho] = \int \rho^{\xi}(\mathbf{r}) d\mathbf{r} \tag{I.8}$$

where ξ is a real number. By applying the definition given in equation (I.1) we calculate

$$\int \frac{\delta F}{\rho(\mathbf{r})} \eta(\mathbf{r}) d\mathbf{r} = \frac{d}{d\epsilon} \left[\int [\rho(\mathbf{r}) + \epsilon\eta(\mathbf{r})]^{\xi} d\mathbf{r} \right]_{\epsilon=0}$$
$$= \xi \int \rho^{\xi-1}(\mathbf{r})\eta(\mathbf{r}) d\mathbf{r}, \tag{I.9}$$

which leads to

$$\frac{\delta F}{\delta\rho(\mathbf{r})} = \xi\rho^{\xi-1}(\mathbf{r}). \tag{I.10}$$

This expression is used in chapter 10 to calculate the derivative of the exchange energy functional in the local density approximation.

References

[1] Parr R G and Yang W 1989 *Density Functional Theory of Atoms and Molecules* (Oxford: Oxford Science Publications)